Insecticides in Pest Control - Impact, Challenges and Strategies

Edited by Sarita Kumar

Published in London, United Kingdom

Insecticides in Pest Control - Impact, Challenges and Strategies
http://dx.doi.org/10.5772/10.5772/intechopen.1003346
Edited by Sarita Kumar

Contributors
A.D. Dileep Kumar, Aarti Sharma, Abhijit Ghosal, Akanksha Singh, Ali Rajabpour , Amitabh Panda, Arman Hakemi, Astrid Jankielsohn, Bibhuti Bhusan Sahoo, Dennis Obonyo Ndolo, Ebenezer Asamani, Fatemeh Yarahmadi, Gaurava Kumar, Grace Kinyanjui, Henry Omoregie Egharevba, Himanshu Pande, Jerry Asalma Nboyine, Job Chemutai Alunga, Jose Barroso, Joseph Adomako, Kahsay Tadesse Mawcha, Mehi Lal, Narasimha Reddy Donthi, Olubukola Oluranti Babalola, Pankaj Verma, Paul Horne, Peter Quandahor, Philip Agrengsore, Rajendra Sangaraju, Rajesh Kumar K, Rebecca Addison, Roopa Rani Samal, Sarita Kumar, Seyed Reza Mousavi, Shubhasri Sahoo, Simon Koma Okwute, Sorabh Chaudhary, Stella W. Simiyu, Sukesh Narayan Sinha, Sunil Kumar Yadav, Sushma Sagar, Swaraj Meher, ThankGod Onuoha, Tien Huynh

First published in London, United Kingdom, 2024 by IntechOpen
IntechOpen is the global imprint of INTECHOPEN LIMITED, registered in England and Wales, registration number: 11086078, 167-169 Great Portland Street, London, W1W 5PF, United Kingdom

For EU product safety concerns: IN TECH d.o.o., Prolaz Marije Krucifikse Kozulić 3, 51000 Rijeka, Croatia, info@intechopen.com or visit our website at intechopen.com.

British Library Cataloguing-in-Publication Data
A catalogue record for this book is available from the British Library

Insecticides in Pest Control - Impact, Challenges and Strategies
Edited by Sarita Kumar
p. cm.

This title is part of the Agricultural Sciences Book Series, Volume 16
Topic: Agronomy and Horticulture
Series Editor: W. James Grichar
Topic Editor: Ibrahim Kahramanoglu
Associate Topic Editors: Murat Helvaci and Olga Panfilova

Print ISBN 978-1-83769-890-5
Online ISBN 978-1-83769-889-9
eBook (PDF) ISBN 978-1-83769-891-2
ISSN 3029-052X

IntechOpen Book Series

Agricultural Sciences

Volume 16

Aims and Scope of the Series

The importance of agriculture cannot be overstated. It helps sustain life, as it gives us the food we need to survive and provides opportunities for economic well-being. Agriculture helps people prosper around the world and combines the creativity, imagination, and skill involved in planting crops and raising animals with modern production methods and new technologies. This series includes two main topics: Agronomy and Horticulture, and Animal Farming. This series will help readers better understand the intricacies of production agriculture and provide the new knowledge that is required to be successful. The success of a farmer in modern agriculture requires knowledge of events happening locally as well as globally that impact input decisions and ultimately determine net profit.

Meet the Series Editor

W. James Grichar has been employed with Texas A&M AgriLife Research for over 45 years with an emphasis on research in agronomy, plant pathology, and weed science. He obtained his BS from Texas A&M in 1972 and his Masters of Plant Protection in 1975. He has published 195 journal articles, over 330 research reports and briefs, 11 book chapters, and over 300 abstracts of profession meetings. He also directs research in many crops including corn, grain sorghum, peanuts, and sesame. He has held various positions in different professional societies including the American Peanut Research and Education Society, Southern Weed Science Society, and Texas Plant Protection Conference in addition to being Associate Editor for Peanut Science and Weed Technology. Significant accomplishments have included spearheading efforts to determine the optimum planting time for soybean production along the upper Texas Gulf Coast. These efforts have shown growers that soybean yields can be improved by 10 to 20% by following a late March to early April plant date. He also has been instrumental in developing a herbicide program for peanut production in the south Texas growing region. Through the development and use of herbicides that are effective against major weed problems in the south Texas region, peanut yields have increased by 25 to 30%.

Meet the Volume Editor

Prof. Sarita Kumar, a distinguished entomologist, works at Acharya Narendra Dev College, University of Delhi, India. In over 32 years of teaching, academics, and administration and 22 years of research, she has contributed immensely to education and research. Working in the area of insect pest and vector control, she is credited with 80+ research publications, 85+ presentations or publications at national and international conferences, 18 e-book chapters, 15 e-modules, and 24 books on biological sciences. A prominent curriculum and educational content developer, she is a member of several national and international committees, a key resource person at renowned organizations, and on the editorial boards of many journals of national and international repute. An academic par excellence, Prof. Kumar has been conferred several prestigious meritorious, and distinguished teacher awards and received recognition from renowned organizations.

Contents

Preface

Insects are the most diverse group of animals, found almost everywhere on the earth. They represent more than half of all animal species, with more than a million described species. Insects are of high economic significance since they help in pollination, soil aeration, dung burial, pest control, and nutrition. Several insects act as pests, damaging crops as they can consume plant foliage, roots, stems, seeds, and fruits, leading to momentous menace to agriculture and substantial economic losses. Many insects act as vectors of diseases in humans and animals, while many are even harmful to our household systems.

These pests have been controlled to some extent by the presence of their predators or parasites in nature and adverse environmental conditions. However, artificial interventions are necessarily required to manage these below the threshold limits. Hence, several interventions have been formulated to combat these insect pests. Among these, chemical insecticide-based approaches are the most preferred among farmers and health workers as the protection that insecticides provide has played a big role in boosting agricultural yield, human health, and environmental safety. Nevertheless, frequent and random use of these chemicals has resulted in several issues causing the exploration of alternate measures. The amalgamation of these measures led to the concept of integrated pest management, which helped to overcome the constraints and challenges associated with single and conventional chemical interventions. It is considered a more efficient, cost-effective, sustainable, and ecologically sound approach for insect pest control.

This book, *Insecticides in Pest Control – Impact, Challenges and Strategies* intends to highlight the features, use, and mode of action of insecticides belonging to different chemical groups; their efficacious use against various insect pests of agricultural importance; challenges associated with their impact on human health, nontarget organisms, and the environment; and recommendations and future perspectives. Apart from these, the eco-safe, innovative, and advanced control interventions in the form of physiological disruptors, and those acting at biochemical and molecular levels, have also been focused upon.

The book includes 17 chapters grouped under four sections, each covering a different aspect of insecticide use. Section 1 deals with the "Basics of Pest Management Strategies", which has two chapters. Section 2 covers "Insecticide in Pest Management: Advantages and Disadvantages", which comprises Chapters 3–7. Section 3 contains the next five chapters (8–12), which highlight "Biorational Strategies in Insect Pest Management", and Section 4 deliberates upon "Sustainable Studies for Efficacious Pest Management", which includes Chapters 13–17. Chapter 1 in Section 1 covers the basics of insect pest management in the agroecosystem, describing the pest-associated problems, potential solutions, and various

agricultural practices to address the

critical problem of pest management. Several management approaches, such as recognition of insect vulnerability, effective monitoring of pests and resistance, increasing insecticide diversity, adopting cultural and biological control interventions, and rotation of insecticides, are discussed. Chapter 2 deals with different kinds of phytophagous insects that damage the crops via their sucking mouth parts, management tactics against them, and a holistic use of these techniques for effective pest management. Section 2 is focused on the types, advantages, and disadvantages of insecticides used in pest management. Chapter 3 illustrates the chemical structure of various insecticides, their historical aspects and sources, and adverse effects on health, environment, climate, and so forth. The chapter also recommends the search for new agents, particularly from natural products of plant origin. The following Chapter 4 deliberates upon the assessment of insecticides, their role in integrated pest management (IPM), ways to select insecticides compatible with other control agents, and associated issues. Chapter 5 highlights the negative impacts of insecticides on agriculture and the environment due to their exorbitant use, development of resistance in insects, and advances in applications resulting in sustainability in agriculture. The next two chapters (6 and 7) describe the adverse effects of insecticides on human health and possible safety measures. Chapter 7 also explores the intrinsic toxicity of pesticides and their adverse environmental effects, contributory factors, suitability of risk mitigation measures, regulatory requirements and failures, and pesticide risk reduction strategy.

Section 3 focuses on the novel biorational measures of insect pest management, which can be safer alternatives to chemical insecticides. Chapter 8 deliberates upon the various classes of botanical insecticides, secondary metabolites, and their characteristics with possible modes of action. Chapters 9 and 10 emphasize the use of biopesticides in agroecosystems. The prospects of integrating emerging biopesticide technologies, such as nano-biopesticides and RNA-based biopesticides, into the existing IPM programs are also discussed in Chapter 9, while Chapter 10 briefly explains the types of biopesticides, their mode of action, and their application in vegetable production with an aim to produce zero-residue vegetables. Chapter 11 recommends the adoption of semiochemical-based behavioral manipulation techniques, while Chapter 12 describes the use of artificial intelligence (AI) to optimize pest control and minimize insecticide use, ensuring a more sustainable and environmentally friendly food production system.

Lastly, Section 4 presents the sustainable studies demonstrating effective pest management. Chapter 13 discusses a protocol developed for mitigating resistance in insects. Investigations displaying the efficacy of neem oil against the pests of the corn crop, Zea mays, have been presented in Chapter 14. Use of bacterial species isolated from tomatoes and beans, their molecular characterization using recent tools, and the potential ability to degrade carbamate pesticides are underscored in Chapter 15, while Chapter 16 discusses ways of mitigating the challenges of insecticide use in agricultural fields. At the end, Chapter 17 addresses the challenges posed by socioeconomic, environmental, and health factors associated with pesticide application and proposes VII a more sustainable approach integrating agroecology into crop protection strategies. It emphasizes the importance of ecological balance, the significance of social and cultural factors, and assessment of IPM effectiveness. I hope and believe that this book will serve as a handbook for students, researchers, and practitioners associated with the field of insects and inspire some future research ideas among them by recognizing potential research directions.

Sarita Kumar
Professor,
Department of Zoology,
Acharya Narendra Dev College,
New Delhi, India

Section 1

Basics of Pest Management Strategies

Chapter 1

Management of Agricultural Pests: Strategies and Constraints

Sarita Kumar, Aarti Sharma and Roopa Rani Samal

Abstract

Agriculture is an intensive activity which is harmed by the attack of multifarious pests. Several measures have been adopted to control these pests among which the use of chemical insecticides is the most frequently used measure. Despite the effectiveness of chemical insecticides, their extensive use has led to pest resistance, their entry in food chain, biomagnification, and adverse effects on human and environmental health. This represents a significant challenge as alternative control methods, such as cultural, biological, microbial, and transgenic approaches, though safer, are often more expensive and time-consuming. Without effective pest management strategies, diverse pest populations continuously and intensively damage crops leading to significant economic losses and threatening agricultural productivity and food security. Therefore, there is a constant search for newer and safer pest control tools, which are implemented after thorough field trials. This chapter highlights the problems associated with insect pests, the potential solutions, and various agricultural practices to address the critical problem of pest management.

Keywords: agro-ecosystem, insect pests, insecticide resistance, resistance mechanism, resistance management

1. Introduction

Crop production is a significant and intensive activity which is important for the economic development of a country and provides food and employment to inhabitants. However, during production, agriculturists face several issues, such as climate adversities, soil erosion, biodiversity loss, change in consumers' preferences, and above all loss of productivity due to occurrence of plant disease and pests. Among these, infestation by insect pests is one of the most concerning issues. Existence of crop pests not only affects the crop productivity but also causes a toll on the farmer's physical and mental health. It has been estimated that the pest attack leads to roughly 20–40% loss of crop production at global level [1].

Farmers use a variety of techniques and measures to save their crops, such as crop rotation, deep plowing, use of fallow lands, intercropping, altering the time of sowing as well as optimizing the time of harvesting among others. However, the application of chemical insecticides, being the quickest and one of the most effective methods, is used most frequently against these pests. Several insecticides have been used for crop pest management, which differ in their chemical nature and mode of action. These include lethal organic insecticides—organochlorines, organophosphates, carbamates, and pyrethroids; botanical insecticides; and the insect growth regulators (IGRs) which control their development by affecting chitin synthesis or hormonal system, etc. Each insecticide is used in specific formulation and concentration optimized based on their application, severity of pest attack, etc. Though agriculturists have also adopted various cultural, biological, and microbial approaches to control crop pests; the chemicals are used at a much larger scale because of their rapid action. This has made insect pests resistant and imposed negative effects on human and environmental health creating a great challenge in agriculture as other control interventions, though effective, are more expensive and slow in action.

Hence, newer and safer tools are continuously explored and implemented after conducting field trials for efficient pest control. In this chapter, we will discuss the problems associated with insect pests, the probable solutions, and various agricultural practices which can be drawn to solve this emerging issue.

2. Pest control in an agro-ecosystem

The efficacious control of agricultural pests is crucial to ensure the growth of healthy crops. Apart from chemical measures, several different approaches can be strategized to manage these pests, such as prevention of the pest introduction in the area, crop rotation, use of mechanical devices (traps, nets, etc.), and boosting natural parasites and predators of pests. In addition, prudent and adequate use of pesticides can not only help us to manage these pests sustainably, safely, and economically but also keep our environment healthy. A brief and comprehensive account of these measures is as follows.

2.1 Assess the pest status: do we need control measures?

The first and foremost requirement for any pest control in an agro-ecosystem is to assess the field and identify pest conditions. The assessment of pest density, knowing about their life cycles along with their stage of attack and damage potential, and the severity of attack, etc., are significant. The computation of economic thresholds (ET) and injury levels (EILs) can assist in estimating economic damage before taking any chemical interventions [2].

Furthermore, maintenance of the field records regarding type and timing of pest attacks, etc., is needed for efficient control. Such data will help us to evaluate the need, timing, and type of pest control method needed in that ecosystem. These assessments will help in selecting appropriate and effective pest control techniques which could be cost-effective, environment-friendly, and sustainable.

2.2 Manage crops: rotate and optimize timings

A few insect pests can be controlled by carrying out crop rotation. The crops with different bionomics, adaptability, and tolerance to insects and diseases can be grown in the region. For example, European corn borer (*Ostrinia nubilalis*) can be managed for several years without using corn crop in the rotation.

Removal of weedicides by timely mowing/tilling can help the provision of adequate nutrients to crops and increase resistance to pests. Likewise, timely sowing and timely harvesting of crops can minimize the productivity loss due to insect attack. However, before adopting such strategies, it is important to assess the agro-ecosystem for the occurrence of pests and their density.

2.3 Use resistant varieties: increase vigor and tolerance

Growing crops in fertile soil and providing crops with suitable and enough nutrients and water can increase their vigor and enhance pest tolerance. Use of resistant varieties of crops with better adaptability and heritable qualities to counter pest attack can also help to increase productivity [3]. These traits include odor, secondary metabolites, waxy layers, trichomes, etc. [4]. Several resistant varieties of crops have been developed which help to alleviate pest damage. Varieties having low to moderate levels of resistance to pests can be used in blend with other pest interventions. Varietal resistance has been demonstrated in minimizing losses because of insect pests in chickpea, pigeon pea, and sorghum.

2.4 Adopt healthy practices: sanitation and cleanliness

Use of healthy and resistant seeds, avoiding the cultivation of host plants of insects, adopting good sanitation, keeping field clean, reducing sources of nutrients by removal of weeds, stubbles, etc., can reduce pest attack.

2.5 Promote biological agents: use of pest's natural enemies

Biological control of pests is one of the highly effective, sustainable, and eco-safe approaches [5]. The parasites and predators of agricultural pests need to be encouraged and protected, which can reduce pest attack without harming our environment. Biopesticides formulated using microorganisms, such as *Bacillus*, *Beauveria*, *Metarhizium*, and *Pseudomonas*, are found to be effective in pest management. In North America, the European corn has been managed successfully, since 1996, by using transgenic crop formulated using *Bacillus thuringiensis* (Bt) [6].

2.6 Biorational volatiles: behavioral modifiers

The biorational and eco-safe volatiles are also used to control insect pests. These include insect pheromones and plant volatiles which alter behavior of insect pests. For example, application of sex pheromones in large amounts can disrupt the mating by impeding the ability of males to find females [7]. Likewise, pheromones can be used for mass trapping of insects or even repelling insects.

2.7 Integrated pest management: a harmonious approach

A holistic approach, called integrated pest management, is considered the best strategy to combat agricultural pests. **Figure 1** shows a pyramid of various pest control strategies—abiotic, ecological, and chemical, which can be used harmoniously as an integrated approach.

a. The base tier represents the *abiotic management* practices (cultural, mechanical, physical, etc.). These interventions can be used before or instantly post-sowing and may continue until the plant is harvested.

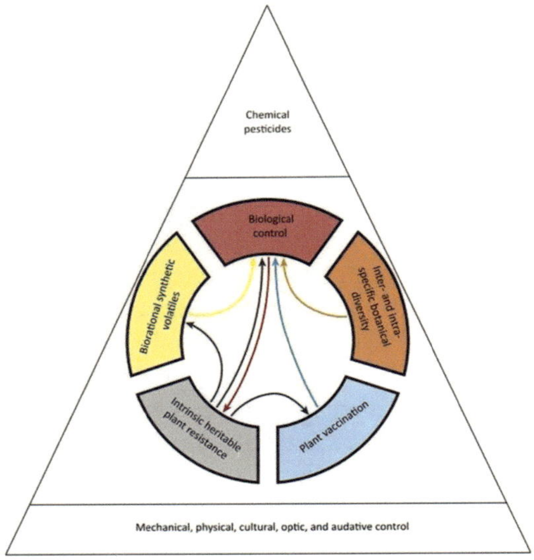

Figure 1.
Various components of pest management [3].

b. The middle circular tier depicts *ecological management* consisting of biological, genetic, eco-safe, and other important ecological elements. The interactions and interrelationships between these components are denoted by the arrows.

c. The top tier defines *chemical management* which should be avoided and only be used in the case of failure of other components, as it may cause harm to the environment and non-target organisms.

3. Chemical-based interventions for pest control

All the methods discussed in the previous section, though effective and eco-friendly, are slow in action. Hence, chemical intervention is regarded as the most effective approach with swift action. Several bioactive chemicals have been utilized for preparation of insecticides against crop pests. These synthetic insecticides contain different chemical constituents based on which they have been grouped as organo-chlorines, organophosphates, carbamates, pyrethroids, etc. Apart from these, insect growth regulators (IGRs) have been developed to control the insect pests. These regulate their development by either inhibiting chitin synthesis (chitin synthesis inhibitors, CSI) or disrupting level of hormones.

Major classes of insecticides and their mode of actions (**Figure 2**) are listed in **Table 1**.

However, frequent application of haphazard doses of chemicals in the field has created havoc in the agro-ecosystem. In addition, development of insecticide resistance has resulted in the over and repeated use of these chemicals in the fields. These chemicals have contaminated not only our water and soil resources, but also their entry in food chain and resultant biomagnification have caused various issues. Negative effects on the life and physiology of beneficial organisms, environmental health, and human well-being are matters of high concerns.

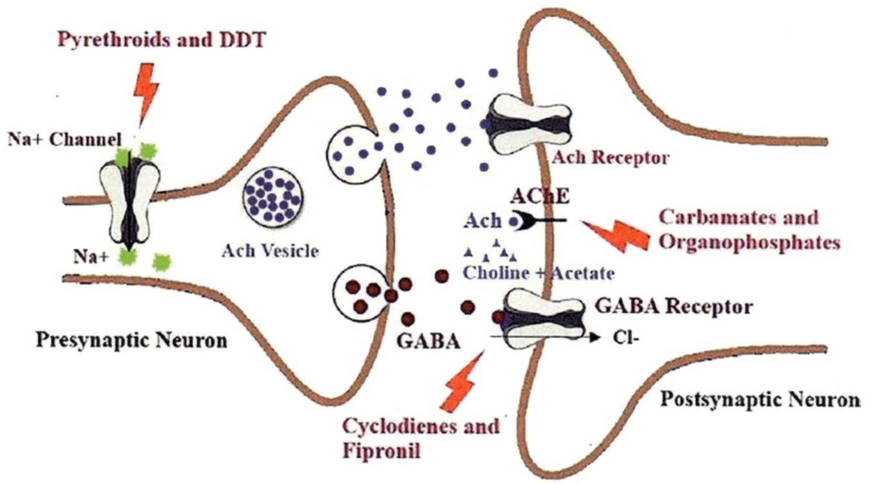

Figure 2.
Mode of action of a few insecticides [8].

Group of insecticides	Mode of action
Organochlorines (OCl^-)	Prevent nerve transmission by (a) affecting Na^+/K^+ pump (DDT) or (b) preventing entry of Cl^- in the neurons by disrupting GABA (γ-aminobutyric acid) receptors (Cyclodienes). Cause repetitive after discharge (RAD), hyperexcitability, ataxia, convulsions, trembles, and death. Broad-spectrum use.
Organophosphates (OPO_4^{3-})	Inhibit acetylcholinesterase (AChE) resulting in the accretion of acetylcholine at the neuromuscular junctions. Cause rapid twitching of muscles, paralysis, and death. Broad-spectrum use.
Carbamates	Same as of organophosphates. Immediate knockdown effect. Broad-spectrum action.
Dinitrophenols	Prevent ATP formation by acting as either uncoupler or inhibitor of oxidative phosphorylation.
Organotins	Prevent ATP formation by inhibiting phosphorylation at the site of uncoupling by dinitrophenols. Used on fruit trees to control mites.
Pyrethroids	Cause hyperexcitability by keeping the Na channels in the neurons open. Lead to ataxia, tremors, confusion, paralysis, and death. Used effectively against most of the crop pests.
Nicotinoids	Neurotoxic compounds which block postsynaptic nicotinic acetylcholine receptors irreversibly. Used against several crop pests, such as Colorado potato beetle, whiteflies, sucking insects, termites, and soil insects.
Spinosyns	Interrupt acetylcholine binding in the nicotinic acetylcholine receptors at the postsynaptic cell. Used to manage caterpillars, leaf miners, and thrips.
Pyrazoles	Inhibit electron transport in mitochondria at the NADH-CoQ reductase site and disrupt the formation of ATP. Effective against aphids, whitefly, psylla, and thrips.
Quinazolines	Chitin synthesis inhibitors. Act on the developing larvae causing rupture of the deformed and distorted cuticle or death by starvation.
Botanicals	a. Pyrethrum—nerve poisons causing repetitive discharges and paralysis. Used against various pests. b. Nicotine—mimics acetylcholine in the ganglia of central nervous system. Causes trembling, convulsions, and mortality. Used against aphids and caterpillars. a. Azadirachtin—antifeedant and anti-moulter. Inhibits metabolism/biosynthesis of ecdysone. Used against caterpillars.

Group of insecticides	Mode of action
Benzoylureas	Chitin synthesis inhibitors, inhibits insect growth. Used to control caterpillars, grubs, gypsy moth, and mushroom fly.

Source: Available from: https://www.epa.gov/caddis/insecticides

Table 1.
Types of insecticides against agricultural pests and their modes of action.

4. Insecticide resistance in crop pests

The insecticide resistance was first demonstrated in the scale insects which became resistant to inorganic insecticides [9]. Subsequently, documentation of more such cases led to the development of DDT and other organic insecticides raising a hope which vanished with the appearance of DDT resistance. Several new insecticides were then formulated, but ironically insects could develop resistance to each and every one, whether slowly or rapidly.

The researches revealed that the major cause of resistance development included intensive use of insecticides and selection of resistance genes through generations. The survival of resistant insects increased their proportion in a population while eliminated the susceptible insects. Consequently, the insects developed immunity against those insecticides making them ineffective. Nevertheless, the speed of resistance development is determined by various factors, such as frequency and dosage of insecticide, reproductive fitness of insect pests, the resistance level, host plants, and chemical nature of the insecticide—degradability, resilience, and specificity.

5. Mechanism of insecticide resistance

The insect pests have developed several ways to develop resistant to agrochemicals (**Figure 3**).

a. *Behavioral resistance*: the pests avoid these chemicals and protect themselves from the toxic chemicals. They may exhibit antifeedant behavior or just move away from the site of application.

b. *Penetration resistance*: frequently, cuticle of resistant insects may become thick and act as barrier to insecticides—prevent their entry or absorb them slowly in their body. This helps the insects to survive the toxins used in the fields. Nevertheless, this mechanism cannot provide sufficient levels of resistance and need to be supplemented by other mechanisms.

c. *Metabolic resistance*: this is the most frequently developed and most efficient mechanism of resistance in insect pests. The resistant insects develop the ability to detoxify the toxins much more rapidly than the susceptible insects. In order to do that, they alter their biochemical machinery and metabolize the insecticides rapidly before experiencing harm.

d. *Altered target-site resistance*: This is another common mechanism of resistance where the modification in the target site of insecticide results in its inefficacy.

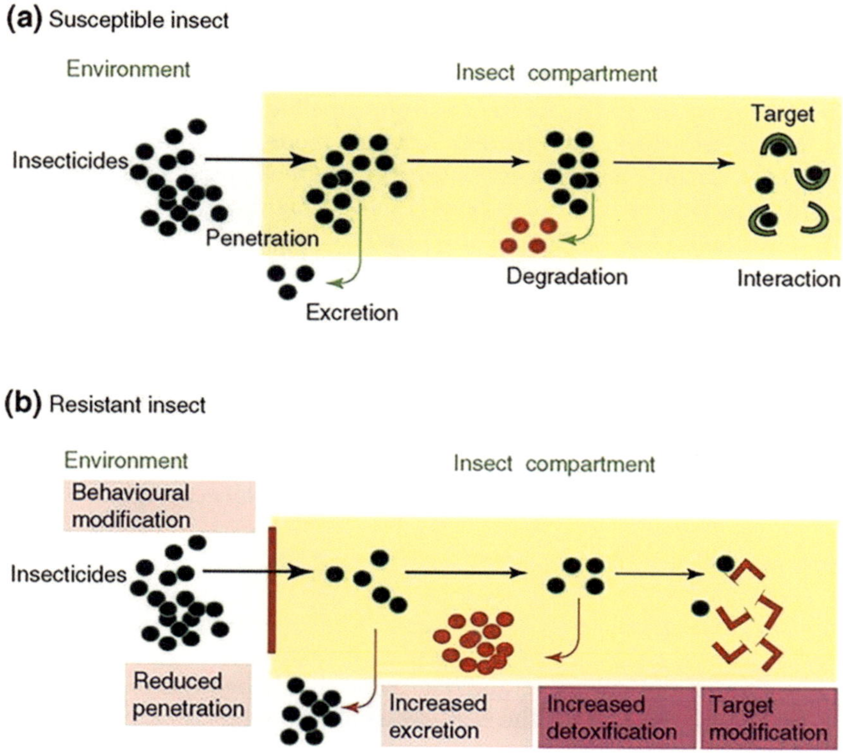

Figure 3.
Different mechanisms of insecticide resistance in insects [10].

6. How can we prevent insecticide resistance?

The effects of resistance to pesticides in agriculture are easy to predict, as insects that survive exposure to insecticides continue to consume crops and lower the productivity. Moreover, underlying evolutionary pressures and resistance mechanisms can often be shared, particularly when mutations within genes, such as the sodium ion channel, are involved for agricultural pests and disease vectors [11].

Prevention of the insecticide resistance or delaying the emergence of resistance is the best approach to manage resistance in agricultural pests. Reducing the amount of pesticide treatments and diversifying the types of insecticide used have been key components of practical accomplishments in resistance management, even though models have revealed a variety of measures with potential to delay resistance. It is evident that for each strategy to be effective, it has to highly rely on the unique characteristics of the pest-crop system. This is in respect to factors such as insect behavior, genetics, dynamics of population, chemical properties of the insecticides, and insecticide formulations.

6.1 Reducing insecticide use

The more insecticides are used, the higher is the pressure for resistance to insecticides. The dose and frequency of insecticide used against any pest is a very important factor that can affect the pace of the development of resistance. The advent of new insecticides, however, poses a serious challenge in agriculture. The large size of the market, with an estimated annual turnover value of US$8 billion, often ushers in new insecticides with multiple uses in the sector of agricultural chemicals, as it enables innovation to take place and guarantees a strong pipeline of insecticides [12]. In fact, many of those insecticides, or with a similar action mode, continue to be used over crops. Evidence has indicated that agriculture undoubtedly plays a major role in the selection for insecticide resistance in pests [13, 14]. Programs in Australia, Israel, and the USA, for instance, have restricted how often and for how long a given insecticide can be applied to control cotton pests [15].

In order to rationally select pesticides, effective resistance management necessitates early problem diagnosis, quick information assimilation about the resistant insect population, innovatory delivery methods, new active ingredients, and non-insecticide-based control measures. Use of integrated pest management that diversifies control measures beyond insecticides, together with surveillance, and monitoring of pests can be a core key approach providing sustainable solution [16–18]. This can reduce the use of pesticides in the environment providing an eco-safe solution.

6.2 Monitor pests

Monitoring and surveillance of pests is one of the crucial strategies in implementing the insecticide resistance management plan. It helps to strategize whether there is a need to apply control measures and if so, what should be the time of insecticide application. The continuation of the pest assessment and interventions would help to manage resistance.

6.3 Assessment of economic thresholds and injury levels

The IPM approach recommends the use of insecticides only when insect population crosses the economic threshold levels and causes economic damage to the agriculture. Thus, conducting assessments of the threshold and injury levels is an important component while applying insecticides and manage resistance.

6.4 Time of insecticide application

Insecticides should be applied when the pests are newly emerged or the most susceptible. The frequency, area covered, and dosage should be appropriate and recommended by experts; otherwise, it can increase resistance in the pests.

6.5 Monitor insecticide resistance

An essential component of pest management is insecticides' resistance monitoring. The monitoring data can give relevant information to direct the application of efficient pest management techniques. This approach provides both proactive and reactive monitoring programs for evaluation of resistance. The proactive monitoring programme identifies initial indications of resistance emergence caused by the utilization of a specific product. It assesses insect sensitivity to active insecticidal ingredients at the population level and tracks product performance over time at a specific site to prevent failure. Such program includes symmetrical field surveys of product performance and/or systematic testing of field-collected populations. Whereas, a reactive monitoring programme first identifies whether resistance evolution is the cause of declining field performance and then puts an appropriate mitigation plan into place. It detects and reports decreasing efficacy of insecticides in the field. To confirm resistance, insects are gathered from the control failure area and examined using a suitable bioassay approach against known sensitivity baselines. Confirmation of resistance may require additional regional management and mitigation actions.

Resistance management techniques can be assigned at random, although these trials are very labor- and time-intensive, expensive, and logistically challenging [19]. Randomized controlled trials (RCTs) are considered gold standard in order to establish the impact of an intercession; nevertheless, it is not practical to indiscriminately assign resistance levels across treatments. These findings point to a significant knowledge vacuum regarding the circumstances, mechanisms, and motivations behind the potential success of these tactics.

6.6 Increasing insecticide diversity

Use of diverse insecticides belonging to the different chemical classes or pesticide families with different modes of action, and application in different localities at different times, can also supplement the approach of reduced reliance on insecticides in order to alleviate the spread of resistance to the insecticide [10]. Of course, having a sufficient supply of pesticides with different modes of action is essential to using alternate methods to prevent or postpone the emergence of resistance in insect populations. While this approach might not be the ideal one, it does allow a nuisance to be contained until more effective management techniques are created and applied to combat the pest. Pesticides are frequently utilized in these tactics; however, they are occasionally and never used at reduced application rates [10].

This may be done by means of using mixtures (two or more insecticides combined with distinct modes of action) or rotation of insecticides (periodic interchanges of different insecticides), use of synergists or additives (enhancing efficacy of insecticides at lower dosages), and application of insecticides as mosaics (use of different insecticides in close by geographical areas). These strategies have received the attention of extensive empirical and theoretical researchers in agricultural contexts [20–23].

It is believed that the probability of developing resistance, in a pest population, to a single insecticide will always be more than the chances of resistance to two or more insecticides containing different chemical constituents and imparting different modes of action. Hence, insecticide mixtures are based on redundancy killing, where if the insect had developed resistance to one of the components, the other one would still be effective. Tank mixing is the process of combining two or more pesticides with distinct mechanisms of action to enhance the effectiveness of each pesticide application and prevent or lessen the emergence of new insect resistance. A review covering xenobiotics showed that combination interventions, such as mixtures of antimicrobial drugs and insecticides, often matched or exceeded alternative approaches in efficacy [24].

Similarly, pesticide rotation is a typical strategy for managing pest resistance when pesticides are the primary or only method of controlling pests. This entails switching between classes of pesticides that have distinct modes of action in order to prevent or lessen the emergence of insect resistance [25]. A pest may respond differently to different pesticide classes. A successful program of resistance management for whiteflies was realized in the USA and Israel by conducting a set rotation program which lessened the insecticide treatment frequency/year and restricted the application of multiple insecticides with similar chemical properties [26, 27]. The basic strategy of specific insecticides being used in response to reported resistance had remained unchanged through time-to-time adjustments. Generally, reactive alternation of insecticides with the resistance tends to be less effective than predetermined rotations [24]. The best strategy may be the use of resistance monitoring as a guide to what insecticides can be rotated, rather than determining when they are used [28].

There may be, however, diverse biological and operational constraints in adopting such strategies. First of all, the development of such co-formulations could be a difficult task, and the advantage of a mixture would be conferred only if both component insecticides remain efficacious. Hence, this limits the usefulness of mixtures in overcoming resistance by being effective only if both the partner insecticides remain effective. Further, for such mixtures to be effective, the resistance alleles in the insects should be low and entirely recessive to avoid high selection pressures for individuals carrying double heterozygous traits, which could initiate rapid resistance evolution against both insecticides [29, 30].

6.7 Adopt integrated approach

The identification of resistance in its nascent stages needs more efficient methods for resistance management. Reducing the selection pressure that a pesticide applies to the pest population can help control pest resistance to that pesticide. Stated differently, the scenario in which a particular chemical eradicates only susceptible bugs and saves the most resilient ones ought to be circumvented. This can be accomplished by employing non-chemical management methods, reducing the needless use of pesticides, and allowing untreated areas as havens for vulnerable bugs to live [31]. In other words, resistance management is a part of integrated pest management, which blends chemical and non-chemical measures in an effort to lower pest populations in a way that is secure, affordable, and long-lasting [32]. In fact, the comprehensive IPM deployment should not be applied only in those agro-systems where the evidence for existing insecticide resistance exists. In other words, a resistance monitoring program should no longer be limited to evaluating an insecticide's reaction with the goal of moving on to a different chemical when resistance levels beyond a threshold that compromises disease management.

It has been stated that developing strategies for resistance management has mostly relied on modeling, as large-scale resistance tests are costly, time-consuming, and have the potential to exacerbate resistance issues [33]. Modeling studies have pointed out the sensitivity of the results toward systemic-specific parameters [34, 35]. This leaves no single strategy supreme, and resistance management plans remain mostly specific for the given pest and crop species. For example, "the strategy may change with pest-crop combinations, such as the potato beetle in the USA, or when dealing with pollen beetles in the oilseed rape fields in Europe" [36].

7. Ensuring effective monitoring of resistance

Currently, the monitoring of pest resistance to insecticides is highly overloaded by uncertainties relating to what needs to be measured and how it should be measured. Molecular markers present an attractive choice for resistance monitoring because of their potential for a faster throughput and finer resolution of resistance patterns. It allows for the further conservation of the insects, to be used for studies, and at the same time, to make molecular tests more easily standardized than the phenotypic bioassays. At times, the association between resistance phenotype and genotype is intricate and poorly understood.

In agricultural pests, knockdown resistance (*kdr*) is one of the extensively investigated target-site mutations. For example, *Leptinotarsa decemlineata* (Colorado potato beetle—CPB) is a severe crop pest that has rapidly developed resistance to most of the insecticides. In these, *kdr* has been recognized as the chief mechanism of resistance to pyrethroids, caused by a polymorphism in single nucleotide (SNP—L1014F) in the voltage-sensitive sodium channel gene (LdVssc1). In order to investigate and find out the occurrence of single or multiple evolutionary origin of *kdr* in these insects, a total of 96 larvae were collected from ten places (nine in the USA and one in Bulgaria). The partial genomic sequences for LdVssc1 were acquired from each organism. The related intron, located in three base pairs downstream from the mutation site, was exploited to identify novel resistant (F1014) and susceptible [37].

Similarly, *kdr* has been found as a primary mechanism against pyrethroids in several other arthropod pests. The onion thrips, *Thrips tabaci*, collected from fields of Washington, USA, have shown significant variance in resistance to λ-cyhalothrin. Researchers found two SNPs in a partial sequence of the *T. tabaci* voltage-gated sodium channel gene. One mutation resulted in a non-synonymous substitution of methionine to leucine (M918L), which is widely known to cause super*kdr* in some pest species. Another mutation caused non-synonymous replacement of valine with alanine at amino acid 1010 (V1010A), which has been discovered and linked to λ-cyhalothrin resistance [38].

Nevertheless, phenotypic bioassays have their own set of challenges and have been pivotal in determining successful resistance management strategies in agriculture [26, 39]. The WHO bioassays are standard, and the results from test to test on the same population have shown large fluctuations [40]. In part due to this, it has been more difficult to identify relationships with bioassay outcomes and other variables.

Another pragmatic constraint is a detectable signal of resistance. Insecticide resistance alleles are likely to be rare, at least in theory. Monitoring will need to be very highly sensitive and able to detect small resistance-allele frequencies. A practical threshold of such monitoring that has been proposed in agricultural literature is that the monitoring should be capable of identifying resistant individuals at frequencies of 1% or less [41]. It is, however, hard to achieve sufficient sensitivity. In many cases, resistance can be detected only when it is too late to implement management measures that could have stopped the spread of resistant strains. Nonetheless, the pitfalls of resistance monitoring should never stand in the way of implementing the insecticide resistance management strategies. They need to be practically applied even more if the strategies to be adopted do not hinge on early resistance detection.

8. Fostering the development of novel control tools

The appearance of insecticide resistance in pest population and adverse effects of chemicals on environment and human health have made it necessary to explore novel tools for pest management. The most recent technology to find its way into commercial crop production is the use of transgenic plants that produce insecticidal proteins to naturally repel insects. Because the pesticidal proteins are specific to a particular target pest, populations of beneficial and other non-target insects are protected.

Though all members of a pest population may be exposed to the insecticidal toxins for an extended period, transgenic plants theoretically have a larger potential than conventional insecticides to promote the development of resistance.

One of the upcoming technologies for crop pest management is the use of sprayable RNAi [42]. It offers a novel way to manage pests by combining the convenience of spray application with a molecular targeting strategy employed in transgenic plants, which opens new modes of action and selectivity. Therefore, sprayable RNAi serves as a conceptual link between GMO crops, traditional chemical control methods, and the more traditional sprayable Bts (*Bacillus thuringiensis*). The goal of each of the strategies is to lower pest pressure on a crop below economic damage levels. However, despite the abundance of more recent methods, chemical pesticides are still the principal instrument for many particular "crop pest locations" situations. As a result, maintaining the long-term efficacy of both new and existing pesticides is crucial because it may be difficult to find new, replacement pesticides due to delays, rising costs, and stricter regulatory requirements [43].

IRM techniques for use with transgenic insect-protected crops are required and approved by local regulatory bodies in nations like the USA and Australia. For instance, a high dosage + refuge strategy has been used for the cultivation of lepidoptera-protected *Bt* maize (which expresses the Cry1 protein from the common soil bacteria *Bacillus thuringiensis*). This approach is predicated on the notion that resistance is a recessive characteristic and that resistance alleles have a low initial frequency. Transgenic crop plantings usually require a refuge area planted with a specific percentage of a traditional, non-GM variety of the same crop. It is very possible that any rare homozygous-resistant individuals from the transgenic variety will mate with any of the far more numerous susceptible individuals from the refuge area. The transgenic variety's high expression level kills the resulting heterozygotes [44]. Recessive inheritance of the resistance alleles, low initial frequencies of those alleles, and the fitness costs linked to resistance are among the factors that have probably contributed to the success of this strategy [45].

However, IRM techniques need support from the entire community. If only a small portion of the society implements the high dose/refuges, they will probably have little effect and will come at a comparatively high cost. This leads to a tragedy of the commons because farmers must put the long-term benefit of protecting the pool of vulnerable insects ahead of the short-term, personal benefit of growing just *Bt* crop. Without any outside pressure, it is challenging for farmers to comply with the management strategy because planting a refuge comes at a cost [46]. In the USA, the Environmental Protection Agency (EPA) requires all *Bt* crops to be grown in refuges, including their size and location, but it is up to the companies that produce and market the products to make sure farmers comply [47, 48].

To prevent future insecticides from rapidly losing their effectiveness due to resistance, it is necessary to assess IRM tactics intended to extend the operational lifespan of insecticides. This is particularly critical considering the novel pesticides that the firms are developing. The multitude of potential IRM tactics, the requirement for replication across various ecological settings, and the extended research durations required to identify phenotypic changes in resistance make evaluation of IRM strategies in the lab and field difficult [49]. While there are many agricultural examples where government regulation has been effective in enforcing good management practices, it is important to remember that non-governmental methods, like community enforcement of sustainable use practices, can also be employed to preserve a public good like susceptibility.

9. Conclusions

Use of insecticides for management of agricultural pests, though most effective, has resulted in several issues. One of the major issues, development of resistance has resulted in use of higher doses and frequent application of these chemicals. The goal of IRM is to limit or postpone the development of pesticide resistance which can help in the effective pest control. Management of resistance requires recognition of insect vulnerability, effective monitoring of pests and resistance, increasing insecticide diversity, and adopting cultural and biological control interventions among various strategies of IPM. Among the pesticide-based strategies, rotation of insecticides with different modes of action is one of the most straightforward and easy-to-perform approaches, requiring the least amount of assumption. It has become necessary to create suitable regulatory frameworks to implement these and other novel upcoming genetics-based control tools.

Author details

Sarita Kumar[1*], Aarti Sharma[2] and Roopa Rani Samal[3]

1 Department of Zoology, Acharya Narendra Dev College, University of Delhi, India

2 School of Biological and Life Sciences, Galgotias University, India

3 Department of Zoology, Dyal Singh College, University of Delhi, India

*Address all correspondence to: saritakumar@andc.du.ac.in

References

[1] USDA-National Institute of Food and Agriculture (USDA). Researchers Helping Protect Crops from Pests 2023. Available from: https://www.nifa.usda.gov/about-nifa/blogs/researchers-helping-protect-crops-pests [Accessed: June 11, 2024]

[2] Weersink A, Deen W, Weaver S. Defining and measuring economic threshold levels. Canadian Journal of Agricultural Economics/Revue Canadienne d'Agroeconomie. 1991;**39**(4):619-625

[3] Stenberg JA. A conceptual framework for integrated pest management. Trends in Plant Science. 2017;**22**(9):759-769

[4] Stenberg JA, Muola A. How should plant resistance to herbivores be measured? Frontiers in Plant Science. 2017;**8**:254269

[5] Eilenberg J, Hajek A, Lomer C. Suggestions for unifying the terminology in biological control. Biological Control. 2001;**46**:387-400

[6] Sappington TW. Critical facets of European corn borer adult movement ecology relevant to mitigating field resistance to Bt-corn. Insects. 2024;**15**(3):160

[7] Witzgall P, Kirsch P, Cork A. Sex pheromones and their impact on pest management. Journal of Chemical Ecology. 2010;**36**:80-100

[8] Shamjana U, Grace T. Review of insecticide resistance and its underlying mechanisms in *Tribolium castaneum*. In: Insecticides-Impact and Benefits of its Use for Humanity. London, UK: IntechOpen; 2021

[9] Melander AL. Can insects become resistant to sprays? Journal of Economic Entomology. 1914;**7**:167

[10] Corbel V, N'Guessan R. Distribution, mechanisms, impact and management of insecticide resistance in malaria vectors: A pragmatic review. In: *Anopheles* Mosquitoes-New Insights into Malaria Vectors. London, UK: IntechOpen; 2013

[11] Soderlund DM, Knipple DC. The molecular biology of knockdown resistance to pyrethroid insecticides. Insect Biochemistry and Molecular Biology. 2003;**33**(6):563-577

[12] Hemingway J. The role of vector control in stopping the transmission of malaria: Threats and opportunities. Philosophical Transactions of the Royal Society B: Biological Sciences. 2014;**369**(1645):20130431

[13] Curtis CF. Control of malaria vectors in Africa and Asia. In: Radcliffe EB, Hutchison WD, Cancelado RE, editors. Radcliffe's IPM World Textbook, 143 Saint Paul. Minnesota: University of Minnesota; 1996. Available from: http://ipmworld.umn.edu/chapters/curtiscf.htm [Accessed: June 12, 2024]

[14] Chouaïbou MS, Fodjo BK, Fokou G, Allassane OF, Koudou BG, David JP, et al. Influence of the agrochemicals used for rice and vegetable cultivation on insecticide resistance in malaria vectors in southern Côte d'Ivoire. Malarial Journal. 2016;**15**(1):426

[15] Zhang J, Wedegaertner T. Genetics and breeding for glandless upland cotton with improved yield potential and disease resistance: A review. Frontiers in Plant Science. 2021;**12**:753426

[16] Huffaker CB, Croft BA. Integrated pest management in the US: Progress and promise. Environmental Health Perspectives. 1976;**14**:167-183

[17] Owen MD, Beckie HJ, Leeson JY, Norsworthy JK, Steckel LE. Integrated pest management and weed management in the United States and Canada. Pest Management Science. 2015;**71**(3):357-376

[18] Rathee M, Singh NV, Dalal PK, Mehra S. Integrated pest management under protected cultivation: A review. Journal of Entomology and Zoology Studies. 2018;**6**(2):1201-1208

[19] Penilla RP, Rodríguez AD, Hemingway J, Trejo A, Lopez AD, Rodríguez MH. Cytochrome P 450-based resistance mechanism and pyrethroid resistance in the field *Anopheles albimanus* resistance management trial. Pesticide Biochemistry and Physiology. 2007;**89**(2):111-117

[20] MacDonald RS, Surgeoner GA, Solomon KR, Harris CR. Development of resistance to permethrin and dichlorvos by the house fly (Diptera: Muscidae) following continuous and alternating insecticide use on four farms. The Canadian Entomologist. 1983;**115**(12):1555-1561

[21] Pimentel D, Burgess M. Effects of single versus combinations of insecticides on the development of resistance. Environmental Entomology. 1985;**14**(5):582-589

[22] Tabashnik BE. Managing resistance with multiple pesticide tactics: Theory, evidence, and recommendations. Journal of Economic Entomology. 1989;**82**(5):1263-1269

[23] Immaraju JA, Morse JG, Gaston LK. Mechanisms of organophosphate, pyrethroid, and DDT resistance in citrus thrips (Thysanoptera: Thripidae). Journal of Economic Entomology. 1990;**83**(5):1723-1732

[24] Rex Consortium. Heterogeneity of selection and the evolution of resistance. Trends in Ecology and Evolution. 2013;**28**(2):110-118

[25] Stehle S, Ovcharova V, Wolfram J, Bub S, Herrmann LZ, Petschick LL, et al. Neonicotinoid insecticides in global agricultural surface waters–exposure, risks and regulatory challenges. Science of the Total Environment. 2023;**867**:161383

[26] Denholm I, Cahill M, Dennehy TJ, Horowitz AR. Challenges with managing insecticide resistance in agricultural pests, exemplisfied by the whitefly *Bemisia tabaci*. Philosophical Transactions of the Royal Society of London. Series B: Biological Sciences. 1998;**353**(1376):1757-1767

[27] Palumbo JC, Horowitz AR, Prabhaker N. Insecticidal control and resistance management for *Bemisia tabaci*. Crop Protection. 2001;**20**(9):739-765

[28] Sparks TC, Nauen R. IRAC: Mode of action classification and insecticide resistance management. Pesticide Biochemistry and Physiology. 2015;**121**:122-128

[29] Curtis CF. Theoretical models of the use of insecticide mixtures for the management of resistance. Bulletin of Entomological Research. 1985;**75**(2):259-266

[30] Tabashnik BE. Modeling and Evaluation of Resistance Management Tactics Bruce E. Tabashnik: Pesticide Resistance in Arthropods; 1990

[31] Siddiqui JA, Fan R, Naz H, Bamisile BS, Hafeez M, Ghani MI, et al. Insights into insecticide-resistance mechanisms in invasive species: Challenges and control strategies. Frontiers in Physiology. 2023;**13**:1112278

[32] Dusfour I, Vontas J, David JP, Weetman D, Fonseca DM, Corbel V, et al. Management of insecticide resistance in the major *Aedes* vectors of arboviruses: Advances and challenges. PLoS Neglected Tropical Diseases. 2019;**13**(10):e0007615

[33] Das TK, Kaur R, Singh R, Shekhawat K, Choudhary AK. Weed Management. In: Division of Agronomy. New Delhi, India: ICAR-Indian Agricultural Research Institute; 2017. p. 44

[34] Lenormand T, Raymond M. Resistance management: The stable zone strategy. Proceedings of the Royal Society of London Series B: Biological Sciences. 1998;**265**(1409):1985-1990

[35] Slater R, Stratonovitch P, Elias J, Semenov MA, Denholm I. Use of an individual-based simulation model to explore and evaluate potential insecticide resistance management strategies. Pest Management Science. 2017;**73**(7):1364-1372

[36] Obermeier C, Mason AS, Meiners T, Petschenka G, Rostás M, Will T, et al. Perspectives for integrated insect pest protection in oilseed rape breeding. Theoretical and Applied Genetics. 2022;**135**(11):3917-3946

[37] Rinkevich FD, Su C, Lazo TA, Hawthorne DJ, Tingey WM, Naimov S, et al. Multiple evolutionary origins of knockdown resistance (kdr) in pyrethroid-resistant Colorado potato beetle, *Leptinotarsa decemlineata*. Pesticide Biochemistry and Physiology. 2012;**104**(3):192-200

[38] Wu M, Gotoh H, Waters T, Walsh DB, Lavine LC. Identification of an alternative knockdown resistance (kdr)-like mutation, M918L, and a novel mutation, V1010A, in the *Thrips tabaci* voltage-gated sodium channel gene. Pest Management Science. 2014;**70**(6):977-981

[39] Alyokhin A, Mota-Sanchez D, Baker M, Snyder WE, Menasha S, Whalon M, et al. The red queen in a potato field: Integrated pest management versus chemical dependency in Colorado potato beetle control. Pest Management Science. 2015;**71**(3):343-356

[40] Badolo A, Traore A, Jones CM, Sanou A, Flood L, Guelbeogo WM, et al. Three years of insecticide resistance monitoring in *Anopheles gambiae* in Burkina Faso: Resistance on the rise? Malaria Journal. 2012;**11**:1-11

[41] Roush RT, Miller GL. Considerations for design of insecticide resistance monitoring programs. Journal of Economic Entomology. 1986;**79**(2):293-298

[42] Xue XY, Mao YB, Tao XY, Huang YP, Chen XY. New approaches to agricultural insect pest control based on RNA interference. Advances in Insect Physiology. 2012;**42**:73-117

[43] Sparks TC. Insecticide discovery: An evaluation and analysis. Pesticide Biochemistry and Physiology. 2013;**107**(1):8-17

[44] IRAC. Prevention and Management of Insecticide Resistance in Vectors of Medical Health Importance. Vol. 1. Manual by Insecticide Research Action Committee; 2011. pp. 1-72

[45] Tabashnik BE, Carrière Y, Dennehy TJ, Morin S, Sisterson MS, Roush RT, et al. Insect resistance to transgenic Bt crops: Lessons from the laboratory and field. Journal of Economic Entomology. 2003;**96**(4):1031-1038

[46] Sternberg ED, Thomas MB. Insights from agriculture for the management of

insecticide resistance in disease vectors. Evolutionary Applications. 2018;**11**(4):404-414

[47] Dove A. Survey raises concerns about Bt resistance management. Nature Biotechnology. 2001;**19**(4):293-295

[48] Bates SL, Zhao JZ, Roush RT, Shelton AM. Insect resistance management in GM crops: Past, present and future. Nature Biotechnology. 2005;**23**(1):57-62

[49] Hobbs NP, Weetman D, Hastings IM. Insecticide resistance management strategies for public health control of mosquitoes exhibiting polygenic resistance: A comparison of sequences, rotations, and mixtures. Evolutionary Applications. 2023;**16**(4):936-959

Chapter 2

Management Guide of Sucking Insect Pest

Abhijit Ghosal

Abstract

Sap feeders have typically piercing sucking mouthparts through which they suck sap from the plant, and due to this feeding, the plant produces specific symptoms. Their role as transmitting plant viruses is well known and considered a very important factor in achieving a desirable yield. Their small size, ability to develop quick resistance and biotypes make them very difficult to manage. Insecticide is considered the most potent weapon to cope with the sap feeders because of its quick knockdown effect and ease of application. At the same time, its indiscriminate use makes the situation harsh, leading towards the disruption of several ecological phenomena. We have seen a significant increase in honey bee colony collapse disorder in recent years, owing to extensive crop coverage with neonicotinoids. Similarly, the emergence of whitefly and jassids in Bt cotton as the dreadest pests is a result of not following the conservation of refuge strategy. As the crop ecosystem is a centre of multiple interactions between several biotic organisms along with abiotic factors, holistic approaches need to be developed combining cultural, mechanical, physical, biological, biopesticidal, and chemical management incorporating the host plant resistance.

Keywords: bio-pesticide, biological control, cultural control, Hemiptera, insecticide, mechanical control, host plant resistance

1. Introduction

Sucking pests, as the name hints, incorporate a wide range of arthropods (mostly members of classes Insecta and Arachnida) that do feed on the cell sap, though pierced with slender, sharp-pointed mouthparts. Among herbivorous insects, those considered as potential crop pests 75% of the known polypagous, oligophagous, and monophagous herbivores are considered sucking insects [1]. Among the different orders of class insecta associated with crop loss hemiptera accounts the largest apart from thysanoptera. The order hemiptera is divided in four sub order Auchenorrhyncha, Sternorrhyncha, Coleorrhyncha and heteroptera [2]; whose members are frequently referred to as "true bugs" (though restricted to the suborder Heteroptera) and are regarded as one of the largest and most important orders. They are tiny to medium-sized, ranging from 1 mm (0.04 in) to around 15 cm (6 in) of true hemimetabolous insects with piercing-sucking mouthparts. They share diverse habitats, while most are terrestrial and some share aquatic habitats (Gerromorpha and the Nepomorpha). Phytophagous hemipterans are found in large numbers in the leaves, stems, or bark of plants and roots also. Some genuses are gall producers; they generally live in galls. Classical examples of hemipteran sap-sucking pests include whiteflies (Aleyrodidae), leafhoppers (Cicadellidae), planthoppers (Delphacidae), aphids (Aphidoidea), mealy bugs (Pseudococcidae), mirid bugs (Miridae), stink or shield bugs (Pentatomidae), psyllids (Psyllidae), scale insects (Coccidae), cicadas (Cicadidae), and many more. Due to feeding from the xylem and phloem, it causes several external symptoms in crop phenology like curling and yellowing/browning of leaves, stunted growth, deformed fruit, wilting in severe cases. Some of the members are important virus vectors responsible for transmitting large numbers of plant viruses. Members of Hemiptera have a unique ability to produce honey dew due to the excretion of sticky sugar-rich residues. This honeydew contains various floras of microbes on leaf surfaces that can lead to a sooty mould appearance [3], The presence of sooty mould is one of the important features in identifying the severity of infestation. The global spread of sucking pests is a major concern for crop production because the development of insecticide resistance in hemiptera is a common phenomenon [4]; while biological control methods are less adopted. Apart from that, resurgence is very well known in hemiptera, associated with sub-lethal doses of insecticides [5]. Not only is insecticidal exposure causing selection pressure, but abiotic factors are also closely linked to sucking pest outbreaks. Thus, it is essential to develop a strategic plan to deal with the insects in a sustainable manner by associating all the management practices.

2. Management of sucking insect pests

Sucking pest management entails integrating several management tactics available to us to protect our crop, which include the following:

2.1 Scouting to know our pest

Monitoring and field scouting is the pre requisite to develop any management strategies in sustainable manner [6]. Unvarying surveys need to be conducted throughout the cropping or pest's breeding season in order to predict future population and forewarn the farmers to initiate the control measures or not. As the pest are associated with have specific crop growth stage thus a clear knowledge about the pest biology, their appearance and breeding potentiality and to know this facts one must have to conduct field scouting or monitoring. Several tactics may be followed to conduct the study like setting yellow sticky trap, rouging survey, population count etc. Sometimes during the off season survival of the pests needs to be checked on wild relatives, alternate and collateral host to understand the bio-ecology of the pests and this will help to predict the level of infestation and sometimes helps to manage judiciously before crop establishment. Visual observation of bagrada bugs at night with a fluorescent lamp is effective [7].

2.2 Cultural management

Cultural management deploy crop management practices related to agronomic practices like adjustment of crop calendar, fertiliser scheduling, irrigation pattern, crop rotation, trap cropping etc. Several practices may be followed against different insect pest based on the bio-ecology factors. Deep summer ploughing is essential to destroy the eggs of several bugs and application of first irrigation at 3–4 weeks after sowing is effective to manage pest population. Immediately ploughed field after harvest is followed to prevent the spread of cabbage aphids to other crops [8] is an important cultural management strategy. Eradication of alternate host plants, like volunteer crucifer plants or other cruciferous weeds and destruction of plant debris at the end of the season can help to kill pest during non-host period [9].

Adjustment of date of sowing exerted potential impact on the population incidence mustard aphid. Early sown brassica crops (late September or early October) have been reported to escape aphid attack from India and Pakistan [10]; while delaying in sowing beyond mid-October or sometimes in November and December results in progressive increase in aphid population on succeeding late sown crops at reproductive stage [11]. But also it is very intricate always to manage the pest without having its bio-ecology knowledge of the target pest as population abundance is highly associated with the abiotic factors [12]. Maintenance of good field sanitation by wiping out and removing the crop residues and weeds in vicinity is an effective practice against the sucking pests like whiteflies and leaf hopper. Cultivation of non-host crops in a crop calendar reduces the population load of aphids and plant hopper. Application of recommended fertiliser dose based on soil testing is always effective to reduce the sap feeders [13] as high dose of N makes the plant succulent and thus succumb to infestation by sap feeders [14]. Increased sulphur fertiliser application exerted negative correlation with aphid population in mustard [15]; while increasing dose of K triggered hostile effects on the reproduction and excretion of the mustard aphid as reported by Bhat and Sidhu [16]. Organic manure like farm yard manure, neem cake, vermicompost reduced the pest population compared to straight fertilised treatments as manures induced the production of phenols and tannins in groundnut [17]. Water stressed condition reduced the fecundity of several sap feeder including aphids when soil water regimes decreased from 8 to 2% [18], tiny sap feeders face difficulties to get concentrated cell sap with low cell turgidity. Afrin et al. [19] reported that intercropping mustard with garlic, onion and coriander significantly reduced aphid population trough acted as repellents. Boarder crop like maize and sorghum may be utilised to reduce the dispersal of the sucking pest like jassid, whitefly, shoot fly etc. Ghosal et al. [20] reported that raising seedlings covered with insect proof net reduces the early infestation of chilli leaf curl viruses vectored by whitefly; he also concluded that border netting with insect proof net at a height of 6 feet help to reduce the dispersal of mite, whitefly and thrips. Planting a nectar plant to attract beneficial insects could also be helpful to build and conserve bio-control agents.

3. Mechanical and physical control

It is a very useful method to check the infestation of pests in any crop; it is an age-old traditional practise now combined with physical management strategies. Removal or clipping of infested parts after field scouting is the quickest and easiest way to reduce the infestation of different insect pests [21]. Most of the sap feeders are attracted to yellow, so it is a well-established fact that the installation of yellow sticky traps at 12 traps/ha for monitoring and 25 traps/ha for mass trapping may be utilised to catch the alate adults [20, 22]. The yellow-coloured water pan trap also exerted a great impact on the reduction of sap-feeding insects. The Nozzle is an important device to check the sucking pest as reported by Nawaz et al. [23]. The hollow cone nozzle with a single orifice (modified) showed the highest percentage of reduction in aphid population by dislodging (90.8%) immediately after spraying of water. Predator or non-target-friendly light traps may be utilised for mass trapping of several sap feeders of cereal crops, pulse crops, vegetable crops, as well as horticultural crops. Light traps of 380–400 lux from energy-saving lamps showed the maximum effect in reducing the pest brown plant hopper [24].

4. Biological control

In nature, there are several insects and pathogens that are associated with the ecosystem as predators, parasitoids, or parasites. Utilisation of this biota as pest management is the main action threshold of natural farming. They may be conserved in the field or may be released to achieve inoculative or inundative control. A clear knowledge of these bio-control agents is the basic pre-requisite to achieving pest management in a judicious way. Ladybird beetles (*Menochilus sexmaculata*, *Coccinella septempunctata*, *Hippodamia variegate*, *Coccinella transversalis*, and *Cheilomones sexmaculata*) are widespread predators that actively predate on soft-bodied sap feeders such as whitefly, aphid, and jassid. Adult ladybird beetles are known to consume an average of 10–15 adult aphids per day. Conservation of natural enemies which predate upon aphids, like ladybird beetles [25], green lacewing (*Chrysoperla zastrowi sillemi*), several species of syrphid/hoverfly (*Syrphus torvus*, *Episyrphus balteatus*, *Sphaerophoria* spp., *Sphaerophoria scripta*, *Xanthogramma* spp., etc.) is the base for in situ management of mustard aphid. The beetles, *Hippodamia convergens* Guerin-Meneville and *Collops vittatus* (Say) fed mostly on whitefly (*Bemisia tabaci*) eggs, but readily and rapidly preyed on all of the whitefly life stages. The true bugs, *Geocoris punctipes* (Say) and *Orius tristicolor* (Say) preyed almost exclusively on adult whiteflies, while *Lygus hesperus* Knight preyed almost exclusively on nymphs. *Cryptolaemus montrouzieri*, often called as mealybug ladybird or mealybug destroyer is well known to check mealybug. Vennila et al. [26] reported that when the population of the green lace wing (*Chrysoperla carnea*) is more (at predatory to prey ratio of greater than or equal to 0.2) the decision to take up insecticidal spray should be abandoned. The braconids are the most dominant bio control agents of an array of hemiptera. *Diaetriella rapae* (Braconidae, Hymenoptera) is copiously present in the mustard ecosystem and naturally parasitize up to 60.0% aphid population. *Acerophagus papayae* Noyes and Schauff (Encyrtidae: Hymenoptera) is one of the efficient parasitoids for the management of papaya mealybug in its native range. In India this parasitoid was introduced from Puerto Rico in 2010 through USDA–APHIS for use against the papaya mealybug as inoculative release [27]. The parasitoids in the genera of *Encarsia* and *Eretmocerus* (Hymenoptera: Aphelinidae) are important biological control agents of whiteflies, and some of them not only parasitize hosts but also kill them with strong host feeding capacity [27]. *Serangium parcesetosum* Sicard (Coleoptera: Coccinellidae) can utilised effectively in biological control of *B. tabaci* [28]. There are known predators and parasitoids that attack *Bagrada hilaris* [29]. 79 species of natural enemies were identified by Chiu [30], comprising 34 species of insect parasites, 1 nematode, 7 pathogens, 21 species of insect predators and 16 species of spider predator. Some show considerable potential for biological control. The predator wolf spider, *Lycosa pseudoannulata* is a promising options in reducing brown plant hopper population successfully in 3 days exposure period indicating its high predation potentiality. Ladybird beetle, *Micraspis discolour, Ophionea indica* is also associated with predation of brown plant hopper [24]. The mirid bug, *Cyrtorhinus lividipennis*, is a plant-feeding and predatory insect, preferring plant and leafhopper eggs and young nymphs [31].

Entomopathogenic fungi are becoming increasingly important in pest management in a sustainable manner. Different workers have been studied on different bio-pesticides to evaluate their efficacy against different sucking pests. A number of entomogenous fungi, *Beauveria bassiana*, *Metarhizium anisopliae*, *Verticillium lecanii* (renamed Lecanicillium), *Cephalosporium* spp. and *Nomuraea rileyi* are also to infect aphids, whiteflies, jassid, mealy bugs etc. [32, 33]. Avoiding spray of chemical pesticides during heavy natural parasitism is quite beneficial to maintain sufficient number of biocontrol agents to suppress this pest. Wild mustard (*Sinapis arvensis*) and white mustard (*Sinapis alba*), Cowpea are very good border crop to conserve biological control agents like lady bird, lace wing etc. Predatory bird *Motacilla cospica* actively feeds on aphids in February–March. Therefore, provision bird perches @ 8–10/acre will boost natural control.

5. Host plant resistance

Host plant resistance is generally based on the principle of antibiosis, antixenosis and tolerance, which has been achieved through hybridization process. Complete resistance to a specific group of insect pest only be achieved through genetic engineering only; but commercial hybridization only can add few resistance trait to the desired cultivar. Breeding programme for resistant variety targets to tap the best source of resistance in this genetically diverse species [34]. For management of several viral diseases vectored by sap feeders modern days hybridization have evolved very successfully. So many resistant varieties are now available in market for commercial cultivation for different cereal, pulse, oilseeds and vegetable crops. But it is difficult to have high degree resistant source which can be exploited in development of aphid resistant varieties in mustard as reported by Kumar and Sangha [35]. Cotton varieties with glabrous leaves are very much susceptible to jassids if compared to the hairy (hirsute) varieties. The vulnerability is correlated with both density & length of hairs on the below surface of the leaf mainly on veins where the jassid feeds & lays eggs. High hair thickness without length is ineffective. Strong positive correlation of jassid population with plant height suggested that improvement in plant height will result increase in jassid population significantly. There was a strong negative correlation of leaf N content, indicated that increasing leaf nitrogen (N) will result negative impact on jassid population. Total sugars, reducing sugars, nitrogen and protein content had a positive correlation with the whitefly population and OYVMV incidence. High resource availability can reduce anti-herbivore resistance in rice, but may also increase tolerance [36]. Development of biotypes in hemiptera is a serious matter of concern to development of resistant varieties. Resistant rice varieties have proved to be highly effective against the brown plant hopper *Nilaparvata lugens*, but their long-term stability is threatened because of the evolution of prolific biotypes [37]. They also reported that Biotype 1 can survive on and damage only those varieties that do not carry any genes for resistance. Biotype 2 can thrive on varieties carrying Bph 1 resistance gene and on those susceptible to biotype 1. Biotype 3 can infest and destroy varieties having bph 2 resistance gene and those susceptible to biotype 1. However, none of these biotypes is capable of damaging varieties with Bph 3 and bph 4 genes for resistance. Several varieties which are resistant in the Philippines are susceptible in India and Sri Lanka because the South Asian biotypes of *N. lugens* are more virulent than the Southeast Asian biotypes. Improved lines of okra have potential to broaden the genetic variability may improve resistance to these pests. Tolerance of BCO-1 genotype to whitefly and that of Arka Anamika to jassid is based on both antixenotic properties of the genotypes as reported by Acharya et al. [38]. In relation host plant, *B. tabaci* is affected mainly by the following features: (i) the external, physical characteristics of the leaf surface, e.g., hairiness vs. glabrousness, sticky glandular trichomes, leaf shape (okra/super okra) and probably the microclimate as a result of foliage density; and (ii) the internal, chemical characteristics of the leaf, e.g., pH of leaf sap [39]. Three different types of host plant resistance mechanisms against Silverleaf whitefly, *B. tabaci* (Gennadius) B Biotype was reported by Miyazaki et al. [40]; (i) okra leaf traits which reduced adult preference, (ii) glabrous leaf traits which reduced oviposition preference and (iii) possible biochemical traits in *Gossypium arboreum* which reduced immature development and/or survival. Although any potential negative effects of these traits on yield need to be carefully evaluated. Incidence of sap feeders are positively correlated with nitrogen content and negatively correlated with phenols and tannins. The organic manures induced the production of phenols and tannins in groundnut plant and thus, the induced resistance played an important role in management of jassid *Empoasca kerri*, aphid *Aphis craccivora* in groundnut [17].

6. Botanicals

The integrated pest management approach strongly encourages the use of botanicals as a component of pest management tools, which are less harmful to the ecosystem, biodegradable, and have a positive effect on a wide range of sap-feeding insects due to their antifeedant and repellent action. The anti-pest philosophy of neem and tobacco is well documented in agricultural history [41]. Apart from that several botanicals may be utilised effectively to reduce the population of sap feeders below ETL. Neem oil (2–5%) and neem seed kernel extract (NSKE-5%) is very effective to manage the several sap suckers like aphid, rice bug, coreid bug, jassid, whiteflies etc. [42]. Concerning susceptibility of different developmental stages, young larvae were the most sensitive. Foliar treatment was the most efficient. The mortality among immature increased in relation to azadirachtin concentrations. Ali et al. [43] reported that Mexican marigold extract @ 1:2.5 g/ml effectively reduced *Lipaphis erysimi* population by 96.4%. Chandel et al. [44] reported varied botanicals like NSKE, leaf extract of *Vitex negundo*, *Parthenium hysterophorus*, *Adhatoda vasica*, *Lantana camara* effective against aphid. Pavela and Herda [45] studied on repellent effects of pongam oil on settlement and oviposition of the common greenhouse whitefly, *Trialeurodes vaporariorum*. Akbar et al. [46] opined that *Calotropis procera* extract can reduce aphid population up to 54.9%. Chattree et al. [47] recorded maximum mortality (92.85%) with 20% neem leaf extract in case of first instar larvae of *L. erysimi* when treated for 18 hours. Castor leaf extract is very effective against aphid [48].

7. Chemical control

Insecticides are the most reliable weapons in pest management strategies. Because of their quick knock down effect and more persistency in the field, they are the most acceptable agents against different pest spectrums. Crop protection with chemicals is a desirable and unavoidable part of integrated pest management. Nowadays, insecticides with varied chemistry are well studied and available in the market. At the same time, development of resistance, resurgence, and residue are three common phenomena associated with chemical management of pests also. Thus, one must be very choosy and selective in the selection of the insecticide effective against the target pest without having hazardous issues. It is very essential to select pesticides for those crops actively visited by an array of pollinators like bees. ETL-based pest management is necessary, but it is awful to desire the insecticide application by the end user following an ETL-based recommendation. One must have sound knowledge of the incidence pattern of insects in relation to weather to protect crops through the adoption of prophylactic spraying [49].

Ghosal et al. [50, 51] reported that imidacloprid 17.8 SL@ 50 g ai/ha was found as a most effective neonicotinoid insecticide against aphid (*Aphis gossypii*) and jassid (*Cestius phycitis*). 84.54% reduction of population over control was achieved. Other two popular neonicotinoids viz., thiamethoxam 25WG@ 50 g ai/ ha and acetamiprid 20SP@ 40 g ai/ha were also found effective and showed better result than acephate 75WP and dimethoate 30EC, those are commonly used against aphid. Dotasara et al. [52] reported that imidacloprid 17.8% SL @ 0.2 g/l reduced the incidence of mustard aphid by 87.5% followed by fipronil 5 SC @ 1.0 ml/l (83.6%) at 7 days after spray. Roy et al. [53] revealed that flonicamid (94.7% mortality) and pyriproxyfen (93.0%) effectively managed *L. erysimi* pest up to 15–18 days. Chau [54] reported that imidacloprid at recommended rates of 28 and 20 g a.i/ha showed very good control of BPH in wet season. Though in India, plant hopper resistance by 35.13, 10.78 and 4.98 fold to imidacloprid, thiamethoxam and clothianidin respectively have been detected during 2006 by Krishnaiah et al. [55]; thus it very pertinent to select the best chemical for the control of plant hoppers. Plant hoppers showed allopatric speciation and showed varied level of resistance depending on exposure. Buprofezin 15% + acephate 35% WP @ 1500 ml/ha is best effective insecticide against brown plant hopper (BPH) and white backed plant hopper (WBPH) as reported by Ghosal et al. [56]. Cotton fields treated with quinalphos, chlorpyriphos induced build-up of jassid population. If Seeds treated with imidacloprid/thiomethoxam - 7 to 10 g/kg of seeds before sowing cotton provides good control against different sap feeders up to 45 days after planting. Ghosal and Chatterjee [5] opined that at sub lethal dose of imidacloprid induced resurgence of whitefly (+2.02% & +9.84%). While, clothianidin induced upsurge rend in whitefly in cotton and tomato. Diafenthiuron 50 WP @ 600 g/ha, imidacloprid 70 WG @ 75 g/ha and thiamethoxam 25 WG @ 100 g/ha reduced the incidence of yellow mosaic viruses in okra through management of its vector whitefly significantly [57, 58]. Spirotetramat could be a better alternative for management of mealy bug instead of chlorpyriphos or any organophosphates as chlorpyriphos shows 100% mortality to vine mealybug parasitoid *Anagyrus* sp. while spirotetramat do not [59]. Cotton mealybug (*Phenacoccus solenopsis*) developed resistance against so many insecticides like pyrethroids, carsulfan, endosulfan, chlorphos, etc. so, restriction of using this insecticides against cotton mealy bug is strictly advisable. Spray volume and nozzle is a very crucial for management of sap feeders as most of them settled down under the leaves. It is preferable to use hollow cone nozzle with spray volume of 500 l/ha for knapsack sprayer and must be greater than 250 l/ha using power sprayers. Proper coverage of the underside of leaves during the insecticidal sprays is very necessary for desired efficacy. Not to spray or avoid spraying during the blooming season for the nectar producing crop and preferably advised to spray the insecticide in the evening hours to avoid mortality of honey bees and other pollinators. Field near bee keeping is strictly instructed to not to use neonicotinoid insecticides during flowering and ethically need to restrict the bee foraging from that field for the following day.

8. Conclusion

Our crops are infested by several sap-feeding insects. They are responsible for a considerable loss in production. Recrudescence of sap-feeding insects due to selection pressure is a common phenomenon. Rapid development of resistance to insecticides is another matter of concern in successful crop production. Sole dependency on insecticides makes the situation more aggravating. A holistic approach involving cultural, mechanical, physical, biological, and botanical boosting of host plant resistance and chemical management is essential to protect the crop and ecosystem.

Author details

Abhijit Ghosal
Sasya Shyamala Krishi Vigyan Kendra, Ramakrishna Mission Vivekananda Educational and Research Institute, West Bengal, Kolkata, India

*Address all correspondence to: ghosalabhijit87@gmail.com

References

[1] Wari D, Okada R, Takagi M, Yaguchi M, Kashima T, Ogawara T. Augmentation and compatibility of *Beauveria bassiana* with pesticides against different growth stages of *Bemisia tabaci* (Gennadius); an in vitro and field approach. Pest Management Science. 2020;**76**:3236-3252

[2] Gerasimos C. True Bugs (Insecta: Hemiptera: Heteroptera): Evolution, Classification, Biodiversity and Biology. Reference Module in Life Sciences; 2019

[3] Wari D, Kabir MA, Mujiono K, Hojo Y, Shinya T, Tani A, et al. Honeydew-associated microbes elicit defence responses against brown planthopper in rice. Journal of Experimental Botany. 2019;**70**:1683-1696

[4] CABI. Available from: https://stateoftheworldsplants.org/2017/plant-health.html [Accessed: March 29, 2021]

[5] Ghosal A, Chatterjee ML. Insecticide induced resurgence study of whitefly in cotton and tomato. University of Sindh Journal of Animal Sciences. 2018;**2**(2):1-6

[6] Yadav S, Rathee M. Sucking pests of rapeseed-Mustarrd. In: Omkar, editor. Sucking Pest of Crops. Singapore: Springer Nature; 2020. pp. 187-233

[7] Huang TI, Reed DA, Perring TM, Palumbo JC. Feeding damage by *Bagrada hilaris* (Hemiptera: Pentatomidae) and impact on growth and chlorophyll content of Brassicaceous plant species. Arthropod-Plant Interactions. 2014;**8**:89-100

[8] Griffin RP, Williamson J. Cabbage, Broccoli and Other Cole Crop Insect Pests. HGIC 2203, Home and Garden Information Center, Clemson Cooperative Extension, Clemson University; 2012

[9] Hines RL, Hutchison WD. Cabbage Aphids. VegEdge, Vegetable IPM Resource for the Midwest. Minneapolis: University of Minnesota; 2013 Available from: https://www.vegedge.umn.edu/pest-profiles/pests/cabbage-aphids

[10] Rahman MA. Effect of Sowing Times and Mustard Varieties on the Incidence of Aphid [M.Sc. thesis]. Dhaka: Sher-e-Bangla Agricultural University Sher-e-Bangla Nagar; 2014. p. 66

[11] Iqbal MF, Hussain M, Ali MA, Nawaz R. Aphid control management as affected by sowing dates of rapeseed. Journal of Agricultural Research. 2015;**53**(3):389-393

[12] Ghosal A, Manna D, Adhikary R. Seasonal incidence of leaf hopper on okra in correlation with abiotic factors and evaluation of some neonicotinoid insecticides against okra leaf hopper *Amrasca biguttula biguttula* (Ishida). Ecology, Environment and Conservation Journal. 2014;**20**(2):169-174

[13] Ghosal A, Das K, Kudu P. Molecular characterization of whitefly *Bemesia tabaci* (Genn.) and development of IPM module against chili leaf curl complex. Indian Journal of Entomology. 2021:e21034

[14] Khattak SU, Khan A, Shah SM, Alam Z, Iqbal M. Effect of nitrogen and phosphorus fertilization on aphid infestation and crop yield of three rapeseed cultivars. Pakistan Journal of Zoology. 1996;**28**(4):335-333

[15] Bakhetia DRC, Rani S, Ahuja KL. Effect of Sulphur nutrition of some

Brassica plants on their resistance response to the mustard aphid, *Lipaphis erysimi* (*Kaltenbach*). In: Calicut Univ Res J (Special Conference Number), 3-5 May. 1982. p. 36

[16] Bhat NS, Sidhu HS. Influence of potassium on amino acids of the host plants and reproduction and excretion of the mustard aphid, *Lipaphis erysimi* (Kalt.). In: Proceedings of the 6th International Rapeseed Conference, 17-19 May 1983. Vol. 6. Paris; 1983. p. 28

[17] Rajasekhara RK. Induced host plant resistance in the management of sucking insect pests of groundnut. Annals of Plant Protection Sciences. 2002;**10**(1):45-50

[18] Sidhu HS, Kaur P. Influence of nitrogen application to the host plant on fecundity of mustard aphid, *Lipapahis erysimi* (Kalt.). Journal of Research: Punjab Agricultural University. 1977;**14**:445-448

[19] Afrin S, Latif A, Banu NMA, Kabir MMM, Haque SS, Emam Ahmed MM, et al. Intercropping empower reduces insect pests and increases biodiversity in agroecosystem. Agricultural Sciences. 2017;**8**:1120-1134

[20] Ghosal A. Chilli Package of Practices and Border Netting Technology in Chilli Leaf Curl Management. India: SSKVK, RKMVERI; 2021. DOI: 10.13140/RG.2.2.10705.45923

[21] Sachan GC, Purwar JP. Integrated pest management in rapeseed mustard. In: Jain PC, Bhargwa MC, editors. Entomology: Novel Approaches. New Delhi: New India Publishing Agency; 2007. pp. 399-423

[22] Ghosal A, Dolai AK. Evaluation of IPM models in managing stem borer and leaf folder of rice and aphid of musterad. Indian Agriculturist. 2017;**61**(3&4):123-130

[23] Nawaz M, Ali S, Abbas Q. High pressure water spray technique for controlling mustard aphid (*Lipaphis Erysimi*) on Brassica crop. American-Eurasian Journal of Agricultural & Environmental Sciences. 2016;**16**(2):224-228

[24] Sardar M, Khatun MR, Islam KS, Haque MT, Das G. Potentiality of light source and predator for controlling brown planthopper. Progressive Agriculture. 2019;**30**(3):275-281

[25] Mishra SK, Kanwat PM. Seasonal incidence of mustard aphid, *Lipaphis erysimi* (Kalt.) and its major predator on mustard and their correlation with abiotic factors. Journal of Entomology and Zoology Studies. 2018;**6**(3):831-836

[26] Vennila S, Biradar VK, Sabesh M, Bambawale OM. Know your cotton insect Pest jassids. Crop Prot Folder Series. 2007;**2**:11

[27] Jothi BD, Surulivelu T, Sonai Rajan T, Valarmathi R. First record on the establishment of the parasitoid (Acerophagus papayae Noyes and Schauff) of papaya mealybug (Paracoccus marginatus Williams and Granara de Willink) on cotton. Karnataka Journal of Agricultural Sciences. 2011;**24**(4):536-537 2

[28] Zyoud FA, Sengonca C, Abbadi SA. Evaluation of *Serangium parcesetosum* (Col.: Coccinellidae) for biological control of *Bemisia tabaci* under greenhouse conditions. Journal of Pest Science. 2007;**80**(2):85-92

[29] Bundy CS, Perring TM, Reed DA, Palumbo JC, Grasswitz TR, Jones WA. *Bagrada hilaris* (Burmeister). In: McPherson JE, editor. Invasive Stink Bugs

and Related Species (Pentatomoidea). Boca Raton, FL: CRC Press; 2018. pp. 205-241

[30] Chiu SC. Biological control of the brown planthopper. In: Brown Planthopper: Threat to Rice Production in Asia. Philippines: International Rice Research Institute; 1979. pp. 335-355

[31] Shepard MB, Barrion AT, Litsinger JA. Friends of Rice Farmer: Helpful Insects, Spiders, and Pathogens. Manila (Philippines): Intl Rice Res Inst; 1987

[32] Ghosal A, Chatterjee ML, Bhattacharyya A. Field bio-efficacy of some new insecticides and tank mixtures against whitefly on cotton in new alluvial zone of West Bengal. Pesticide Research Journal. 2018;**30**(1):31-36

[33] Deka AC, Goswami NK, Sharma I. Biocontrol prospects of entomopathogenic fungi for management of mustard aphid (*Lipaphis erysimi* Kalt.) on rapeseed-mustard. Advances in Applied Science Research. 2017;**8**(4):21-29

[34] Rohilla HR, Singh H, Kumar PR. Strategies for the identification of the sources of resistance in oilseeds Brassica against *Lipaphis erysimi* (Kalt.). Annals of Biology. 1993;**9**:174

[35] Kumar S, Sangha KS. Biochemical mechanism of resistance in some Brassica genotypes against *Lipaphis erysimi* (Kaltenbach) (Homoptera: Aphidiae). Vegetos. 2013;**26**(2):387-339

[36] Horgana FG, Cruz AP, Berna CC, Rama AF, Almazan MLP, Wilby A. Resistance and tolerance to the brown planthopper, *Nilaparvata lugens* (Stal), in rice infested at different growth stages across a gradient of nitrogen applications. Field Crops Research. 2018;**217**:53-65

[37] Saxena R, Barrion A. Biotypes of the brown planthopper *Nilaparvata lugens* (Stal) and strategies in deployment of host plant resistance. Insect Science and Its Application. 1985;**6**(3):271-289

[38] Acharya J, Chatterjee S, Konar A, Chattopadhyay A, Mandal AK, Dutta S. Host plant resistance through physico-chemical characters against major insect pests of okra occurring in the Gangetic plains of eastern India. International Journal of Pest Management. 2019;**65**:137-146

[39] Berlinger MJ. Host plant resistance to *Bemisia tabaci*. Agriculture, Ecosystems and Environment. 1986;**17**(1-2):69-82

[40] Miyazaki J, Stiller WN, Wilson JL. Identification of host plant resistance to silverleaf whitefly in cotton: Implications for breeding. Field Crops Research. 2013;**154**:145-152

[41] Sable M, Kushwaha RK. Efficacy of different plant leaf extract against mustard aphid *Lipaphis erysimi* (Kalt). Journal of Industrial Pollution Control. 2014;**30**(2):231-233

[42] Ghosal A. Studies on the potency of some bio pesticides against whitefly in cotton and tomato. Journal of Biopesticides. 2018;**11**(1):64-72

[43] Ali A, Rizvi PQ, Khan FR. Bio-efficacy of some plant leaf extracts against mustard aphid, *Lipaphis erysimi* Kalt. On Indian mustard, *Brassica juncea*. Journal of Plant Protection Research. 2010;**50**(2):130-132

[44] Chandel BS, Singh V, Trevedi SS. Aphidicidal potential of *Azadirachta indica*, *Adhatoda vasica*, *Vitex negundo*, *Parthenium hysterophorus* and *Lantana camara* against mustard aphid, *Lipaphis erysimi* Kaltenbach (Hemiptera: Aphididae). Journal of Natural and Applied Sciences. 2012;**4**(2):181-186

[45] Pavela R, Herda G. Repellent effects of pongam oil on settlement and oviposition of the common greenhouse whitefly *Trialeurodes vaporariorum* on chrysanthemum. Insect Sci. 2007;**14**:219-224

[46] Akbar W, Asif MU, Muhammad R, Tofique M. Bio-efficacy of different plant extracts against mustard aphid (*Lipaphis erysimi*) on canola. Pakistan Journal of Entomology. 2016;**31**(2):189-196

[47] Chattree PV, Mishra MS, Srivastava AN. Residual effect of neem leaf extract on the mortality of *Lipaphis erysimi* and its larvae. IOSR Journal of Environmental Science, Toxicology and Food Technology. 2016;**10**(10):115-119

[48] Khan S. Management of Mustard Aphid [*Lipaphis erysimi* (Kalt.)] by Different Eco-Friendly Tactics [Ph.D. thesis]. Gwalior, MP: Rajmata Vijayaraje Scindia Krishi Vishwa Vidyalaya; 2017. p. 14

[49] Ghosal A. Influence of Abiotic Factors on Whitefly Population Abundance in Cotton. London, UK, London, UK: IntechOpen; 2022. DOI: 10.5772/intechopen.103006

[50] Ghosal A, Chatterjee ML, Bhattacharyya A. Bio-efficacy of neonicotinoids against *Aphis gossypii* glover of okra. Journal of Crop and Weed. 2013;**9**(2):181-184

[51] Ghosal A, Chatterjee ML. Evaluation of imidacloprid 17.8 SL against brinjal jassid, *Cestius phycitis*. Journal of the Entomological Research. 2013;**37**(4):319-323

[52] Dotasara SK, Agrawal N, Singh N, Swami D. Efficacy of some newer insecticides against mustard aphid Lipaphis erysimi Kalt. in cauliflower. Journal of Entomology and Zoology Studies. 2017;**5**(2):654-656

[53] Roy DP, Arkar KS, Sarkar S. Field efficacy, non-target toxicity and economics of novel systemic molecules against *Lipaphis erysimi* K. and seasonal incidence in mustard. Indian Journal of Entomology. 2018;**80**(2):217-222

[54] Chau LM. State of insecticide resistance of brown plant hopper in Mekong delta, Vietnam. Omonrice. 2007;**15**:185-190

[55] Krishnaiah NV, Jhansi Lakshmi V, Pasalu IC, Katti G, Kondala Rao Y. Insecticide resistance development in brown planthopper (*Nilaparvata lugens* Stål) populations in east Godavari district of Andhra Pradesh. In: 26th Int. Rice Res. Conf., ICAR, New Delhi. 2006. pp. 424-425

[56] Ghosal A, Dolai AK, Chatterjee ML. Evaluation of a new ready mixed insecticide (buprofezin 15% + acephate 35% WP) against BPH (*Nilaparvata lugens* (Stal)) and WBPH (*Sogatella furcifera* (Horvath)) of rice in West Bengal. Journal of Crop and Weed. 2018;**14**(1):205-211

[57] Ghosal A, Chatterjee ML. Management of okra yellow vein mosaic virus by chemical control of *Bemisia tabaci* Gennadius. Indian Journal of Entomology. 2013;**75**(3):236-238

[58] Ghosal A, Chatterjee ML. Bioefficacy of imidacloprid 17.8 SL against whitefly, *Bemisia tabaci* Gennadius in brinjal. The Journal of Plant Protection Science. 2013;**5**(1):27-31

[59] Mansour R, Suma P, Mazzeo G, et al. Evaluating side effects of newer insecticides on the vine mealybug parasitoid *Anagyrus* sp. near *pseudococci*, with implications for integrated pest management in vineyards. Phytoparasitica. 2011;**39**:369-376

Section 2

Insecticide in Pest Management: Advantages and Disadvantages

Chapter 3

Insecticidal Agents in Pest Control: Sources, Challenges, and Advantages

Simon Koma Okwute and Henry Omoregie Egharevba

Abstract

Insect pests are found commonly in two critical areas of human life: the farms and crop storage facilities and the home, where they cause a number of problems, including the destruction of various forms of materials such as clothes and cellulose-based items by ants, as well as nuisances and disease-transmitting activities perpetrated by mosquitoes, flies, cockroaches, and bugs. For ages, man has tried to fight the menace of insects using traditional and scientific methods, including the use of chemicals. In this work, the historical aspect of the use of insecticides for pest control, the challenge of the development of insect resistance, the potential for and incidences of environmental and health hazards, adverse effects on climate change, and the search for new agents, particularly from natural products of plant origin, are discussed. The challenges and the strategic advantages of the use of various classes of insecticides are also presented. The need for the application of lessons learned from human pharmaceutical science, the deployment of emerging technologies in the search for new insecticidal moieties and biopesticides, and the development of new and more efficient insecticide application tools and technologies are also discussed.

Keywords: pest control, insecticides, sources, classification, challenges, advantages

1. Introduction

Pests have been defined in various ways by various researchers, but may be summarized as any form of plant or animal or any pathogenic agent that is injurious or has the potential to be injurious to plants, plant products, livestock, and man. They include insects and the other arthropods, vertebrates, nematodes, weeds, and pathogens such as fungi, bacteria, viruses etc. [1, 2]. An estimated 67,000 species of vertebrate and invertebrate pests exist globally, destroying about 40% of agricultural produce annually [3].

Pests have been in existence since the creation of the universe, and the threat of pests such as mosquitoes, cockroaches, rodents, parasitic worms, pathogens, and snails has been well known and challenged by man. Both vertebrates and invertebrate pests cause huge damage to crops annually in the fields, gardens, and homes. Of the invertebrates the arachnids and in particular the insects constitute the most important crop pests. They cause damage by their eating habits, leading to defoliation and damage to the roots of plants by insects and larvae. Insects such as the aphids suck the plant sap and therefore cause physical damage as well as act as vectors for infective agents. The resultant effects are decreased agricultural yields, food quality and quantity, and increased hunger among the population [4]. Ancient man had used his ingenuity to deploy various methods to control their menace, including smoky fire, the spread of mud on the body, prayers, magic spells, cultivation systems, mechanical practice, as well as application of organic and inorganic substances to protect his crops from attack of weeds, diseases, and insect pests [5]. While agricultural yield has greatly improved to 70–100% in the industrialized nations due to better pest management and improved plant species, the challenge of poor yield due primarily to pests persists in the developing countries of Africa [3].

In recent years, the global concern is the impact of the activities of pests on agricultural outputs in terms of quantity and quality of food production by farmers all over the world, and disease transmission [6–10]. Annual crop losses due to pests is 35–45% which amounts to about $470 billion USA [11]. The world population is rapidly increasing, and as of today it is about 7.95 billion [12]. Thus, population growth pressure and limited options have put an enormous burden on the need to develop measures to meet food demands, which is a serious challenge in developing countries of Asia and Africa [13, 14]. To minimize the devastating impact of pests, particularly on agricultural production, as well as eliminate disease-borne vectors as part of primary healthcare measure, pesticides have been developed. Some reports have it that as of today, the pesticide consumption has reached 4 million tons of which 50% are herbicides, 30% are insecticides, and 18% are fungicides [15]. Also, the world market in 2019 was US$84.5 billion, with the prospect of reaching US$130.7 billion in 2023 [16]. As our knowledge of pests and their activities has expanded, the definition of pesticides has also been expanded in terms of actions on the pests. A version by Maton et al. [1] defines a pesticide as "any natural or synthetic chemical that causes directly the death, impulsion, attraction, deterrence of pests or in other ways influences pests for useful purposes". However, Lopes et al. [11] while not disagreeing with the above definition defined a pesticide as any compound or mixture of components intended for preventing, destroying, repelling, or mitigating any pest. In any case, they include herbicides, insecticides, fungicides, rodenticides, and nematicides [17].

Among the pests, perhaps one of the most devastating agents are the insects, which destroy plant crops on the farmlands, crops in storage, and home furniture and act as disease vectors in aquatic and terrestrial environments. The chemicals for killing or inhibiting insect pests worldwide are called insecticides and probably were the earliest chemicals to be used by man in agriculture and in the home, to prevent vector-borne diseases, for crop protection and food preservation, and in aquaculture [18].

This work focuses on insecticidal agents in pest control and addresses issues relating only to insects and chemicals in use to control them. Thus, as much as possible chemicals used as antimicrobial (antibacterial and fungicidal), anthelmintic, herbicidal, molluscicidal, or piscicidal agents will not be discussed. However, associated chemicals such as repellents, attractants, and synergistic agents are included. The history and sources of insecticidal agents and their contribution to agricultural productivity and healthcare as well as the challenges and advantages in their applications will be presented.

2. History and sources of insecticides

A number of substances were used in pest control between 500 BC and World War II in the 19th century, and were considered as pesticides. A pesticide is "any substance or mixture of substances intended for preventing, destroying, repelling or mitigating any pest". They included brimstone (sulfur), arsenic, lead, and mercury [13, 19]. Others were cryolite (sodium hexafluoroaluminate), petroleum oil, hydrogen cyanide gas, nicotine, pyrethrum, and rotenone [20]. In 1874 DDT (dichlorodiphenyltrichloroethane) was synthesized, and during the second half of World War II, its insecticidal activity was discovered and effectively used to control malaria and typhus diseases among the troops. It became the first synthetic organic pesticide used after the war for agricultural purposes [21]. However, it was established that synthetic organic pesticides, particularly chlorinated hydrocarbons such as DDT and derivatives, led to serious environmental pollution (water, air, and soil), affecting human health and causing deaths of non-target organisms (animals, plants, and fish). This led to the Stockholm Convention in 2001 and the eventual ban of DDT in 2004 [22]. Before the ban, efforts were already made by researchers for alternative sources of pesticides due to other reasons including (a) non-selectivity/specificity, (b) ineffectiveness, (c) not many of the synthetic compounds having been successfully marketed due to lack of interest by potential users, (d) high cost of synthetic chemicals, and (e) development of resistance [23, 24].

The development of pesticide resistance has led to rising number of insect species, and today there are over 10,000 species of insects. The "green revolution" in the 1960s by a number of countries led to the heavy use of pesticidal agrochemicals, with Argentina, Brazil, China, Russia, Canada, Australia, Columbia, France, and India as top consumers [25].

3. Classification of insecticides

Based on these earlier uses, insecticides were generally classified according to their mode of action as they have very diverse chemical structures. Thus, they included fumigants that were delivered as smokes, aerosol sprays, volatile liquids, and gases and they enter the insects through the breathing pores called spiracles. Other classes included stomach poisons, which were inorganic in nature and need to be eaten, and their organic counterparts are contact poisons, which penetrate the insect cuticle into the nervous system to cause paralysis. More recently, the systemics were introduced and they act by entering the plant tissues and then being absorbed by the insects. Thus, they are similar to stomach poisons [4]. Today, we have broader and sometimes micro-classifications such as contact (with no residual effect), systemic (with long-term activity), repellent and non-repellent, deterrents, and botanicals [1, 26, 27]. The broad classes are presented as follows:

3.1 Chemical insecticides

Chemical insecticides are synthetic compounds designed to target and eliminate specific insect pests. These agents are categorized based on their mode of action, including neurotoxins, growth regulators, and stomach poisons. Widely used in agriculture, chemical insecticides provide rapid and effective control of pests but are also associated with concerns such as resistance development and environmental impacts. As earlier emphasized, sulfur and a number of heavy metals such as arsenic, copper, lead, mercury, etc., were used as insecticides by the ancient man. However, since World War II (1940), which opened the "*Modern Era of Chemical Control*", a number of organic chemical insecticides have been developed and used, including naphthalene, carbon tetrachloride, tetrachloroethane, 1,1-dichloro-1-nitroethane, nicotine, azobenzene, sulfotep, fumazone, telone, dazomet, metham sodium, triphenyl phosphine, phenothiazine, DNC (4,4′-dinitrocarbanilide), dinex, DDT, gammexane, carbamates, the cyclodienes such as aldrien, and dieldrin, and the organophosphorus compounds dimefox, TEPP (Tetraethyl pyrophosphate), schraden, malathion, parathion, and dichlorvos as well as the bridged diphenyl compounds, dicofol and tedion (tetradifon) [10, 19, 20]. These chemicals can be further classified as follows.

3.1.1 Organochlorines

This group includes the diphenyl aliphatics such as chlorobenzilate (1), DDT (2), DDD (Dichlorodiphenyldichloroethane, 3), dicofol (4), ethylan (5), and methoxychlor (6). Most of these have been banned because of environmental toxicity [20].

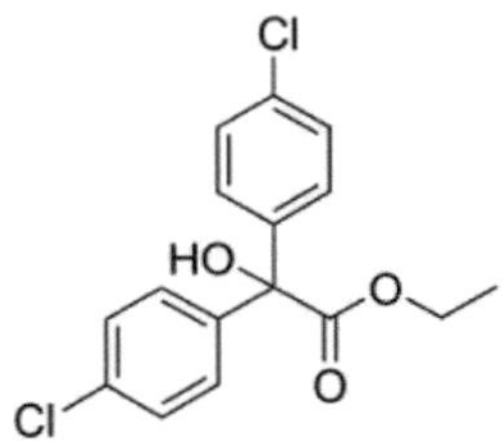

(1) Chlorobenzilate

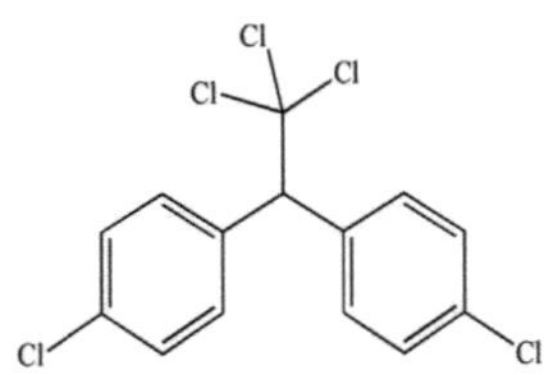

(2) DDT

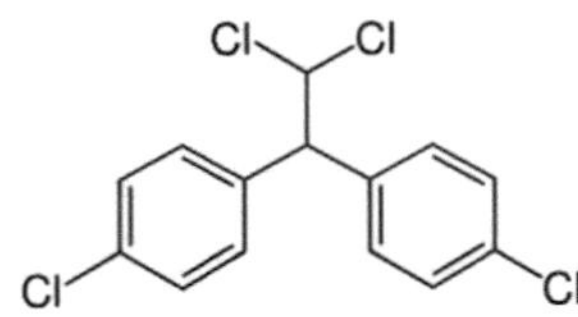

(3) DDD

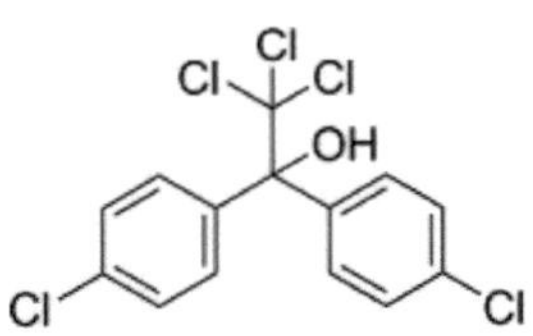

(4) Dicofol

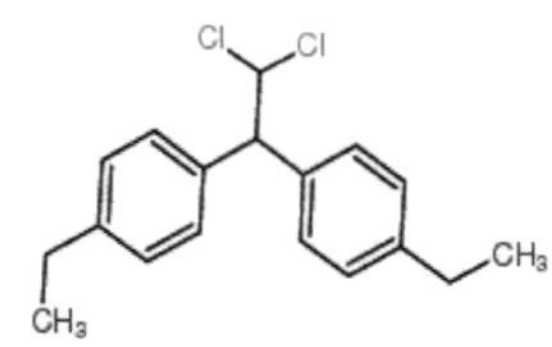

(5) Ethylan

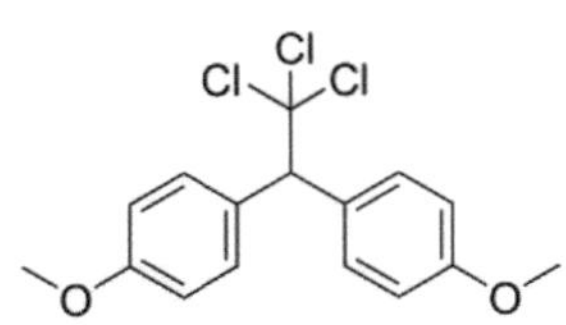

(6) Methoxychlor

Of the above organochlorides, DDT has been the most widely used from 1940 up till 1973 when its use was prohibited in the United States by the United States Environmental Protection Agency (USEPA), though it is still being widely used in many developing countries for the control of insect-vectored diseases such as malaria and yellow fever. For its high toxicity, DDT was eventually banned in 2004 after the Stockholm Convention in 2001 [20, 22]. The mode of action of DDT is suspected to be by disrupting the balance of Na^+ and K^+ ions in the axon of the neurons blocking normal transmission of nerve impulses resulting in spontaneous convulsion and death [20].

Hexachlorohexane, also known as benzene hexachloride (BHC), was discovered in 1940. Of the five isomers, *alpha, beta, gamma, delta,* and *epsilon*, only the *gamma* isomer has insecticidal properties and is sold under the trade name *lindane*. It was banned in the US in 2002, although it is still used in some developing countries. Like DDT it is a neurotoxicant [20].

The cyclodienes class of organochlorines was developed after World War II between 1945 and 1958. They include chlordane (7), aldrin (8), dieldrin (9), heptachlor, endrin, mirex, endosulfan, and chlordecone. They were mostly used as termiticides and larvicides against soil-borne insects. Cyclodienes are relatively stable to ultraviolet light from the sun, and persistent in the soil. Chlordane, aldrin, and dieldrin are able to protect treated wood for over 60 years. Due to its environmental persistence, bioaccumulation in the wildlife food chain, and causing the development of insect-pest resistance, it was banned in the US in the 1980s. Unlike DDT and BHC, cyclodienes have a positive temperature correlation. Their activities increase as the temperature increases. Cyclodienes are believed to be neuro-active like DDT and BHC, but act by inhibiting GABA (g-aminobutyric acid) receptor [19, 20].

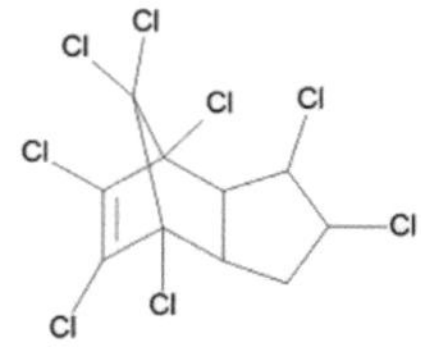

(7) Chlordane

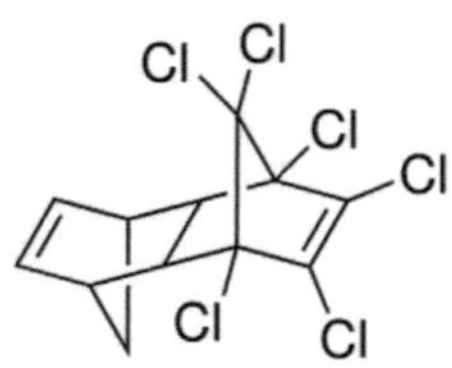

(8) Aldrin

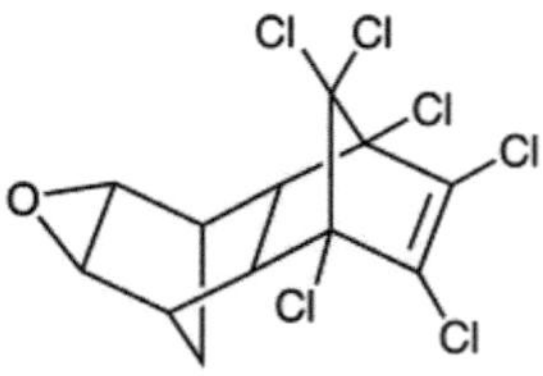

(9) Dieldrin

The polychloroterpenes include the toxaphene (10) and strobane (11), developed in 1947 and 1951, respectively. Toxaphene has been widely used alone or as a synergist with DDT and organophosphorus insecticides such as methyl parathion. Its mode of action is similar to that of the cyclodienes. Its use was discontinued in the US in 1983 by the USEPA due to its high aquatic toxicity [20].

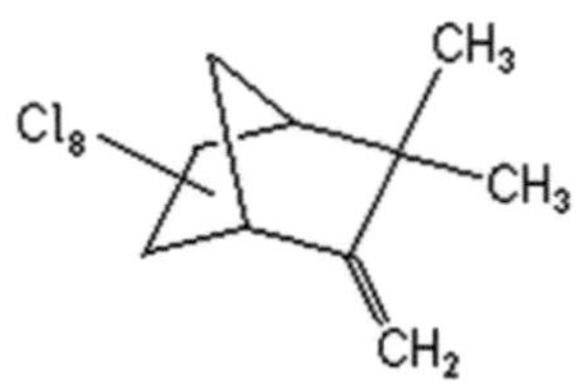

(10) Toxaphene

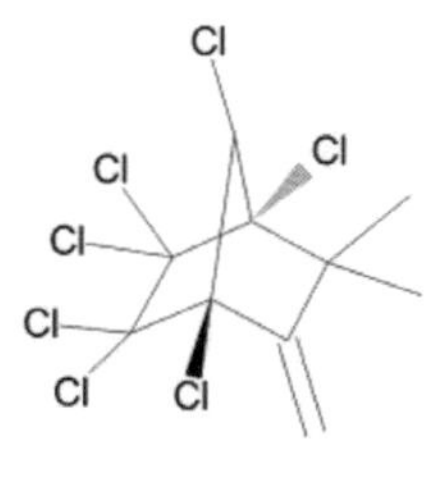

(11) Strobane

3.1.2 Organophosphates

These are phosphorus-containing organic compounds used as insecticides. Their insecticidal property was first noted during World War II when the extremely toxic organophosphate nerve gases, sarin, soman, and tabun, were studied for their structural similarities and as substitutes for nicotine, which was in short supply as insecticide in Germany. Organophosphates are generally more toxic to vertebrates than other classes of insecticides, and most are chemically unstable or nonpersistent. Their non-persistence made them more environmentally friendly and constituted a suitable substitute for organochlorines. Organophosphates inhibit cholinesterase, an enzyme of the nervous system, resulting in the accumulation of acetylcholine at the neuron/neuron and neuron/muscle (neuromuscular) junctions or synapses. Most organophosphates were discontinued for use in the 1990s for food-safety reasons [19, 20].

All organophosphates are esters of phosphorus having varying combinations of oxygen, carbon, sulfur, and nitrogen attached, resulting in six different subclasses: phosphates, phosphonates, phosphorothioates, phosphorodithioates, phosphorothiolates, and phosphoramidates. Generally, organophosphates are divided into three groups, viz.: aliphatic, phenyl, and heterocyclic derivatives. Tetraethyl pyrophosphate (TEPP, 12), and dichlorvos (13) are aliphatic organophosphate; ethyl parathion (14) and famphur (15) are phenyl derivatives, while diazinon (16) and chlorpyrifos-methyl (17) are heterocyclic derivatives [20].

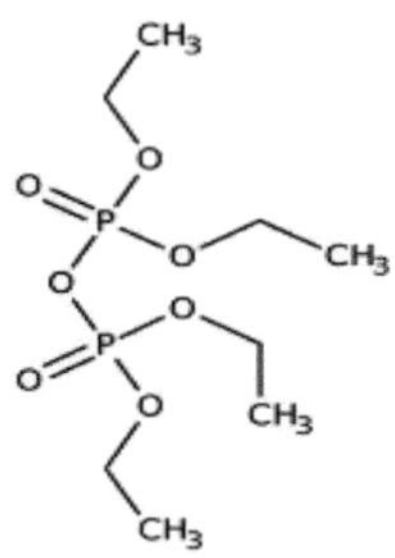

(12) TEPP

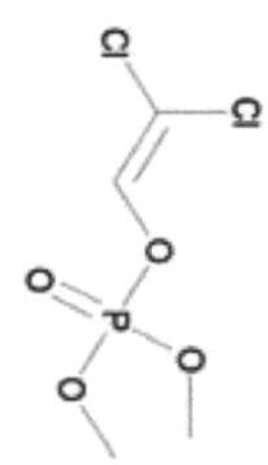

(13) Dichlorvos

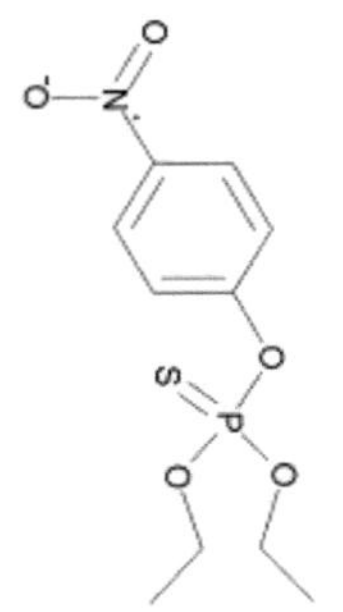

(14) Ethyl parathion

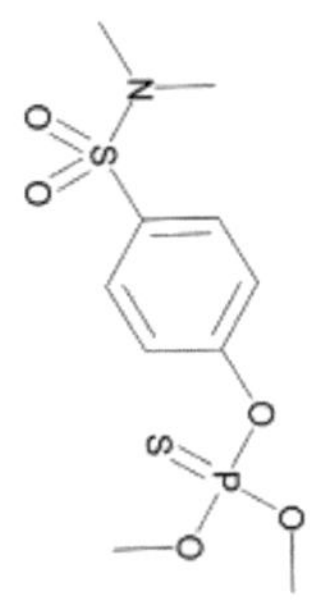

(15) Famphur

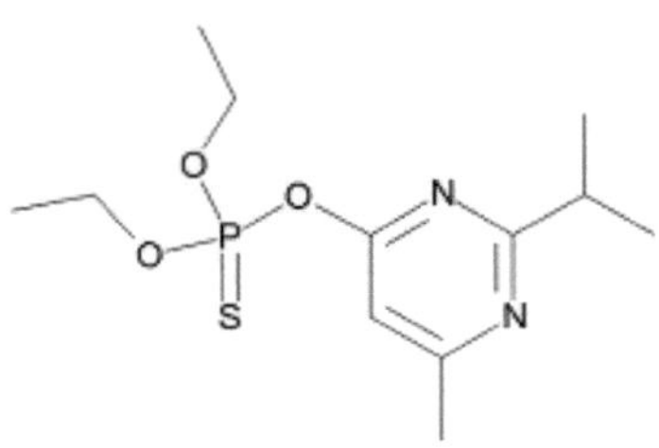

(16) Diazinon

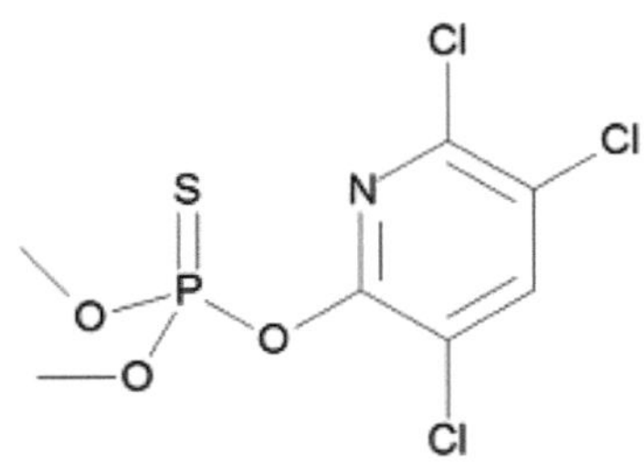

(17) chlorpyrifos-methyl

3.1.3 Organosulfurs

This class has two phenyl rings like DDT but has sulfur in place of carbon as the central atom. They are used as acaricides (miticides) for their low toxicity. Examples include tetradifon (18), propargite (19), and ovex (20) [20].

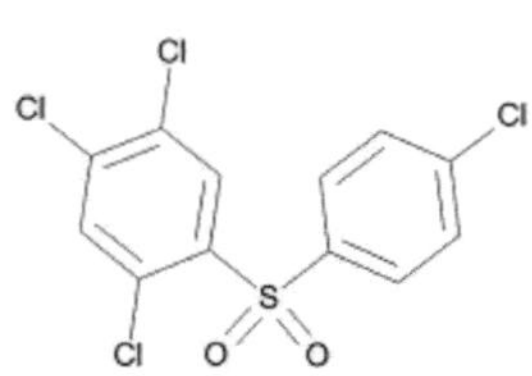

(18) Tetradifon

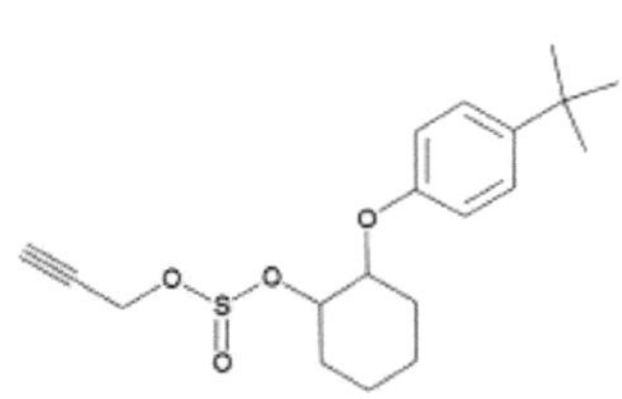

(19) Propargite

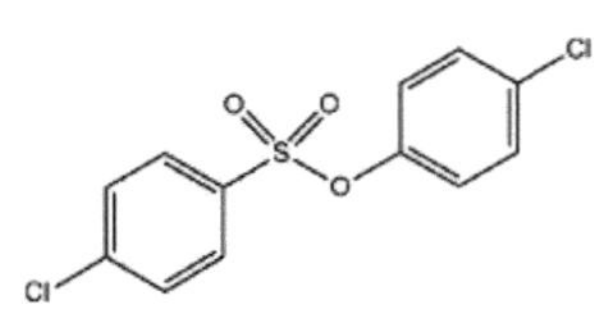

(20) Ovex

3.1.4 Carbamates

The carbamates are derivatives of carbamic acid just as organophosphates are from phosphoric acid. They act by inhibiting cholinesterase enzyme. Members of the class are carbaryl (21), methomyl (22), carbofuran, and primicarb (23), etc. [19, 20].

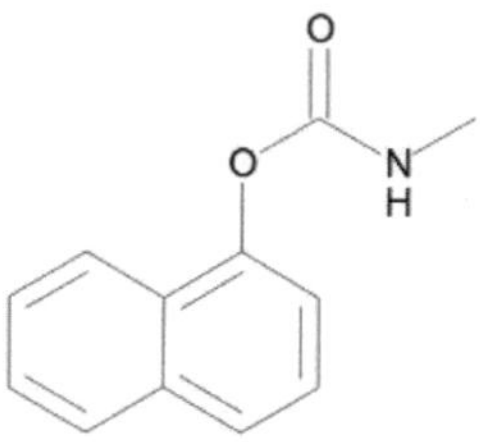

(21) Carbaryl

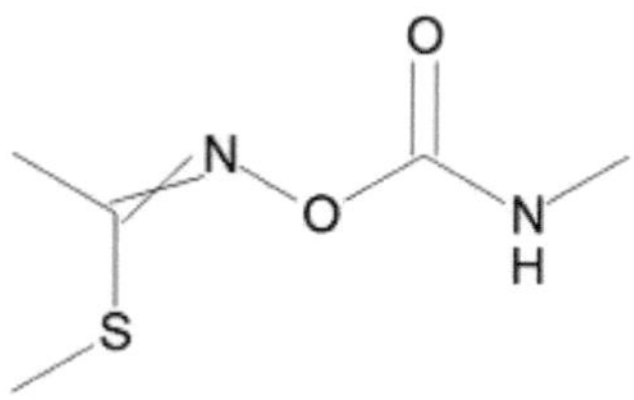

(22) Methomyl

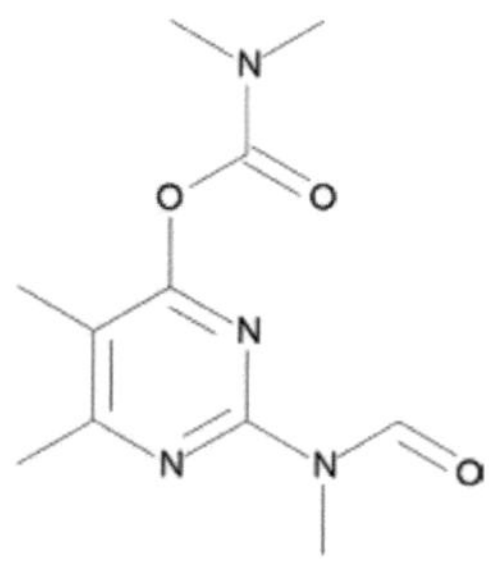

(23) Primicarb

Most carbamates exhibit an irreversible inhibition of cholinesterase, poisoning the insect central nervous system (CNS) [28].

3.1.5 Formamidines

Examples of this group include chlordimeform (24), formetanate (25), and amitraz (26). They are now mainly used in the control of organophosphate and carbamate-resistant pests. Formamidines inhibit monoamine oxidase responsible for degrading the neurotransmitters norepinephrine and serotonin leading to their accumulation. The high concentration of the neurotransmitters causes paralysis and death of the insect [20].

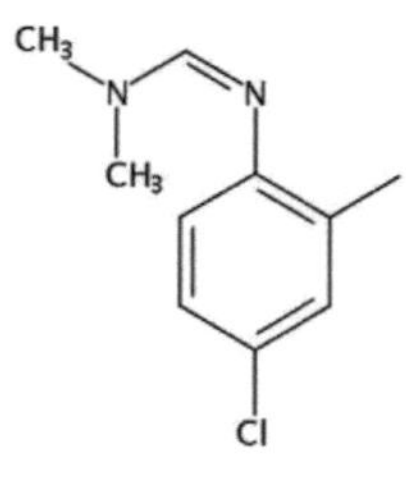

(24) Chlordimeform

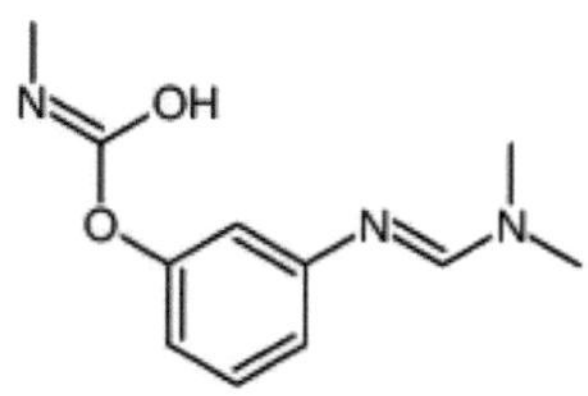

(25) Formetanate

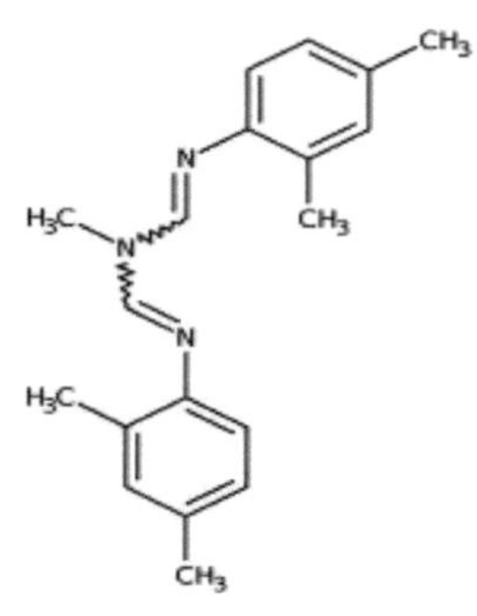

(26) Amitraz

3.1.6 Dinitrophenols

The dinitrophenols have a broad range of uses as insecticides, ovicides, herbicides, and fungicides. They act by inhibiting oxidative phosphorylation, preventing the generation of adenosine triphosphate (ATP). They have all been withdrawn from industrial use due to toxicity. Examples include binapacryl (27) and dinocap (28) [20].

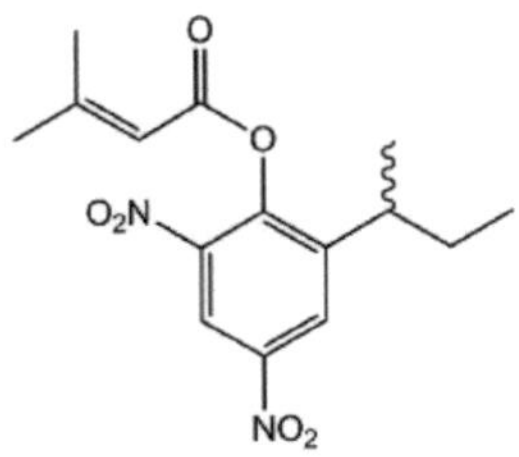

(27) Binapacryl

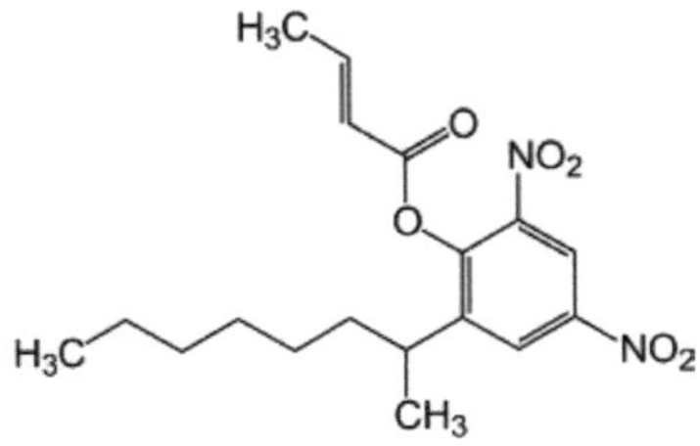

Dinocap (28)

3.1.7 Organotins

These are compounds containing tin (Sn) and used mainly to eliminate ticks and mites, although they are also good fungicides. They also act by preventing oxidative phosphorylation and the production of ATP. Examples are cyhexatin (29) and fenbutatin oxide (30) [20].

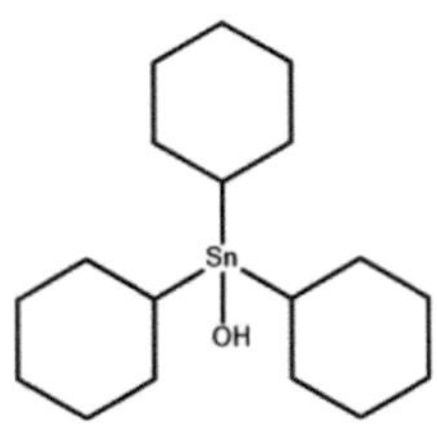

(29) Cyhexatin

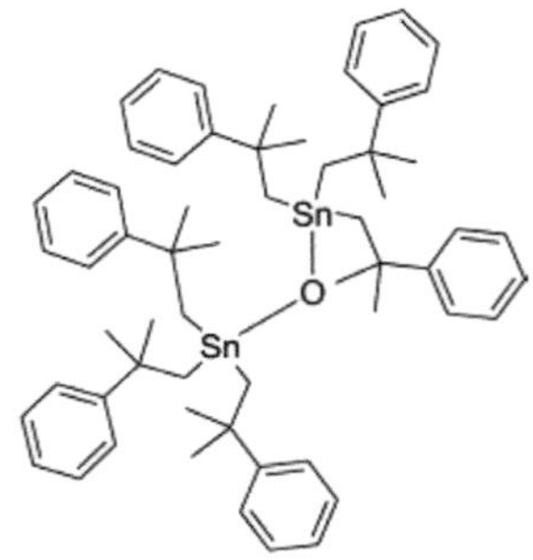

(30) Fenbutatin-oxide

3.1.8 Pyrethroids

These are synthetic pyrethrin-like substances created in recent decades to replace natural pyrethrum for agricultural purposes because of their lower cost and better stability in sunlight. Natural pyrethrums are unstable in sunlight and expensive to produce. They are generally effective against several agricultural insect pests at low doses of 0.0045 to 0.045 kg per acre. Pyrethriods act by attacking the sodium channel of the axon of the neurons like DDT, affecting both the peripheral and CNS of the insect. Examples include allethrin, tetramethrin, phonothrin, permethrin, cypermethrin (31), acrinathrin, imiprothrin (32), and *gamma*-cyhalothrin (33), etc. [19, 20].

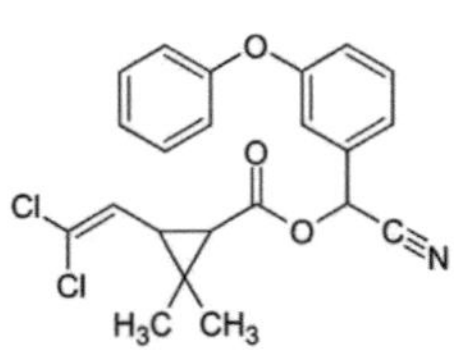

(31) Cypermethrin

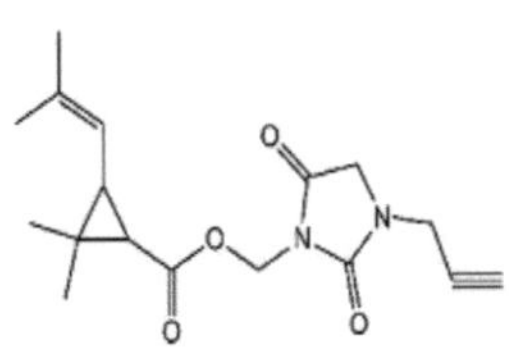

(32) imiprothrin

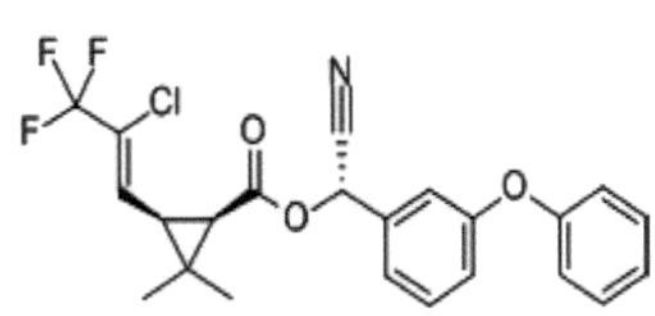

(33) *gamma*-Cyhalothrin

3.1.9 Nicotinoids

These are a class of synthetic insecticides that are similar to or analogs of the natural nicotine. They were previously referred to as nitro-quanidines, neonicotinyls, neonicotinoids, chloronicotines, and chloronicotinyls. An example is imidacloprid (34), which has root-systemic profile and notable contact and stomach action. It works by irreversibly blocking the postsynaptic nicotinergic acetylcholine receptors causing the collapse of the CNS and death of the insect. It is used as a soil, seed, or leaf treatment for the elimination and long residual control of sucking insects, soil insects, whiteflies, termites, turf insects, etc. Imidacloprid does not affect mites or nematodes [19, 20, 28–30]. Other examples include acetamiprid, thiamethoxam (35), thiacloprid, nitenpyram, clothianidin, dinotefuran (36) and clothianidin.

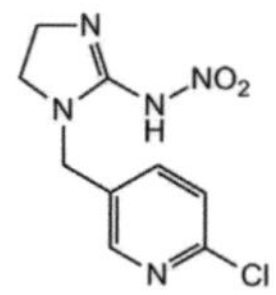

(34) Imidacloprid

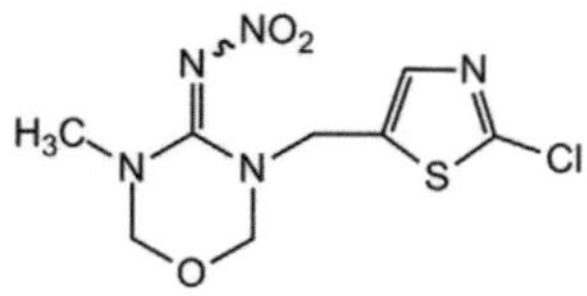

(35) thiamethoxam

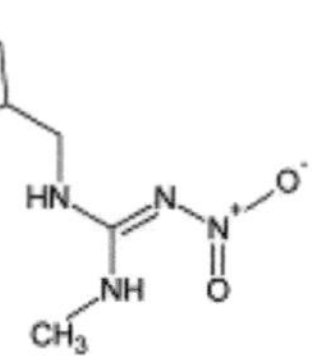

(36) dinotefuran

3.1.10 Fiproles (or phenylpyrazoles)

The only example is fipronil (37). It is effective against insects resistant or tolerant to pyrethroid, organophosphate, and carbamate insecticides. It acts by blocking the *g*-aminobutyric acid (GABA) regulated chloride channel in neurons, thus antagonizing the "calming" effects of GABA, similar to the action of the cyclodienes [20].

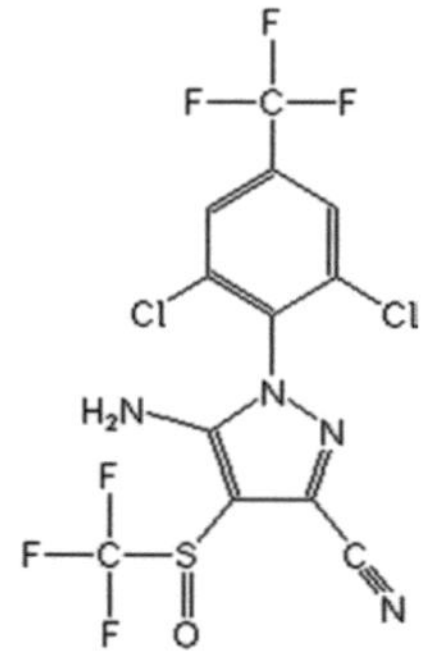

(37) Fipronil

3.1.11 Pyrroles

The only available member of this group is the chlorfenapyr (38). It acts by inhibiting oxidative phosphorylation and the production of ATP, mediated by the mitochondrial electron transport at the NADH-CoQ reductase site. They are effective against aphids, psylla, thrips, and whitefly [19, 20].

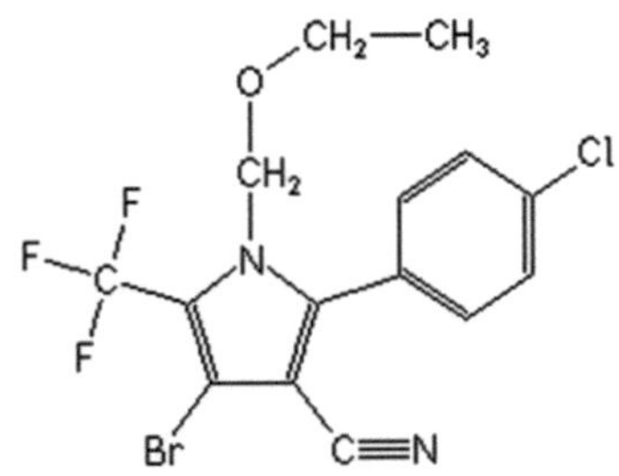

(38) Chlorfenapyr

3.1.12 Pyrazoles

Examples of this class include tebufenpyrad (39), fenpyroximate (40), ethiprole (41) and tolfenpyrad (42). They have broad activity against a wide range of chewing and sucking insects. They are contact and stomach active. Their mode of action is by inhibiting mitochondrial electron transport at the NADH-CoQ reductase site, leading to the disruption of adenosine triphosphate (ATP) formation [19, 20].

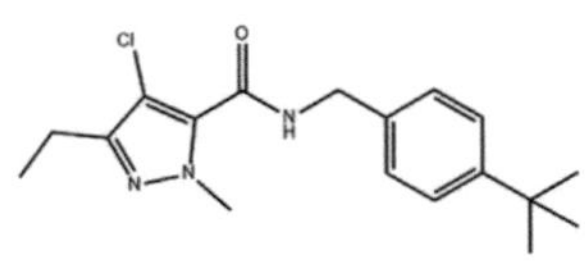

(39) Tebufenpyrad

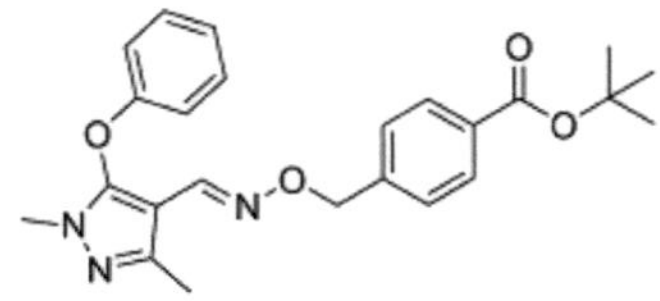

(40) Fenpyroximate

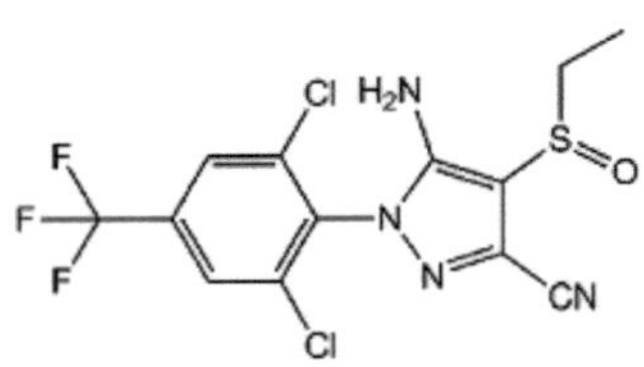

(41) Ethiprole

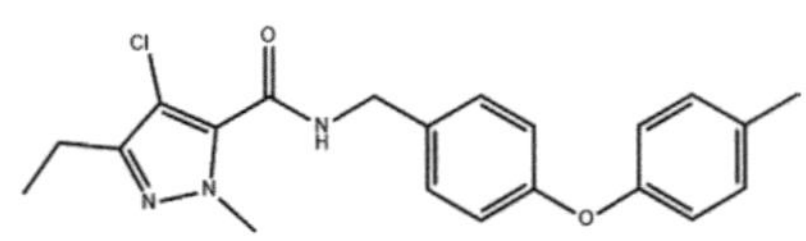

(42) Tolfenpyrad

3.1.13 Pyridazinones

Pyridaben (43) is the only member of this class. It is a selective contact insecticide and miticide, effective against thrips, aphids, whiteflies, and leafhoppers. It is a metabolic inhibitor that interrupts mitochondrial electron transport at site 1, similar to the quinazolines [19, 20].

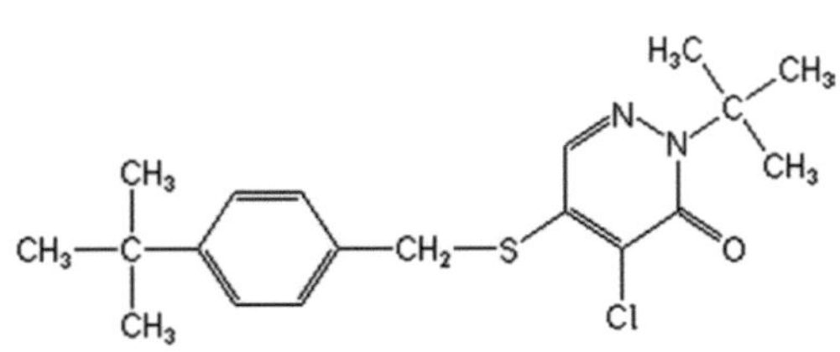

(43) Pyridaben

3.1.14 Quinazolines

The only example is fenazaquin (44), which is a contact and stomach active agent. It inhibits mitochondrial electron transport at Site 1, similar to the pyridazinones [20]. Quinazolines also act on the larval stages of most insects by inhibiting or blocking the synthesis of chitin in the exoskeleton. Developing larvae exhibit rupture of the malformed cuticle or death by starvation [19].

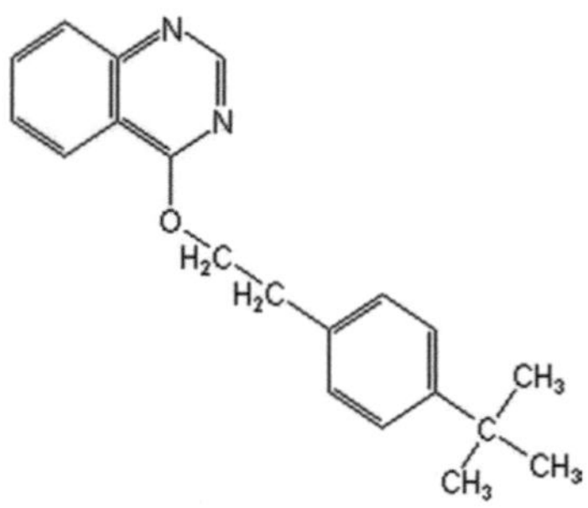

(44) Fenazaquin

3.1.15 Benzoylureas

Benzoylureas are quite a different class of insecticides that act as growth regulators in the insects by regulating the synthesis of chitin, rather than attacking the insect nervous system like the other poisons. They are taken up more by ingestion than by contact. Their greatest value is in the control of caterpillars and beetle larvae. Examples are triflumuron, chlorfluazuron, teflubenzuron, hexaflumuron, flufenoxuron, flucycloxuron, flurazuron, novaluron (45), lufenuron (46), diafenthiuron (47), bistrifluron, noviflumuron, diflubenzuron (48). Though not a benzoylurea, cyromazine (49), a triazine, is also a potent chitin synthesis inhibitor. Chitin synthesis inhibitors are very effective against the larval stages of most insects since chitin is the vital and almost indestructible part of the insect exoskeleton. The developing larvae are eliminated by the rupture of malformed cuticles or death by starvation. Adult female boll weevils exposed to diflubenzuron lay eggs that do not hatch. Thus, mosquito larvae control can be achieved with as little as 1.0 gram of diflubenzuron per acre of surface water [19, 20].

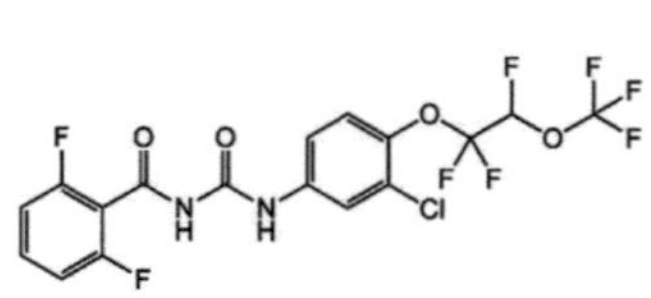

(45) Novaluron

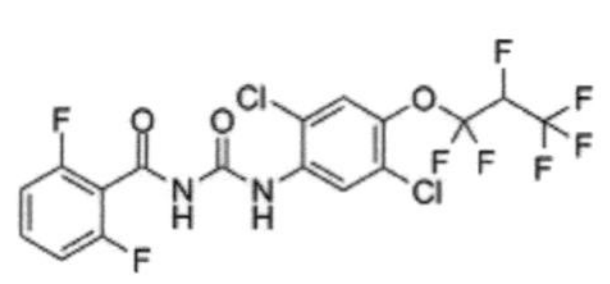

(46) Lufenuron

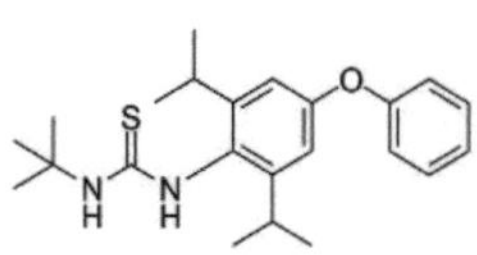

(47) Diafenthiuron

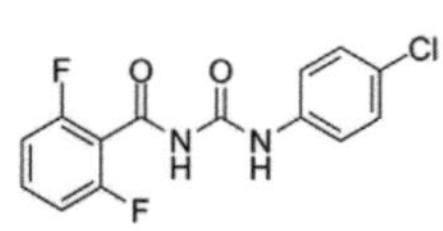

(48) diflubenzuron

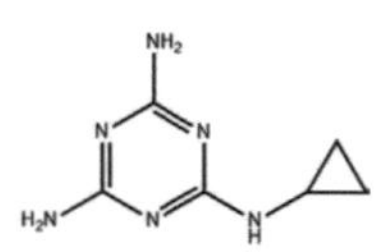

(49) Cyromazine

In the last decade, more than 34 insecticides have been launched with a shift from organophosphorus, carbamates, and synthetic pyrethroids to nicotinic and diamide insecticides known as neonicotinoids, which were used in Europe in 2005–2013 to protect wildlife such as pollinators (honey bees), mammals, birds, fish, amphibians, and reptiles [11]. Another advance in the development and use of insecticides to control pests was the introduction of groups of chemicals known as attractants and repellents to minimize the use of highly toxic chemical insecticides [27].

3.2 Biological control agents

Biological control agents encompass a range of living organisms that can regulate pest populations. Beneficial insects, such as parasitoids and predators, nematodes, fungi, and bacteria, are examples of biological control agents that offer a sustainable and eco-friendly approach to pest management. These agents can be integrated into pest control programs to reduce reliance on chemical insecticides and promote natural pest suppression mechanisms. A new class of biological insecticides has recently been discovered, and its use is gradually gaining importance, particularly in developed countries. They are the microbial insecticides, which use microbial metabolites from microorganisms to kill insect larvae. For example, *Bacillus thuringiensis* produces delta-endotoxin, which is active against lepidopterous and dipterous larvae [29]. Another microbial insecticide is beauvericin, a toxic substance from the fungus, *Beauveria bassiana* [31].

Also included among the biological insecticides is the spinosad, derived from naturally occurring soil bacteria, *Saccharopolyspora spinosa*. Spinosad is a mixture of two compounds, spinosyn A (50) and spinosyn D (51) (thus its name, spinosAD), and is used in organic farming. It is obtained as a fermentation metabolite of the soil-inhabiting microorganism *Saccharopolyspora spinosa* and, thus, can rightly be classified under biological insecticides. It is particularly effective as a broad-spectrum material for most caterpillar pests at the low rate of about 18 to 40 grams per acre. It has both contact and stomach activity against lepidopteran larvae, leaf miners, thrips, and termites, with long residual activity. It acts on the nervous system of insects by disruptively binding acetylcholine in nicotinic acetylcholine receptors at the postsynaptic cell [19, 20, 32].

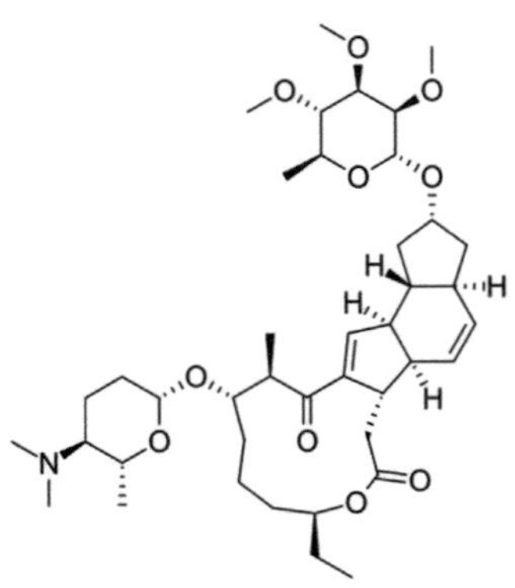

(50) Spinosyn A

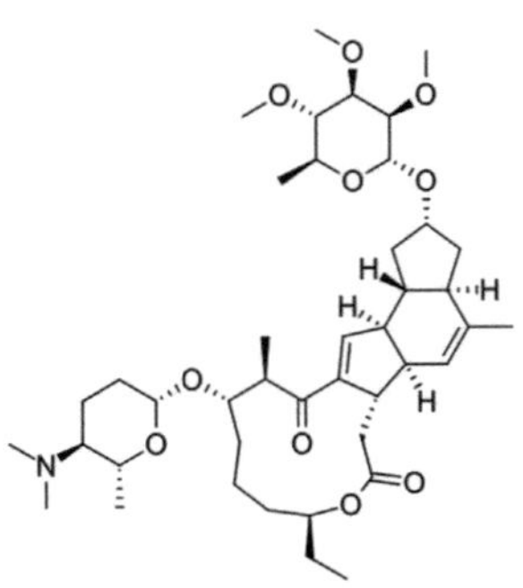

(51) Spinosyn D

Of great interest today are the insect growth regulators (IGRs), which have the capacity to disrupt the growth and development of insects by interfering with their hormonal regulation, the molting process, or reproduction. They include juvenile hormone analogs, chitin synthesis inhibitors, and ecdysone receptor agonists. Some natural chemical compounds such as cinnamyl phenols, though not insecticides themselves, possess sterilizing properties against adult mosquitoes and houseflies, and therefore help to control the insect population. In this way, they act remotely as insecticides and do not impact the environment or human health [33–35].

3.3 Botanicals

Natural insecticides derived from plants or plant products are perhaps the earliest insecticides used by man. Botanical extracts derived from plants contain compounds with insecticidal properties. Neem oil, camphor, pyrethrum, hellebore, quassia, turpentine, tobacco, and citrus extracts are examples of botanical insecticides that can repel or kill pests [20]. Other plant-derived insecticides such as cinnamaldehyde (52), eugenol (53), and limonene (54) are referred to as floral or scented plant chemicals.

Of particular economic importance among the plants in common use today are the pyrethrum and the tropical plant, *Azadirachta indica* A. Juss., popularly known as the neem tree. The neem oil is known to control over 25 species of insect pests, and the activity has been linked to the natural compound, azadirachtin (55), a tetranortriterpenoid found in the kernel. Azadirachtin is used on ornaments and in greenhouses. Azadirachtin is a powerful antifeedant that acts by disrupting insect growth regulators. It causes growth malformation and mortality in insect larvae. It disturbs insect development apparently by interfering with the biosynthesis, and release or action of juvenile and ecdysteroids such as the prothoracicotropic hormone (PTTH), allatotropins, and other growth regulators of insect molt [35, 36].

Natural nicotine (56) is an alkaloid derived from the leaves of *Nicotiana spp.,* especially *Nicotiana tabacum,* the tobacco plant [20]. Nicotine is an autonomic blocking agent acting like acetylcholine at the ganglia and neuromuscular junctions. Like the carbamates, nicotine acts by inhibiting cholinesterase activity responsible for the breakdown of acetylcholine resulting in acute cholinergic toxicity, and CNS disruption [28].

Rotenone (57) naturally occurs in *Derris, Lonchocarpus, Tephrosia,* and *Mundulea* species, particularly in the roots of *Lonchocarpus* species. Rotenone is a colorless and odorless crystalline isoflavonoid that is highly toxic to insects and fish, but moderately hazardous to humans and other mammals [37, 38]. Rotenone is used to eliminate ticks and lice on cats, dogs, and horses and mites on poultry. Rotenone acts by inhibiting the activity of mitochondrial respiratory complex I, generating excess free radicals resulting in cell death by apoptosis [39, 40].

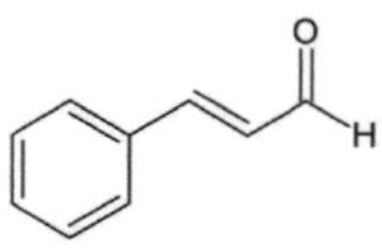

(52) Cinnamaldehyde

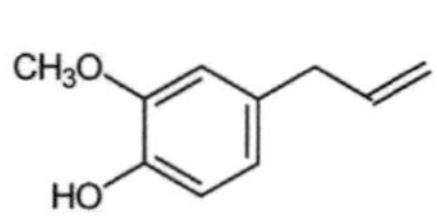

(53) Eugenol

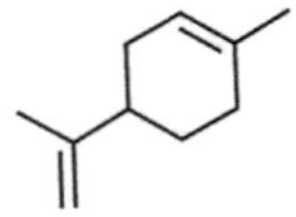

(54) Limonene

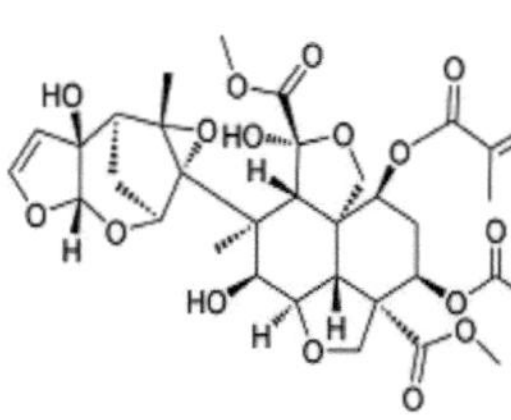

(55) Azadirachtin

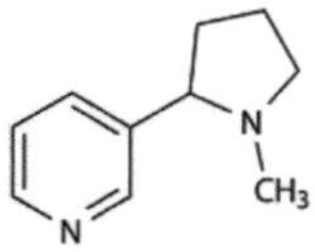

(56) Nicotine

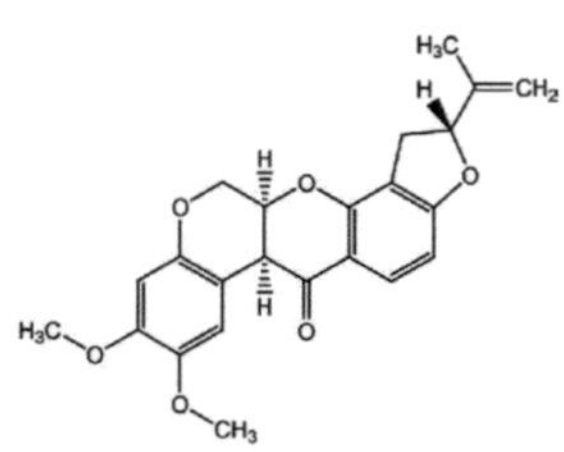

(57) Rotenone

Pyrethrins are a mixture of 6 compounds (pyrethrin I & II (58 & 59), cinerin I & II (60 & 61), and jasmolin I & II (62 & 63)) and are found in over 2000 commercial pesticides, as the active ingredient along with synergists, commonly used against many insect pests including ants, fleas, flies, moths, and mosquitoes. It excites the nervous system of the insects that feed on it causing paralysis and death [41, 42].

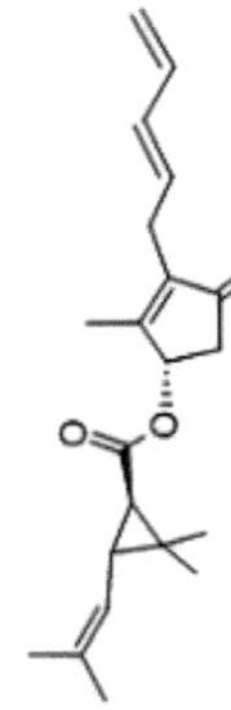

(58) Pyrethrin I

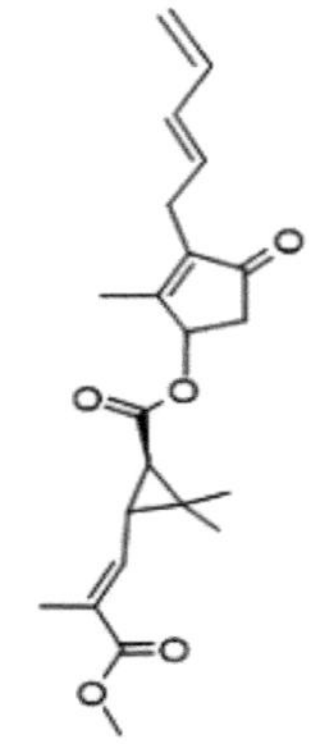

(59) Pyrethrin II

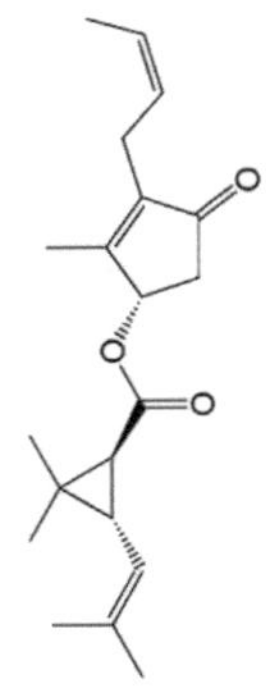

(60) Cinerin I

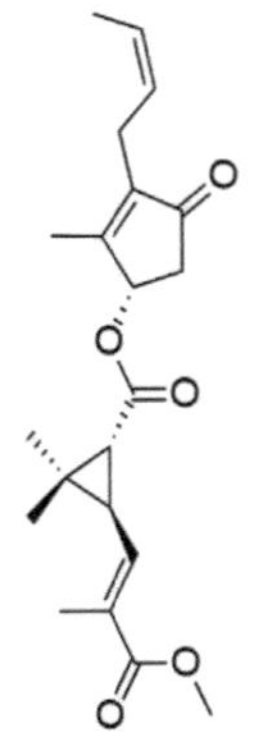

(61) Cinerin II

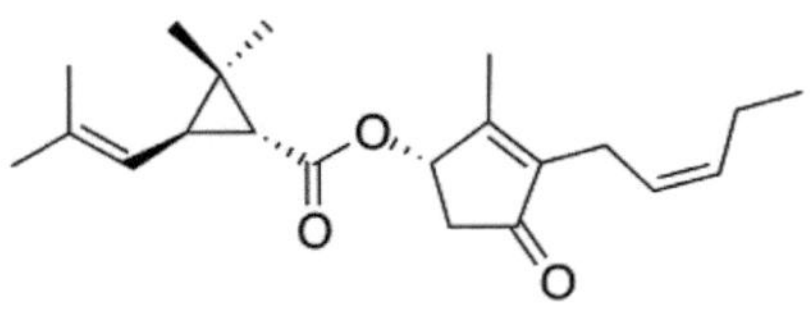

(62) Jasmolin I

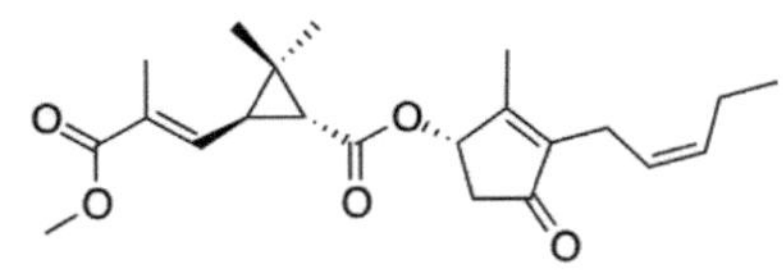

(63) Jasmolin II

These natural alternatives are favored for their lower environmental impact and reduced toxicity to non-target organisms, making them suitable for organic farming and integrated pest management practices. The number of plants that have been investigated for their efficacy and constituents has continued to increase in the past decade [43]. Also, the effectiveness of nine Chinese plant species has been compared with synthetic insecticides against 40 species of insects [43–46].

Some Nigerian plants such as *Piper guineense*, certain *Dalbergia spp.*, such as *Dalbergia saxatilis* and *Dalbergia lactea*, and essential oils from *Eucalyptus spp.*, like *E. citriodora* and *E. globulus*, and *Cymbopogon spp* (lemon grass) have been studied in some detail and found to exhibit attractant, repellent, and/or insecticidal activities and are worthy of exploitation for commercial purposes. Of particular interest is guineensine (64) which was isolated from the fruit of *Piper guineense* and has been synthesized [45–50]. The piperamide, guineensine, is an N-isobutylamide alkaloid that is insecticidal to many insects such as bean weevil, and mosquitoes (*Culex pipiens pallens* and *Aedes aegypti*) and thus often used as an insect repellant and antifeedant [51, 52]. Piperamides act as neurotoxins to insects through a pathway that does not interfere with the neurotransmitter system but through the inactivation of voltage-gated sodium ion channels that regulate neuronal intracellular calcium ion release creating Ca^{2+} toxicity. Piperamides also inhibit cytochrome P-450 dependent polysubstrate monooxygenase (PSMO) in insects preventing the detoxification of toxicants, and enhancing the effect of the toxic insecticide(s) [19, 52]. Piperamides are good synergists with additive effects to insecticides like pyrethrum. Similar insecticidal piperamides from *Piper nigrum* include pipercide, pellitorine, and piperine [52]. Most piperamides are easily degraded under sunlight. Evidence has shown that they possess a half-life of about 40–50 minutes under ultraviolet light [52].

Citronella oil from *Cymbopogun citratus* (D.C.) Stapf, and eucalyptus oil are strong repellants of some species of *Anopheles* mosquitoes, especially *An. Stephensi*. Citronellal (65) and eucalyptol (or 1,8-cineole) (66) are the major constituents of citronella oil and eucalyptus oil, respectively [49, 53, 54]. Linalool, a monoterpenoid and constituent of many essential oils, influences the ion transport and acetylcholine esterase release in insects. Some other insecticidal constituents of essential oils have been demonstrated to act through octopamine-mediated modification of insect physiology by activating the octopaminergic receptor [55–57]. Limonene affects the sensory nerves of the peripheral nervous system [19].

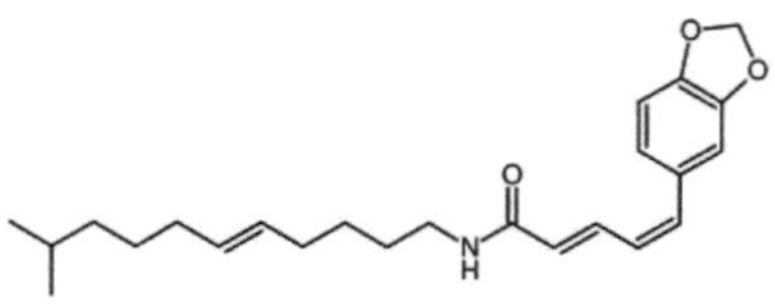

(64) Guineensine

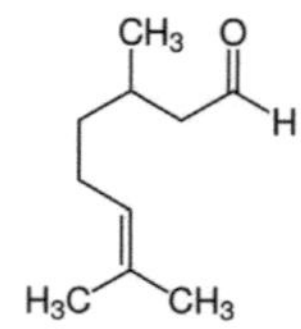

(65) Citronellal

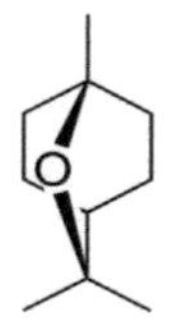

(66) Eucalyptol

The list of some commercial botanical insecticides, the constituents, mode of action, and targets is shown in **Table 1**.

Plant Name	Product/trade name	Active compound	Group/mode of action	Targets
Adenium obesum (Forssk.) Roem. & Schult.	Water extract of Chacals Baobab (Senegal)	Chacal Baobab extract	Insecticidal	Cotton pests, particularly the larvae of ballworm (*Heliotis spp.*), spiny ballworm (*Earias sp.*), and Sudan ballworm (*Diparopsis watersi*)
Azadirachta indica L.	Neem oil, Neem cake, Neem powder, Bionimbecidine (GreenGold)	Azadirachtin	Repellent Antifeedant Nematicide Sterilant Anti-fungal	Dandruff, Eczema, Nematodes, Sucking and chewing insects (caterpillars, aphids, thrips and maize weevils)
Chrysanthemum cinerariaefolium (Trevir.) Vis.	Pyrethrum	Pyrethrins	Insecticidal	Crawling and flying insects such as cockroaches, ants, mosquitoes, termites, bees
Citrus trees	Essential oil	δ-Limonene Linalool	Contact poison	Fleas, aphids, mites, paper wasps, house crickets, dips for pets, bugs, blister beetles
Derris elliptica (Wall.) Benth.	Rotenone	Rotenone	Insecticidal	Aphids, bean leaf beetle, cucumber beetles, leafhopper, red spider mite
Lonchocarpus spp	Rotenone	Rotenone		

Plant Name	Product/trade name	Active compound	Group/mode of action	Targets
Nicotiana tabacum L.	Tobacco	Nicotine	Insecticidal Antifungal	Aphids, thrips, mites, bugs, fungi, gnats, leafhoppers
Ryania speciose	Ryania	Ryanodine	Insecticidal	Caterpillars, thrips, beetles, bugs, aphids
Schoenocaulon officinale (Schltdl. & Cham.) A.Gray	Sabadilla dust from the seed	Veratran D and Cevadine	Insecticidal	Flies, caterpillars, potato leafhopper

Table 1.
List of some commercial botanical insecticides (From: Okwute [43]).

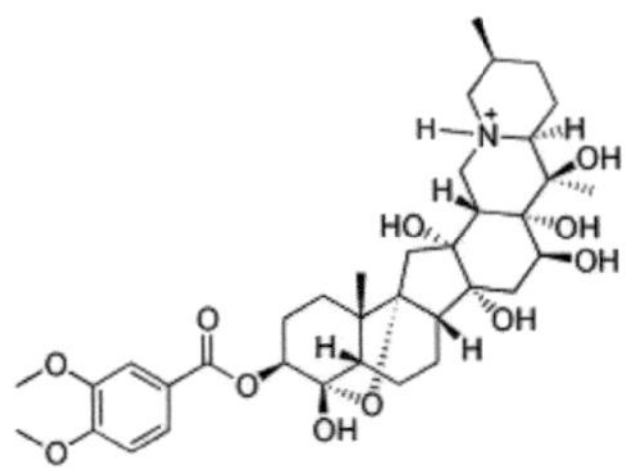

(67) Veratran D (or Veratridine) [42, 56]

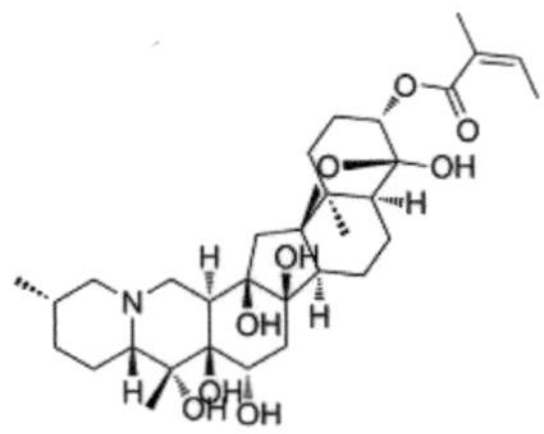

(68) Cevadine

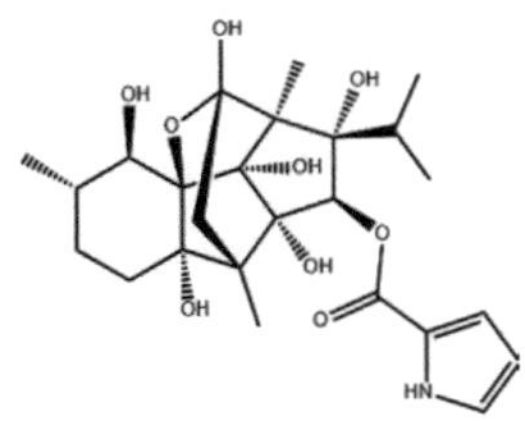

(69) Ryanodine [57]

4. Challenges of insecticidal agents in pest control

There is no doubt that the use of insecticidal agents has contributed enormously to agricultural productivity and global healthcare, but it is evident that over time because of intensive use of insecticides the environment has been saturated with toxic chemicals and some of these have impacted negatively on non-intended targets. Thus, indiscriminate use of insecticidal agents has the potential to upset the ecosystem [58]. One of the primary challenges associated with insecticidal agents is the development of resistance in pest populations. Prolonged exposure to the same insecticide can lead to genetic adaptations in pests, rendering the agent less effective over time. To combat resistance, integrated pest management strategies that combine multiple control methods have been recommended [59].

Environmental impacts are another significant concern related to insecticidal agents, as shown in **Figure 1**. Chemical insecticides can persist in the environment, contaminating soil, water sources, and non-target organisms. This can disrupt ecosystem dynamics, harm beneficial insects, and pose risk to human health. For example, some common household insecticides used in Nigeria have been shown to induce oxidative stress in Wistar rats [60]. Also, insecticide poisoning fatalities in children have been reported in some countries including South Korea, South Africa, Canada, Turkey, India, and Tanzania [3, 9, 18, 19]. Carbamate and organophosphate insecticides are very toxic to humans. The potential hazards of insecticides from environmental pollution, ozone depletion, persistence, and bioaccumulation, have made the search for new and environmentally friendly insecticides an imperative. This has

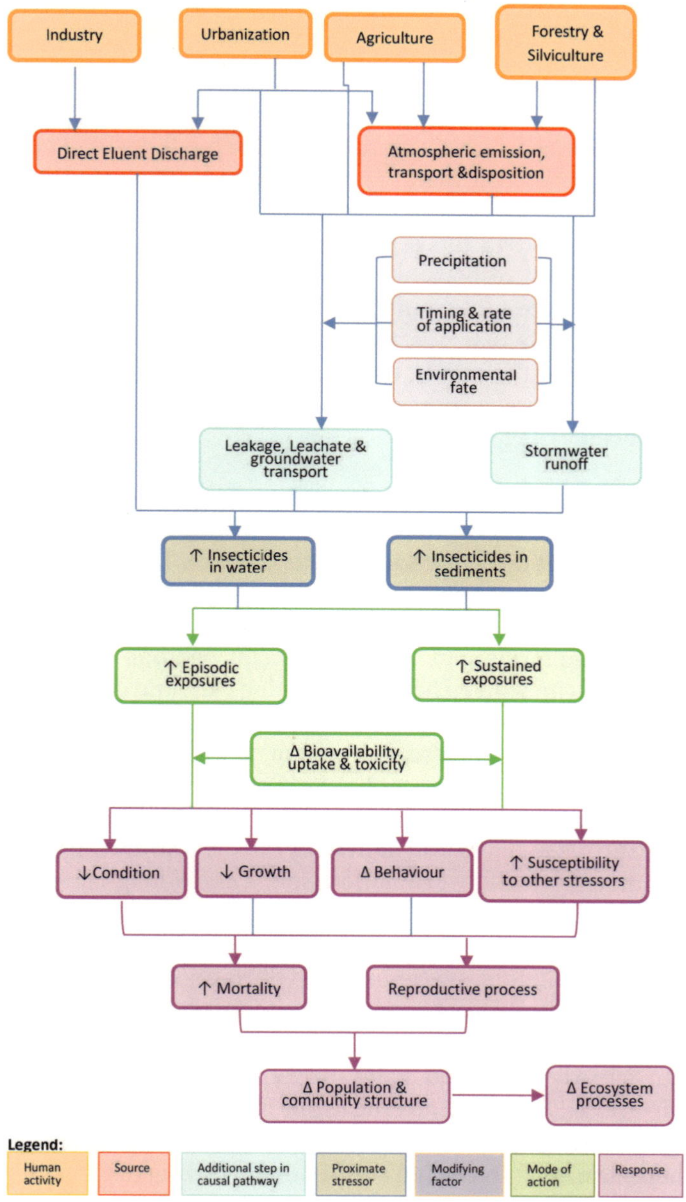

Figure 1.
Environmental pollution pathway for insecticides' use [19].

led to a renewed desire to develop new and environmentally friendly biological and botanical insecticides for agricultural and healthcare purposes [3, 19, 55, 61, 62]. However, the short half-life of many biological and botanical insecticides is also a concern. Many natural insecticides are degraded under sunlight and in the soil within 48 and 240 hours, respectively, requiring frequent and repeated application for optimum desired effect [52].

5. Remedial approaches to the challenges

The desire to improve agricultural yield and eliminate insect-vector diseases has led to the increasing use of insecticides over the decades, despite growing health and environmental hazards, and the destruction of global biodiversity. The global use of pesticidal agents, particularly insecticides, increased between 1990 and 2017 by about 80% leading to increased yields in agricultural products and sharp drops in their market prices [63]. Safety concerns surrounding insecticidal agents therefore highlight the importance of responsible application and handling. Improper use of chemical insecticides can result in human exposure, pesticide residues in food products, and toxicity to wildlife [63]. Sustainable pest control practices aim to minimize environmental impacts by promoting the judicious use of insecticidal agents and exploring alternative solutions. Some of the steps that need to be taken to remediate these challenges include the following:

5.1 Responsible handling of insecticidal agents and regulatory oversight

In the home, to combat disease-vector insects there has been extremely careless application of very toxic insecticides usually as sprays in very high concentrations. For most African dwellers, in most cases personal protective equipment (PPE) are not worn when spraying farms, and homes (especially bedrooms and living rooms), and perhaps not much gap is given between spraying time and cultivation or sleeping time. This practice, which is a gross display of ignorance of the hazardous effects of chemical insecticides, is very dangerous to human, animal, and environmental health. It is perhaps the most hazardous aspect of our effort to eradicate the destructive effect of insect invasion in agricultural practices, and the spread of insect-vectored diseases like malaria and trypanosomiasis transmitted by mosquitoes and tsetse flies. For instance, non-adherence to safety precautions in the use of long-lasting insecticidal nets (LLINs), and failure to use PPE in indoor residual sprayings (IRS), could be potential sources of insecticide hazards [59]. The use of facemask, chemical-resistant gloves, goggles, and respirators is recommended during spraying of farms, homes, and environment to minimize chemical hazards. Care should also be taken to avoid contact with non-targeted animals and plants. For highly toxic chemical insecticides, a number of countries, particularly the United States, have enacted laws to ban some insecticides. For example, the United States banned the use of insecticides of the group chlorpyrifos in 2021, while Brazil, China, and Europe have planned to do the same [18, 64, 65]. Some researchers have also established a list of about 18 insecticides considered safe to comparatively natural enemies [14]. This is expected to limit the indiscriminate use of very toxic chemical insecticides in the environment and target specific crops and pests.

Regulatory oversights and education programs are also essential to ensure insecticidal agents' safe and effective use while minimizing risks to human health and the environment. Such policies should control the indiscriminate sale, purchase, and use of chemical insecticides that have become common tools for committing murder and suicide by food and drink poisoning, particularly in the developing countries where insecticidal products are displayed in the open markets. Thus, in many Sub-Saharan African and South Asian countries, while there may be registration of agrochemical businesses, the pesticides are unlabeled and sold in the open markets [17]. Thus, treaties such as the Stockholm Convention, and local national regulatory agencies should be strengthened or empowered to perform their oversight functions [65].

5.2 Use of combined intervention methods

Another approach to the reduction of insecticide concentration in the environment is perhaps to use a combination of small quantities of insecticides in combination with non-toxic pesticides such as insect attractants or sex pheromones, which have the capability of bringing insects in mass to a location where the insecticide is applied. During the mating season, the male insects are attracted by the pheromones produced and released by their female counterparts. Hence, the use of pheromones as additives helps attract the insects for insecticidal action of applied insecticide(s), and also to disrupt their natural mating cycle. Undesired biodiversity destruction can be avoided by the use of selective pesticidal natural biomolecules, also called biorational products. Biorationals exhibit selective toxicity and have relatively low or no direct toxicity to non-targeted organisms [66]. Some secondary metabolites of plants act as allelochemicals stimulating antixenotic intrinsic and extrinsic defense against herbivorous insects. Thus, allelochemicals can act as kairomones (defensive stimulus for adaptive benefit toward another species), allomones (defensive stimulus for adaptive benefit toward the same species), or synomones (defensive stimulus for adaptive benefit toward same species and other species) depending on the effect they have on the specific insect [66]. The use of pheromones (or allelochemicals) can reduce the quantity and the applicable space for the application of the insecticide, especially in cases involving sprays, promoting environmental safety and cost-effectiveness in insect pest control. On the other hand, use of insect repellents, antifeedants, and deterrents that do not primarily aim to kill the insects ensures that no real toxic chemical insecticide will be applied, as toxic and staining repellents are not suitable for human use, especially against mosquitoes and flies [67, 68].

5.3 Use of botanicals

In the face of insect resistance to synthetic insecticides, their high toxicity to non-target organisms, and environmental persistence, attention has been shifted to plant-based pesticidal plants in the past few decades. The folkloric use of higher terrestrial plants by the natives of various parts of the world as pesticidal materials has been well known [69, 70]. Perhaps, one of the early plants so recorded as pesticidal material was tobacco (*Nicotiana tabacum*). The use of tobacco leaf infusion to kill aphids led to the isolation of the alkaloid, *nicotine*, while the chemical investigation of the Japanese plant, *Roh-ten* (*Rhododendron hortense*) in 1902 showed *rotenone* as the active constituent [4]. In this class of age-old pesticidal plants are species belonging to the genus *Chrysanthemum* found in Kenya and other highlands in Africa, which are the sources of the all-purpose and very successful insecticidal extract, *pyrethrum*, and the active constituents, the *pyrethrins* [71]. The relative success of these early botanical insecticides gives impetus to the current search for new and eco-friendly insecticides from plant sources.

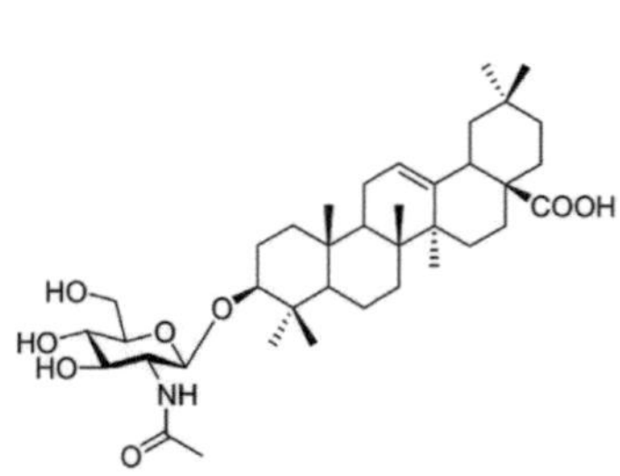

(70) Aridanin

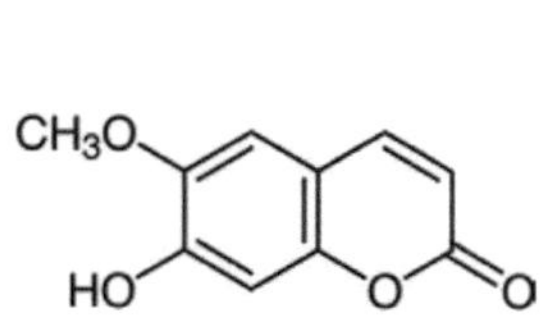

(71) Scopoletin

Family	Species	Parts screened	Form	Activity	Organisms	Compounds
Annonaceae	*Dennettia tripetala*	Leaf	Extract	Protectant	Weevils	β-phenyl-nitroethane Terpenes
Fabaceae	*Dalbergia saxatilis*	Stem bark & leaf	Extract & Powder	Insecticidal	Mosquitoes Housefly	Unidentified
Melliceae	*Azadirachta indica*	Seeds	Powder	Protectant	25 species of plant pests	Limonoids
Mimosoidae	*Tetrapleura tetraptera*	Root & stem bark LeafFruit	Extracts Seed oil	Insect repellant Insecticidal Larvaticidal	*Callosobruchus maculatus* (cowpea weevil) *Anopheles Gambiae*	Aridanin (70) transgeranylgeraniol Scopoletin (71) [72–74]
Amaranthaceae	*Alternanthera sessilis* L.	Leaf	Extracts	Larvaticidal	Culex quinquefasciatus (mosquito larvae)	Phytosterols, flavonoids, and terpenes [75–77]
Piperaceae	*Piper guineense*	Fruits	Powder Extract	Insecticidal	Weevils Grasshoppers	Alkaloids
Rutaceae	*Clausena anisata*	Leaf Root	Extract	Insecticidal Antifeedant	Grasshopper Armyworm	Alkaloids Coumarins

Table 2.
Biological and phytochemical screening results on some Nigerian pesticidal plants (Source: [45]).

Sample	% of insects (dead or moribund)	Effective concentration (ppm)
Piperine	50.0	5000
Guineensine	35.0	500
Chloroform extract	30.0	1000
Petroleum ether extract	20.0	1000

Table 3.
Percentage of insects dead or moribund in 1 hour of treatment with extracts and extractives from Piper guineense (Source: [45]).

The search for insecticidal plants, including repellents, antifeedants, and deterrents, has been extensive and successful particularly in Africa and Asia, including some Nigerian plants (**Table 2**) [45, 78, 79].

Prominent among the plants in common use today is the tropical plant *Azadirachta indica*, popularly known as the neem tree. In India as well as in Nigeria the plant is effectively used to control over 25 different species of insect pests. The activity has been associated with the presence of azadirachtin, which is said to be eco-friendly and safe and the concentration is highest in the kernel than in the leaves and other tissues of the plant [5, 44].

5.4 Use of synergists

Another approach to avoiding the application of high concentrations of highly toxic chemical insecticides is the use of synergists along with moderately toxic insecticides. Synergists have the capability to potentiate moderately active insecticides. For example, piperine (72) has been shown to act synergistically with guineensine in its activity to control the garden insect, *Zonocerus variegatus* [45]. **Table 3** shows the comparative activities of the crude petroleum and chloroform extracts of *Piper guineense* and the insecticidal activities of its constituents, guineensine and piperine. The presence of a synergist provides for use of a limited quantity of the toxic insecticide though the exact mechanism is not quite understood.

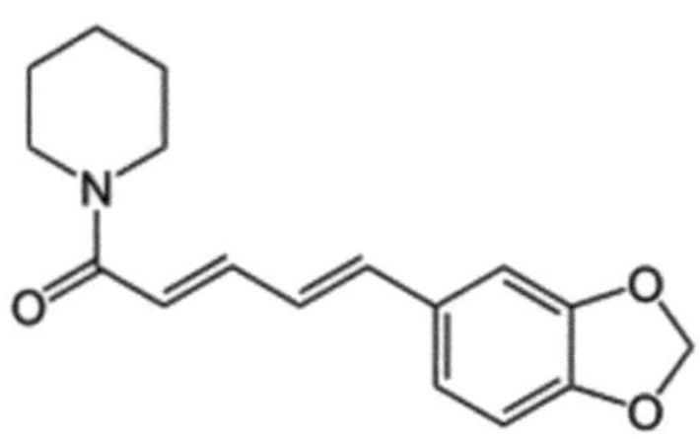

(72) Piperine

6. Advantages of insecticidal agents in pest control

Despite the associated challenges, insecticidal agents offer several advantages in pest control. These agents provide efficient and targeted control of pest populations, helping to protect crops, reduce economic losses, and safeguard public health. By selectively targeting pests, insecticidal agents can minimize damage while preserving beneficial organisms in the ecosystem. Thus, they have contributed to increased food production, profits for farmers, and prevention of diseases. Globally, pesticide use has increased since the 1940s just as the average life expectancy. Data in the United States showed that the life span increased from 68 years in 1950 to 70.5 years in 1970 and was projected to increase to 75 in 1990 [13]. It has been estimated that between 1945 and 2018, the use of pesticides has prevented the death of about seven million people by killing pests that transmit diseases such as malaria caused by mosquitoes, bubonic plague carried by rat fleas, and typhus carried by body lice [80]. It has been reported that 17% of human infections are vector-borne, causing about 700,000 deaths annually, with 400,000 deaths attributable to malaria, of which Sub-Saharan Africa accounts for over 90%. However, a substantial achievement has been recorded in malaria control between 2000 and 2015 using a combination of interventions involving long-lasting insecticide-treated nets (LLINs), indoor residual spraying (IRS), and prompt management with ACTS therapies [81]. In Nigeria, a plant disease sarcastically labeled *tomato ebola*, which was quickly stopped from spreading in 2016 by the use of insecticides, is a strong demonstration of the importance of pesticides in modern agriculture. For the period it lasted, there was a serious scarcity of tomatoes leading to the collapse of some tomato paste–producing companies, and thus affecting the national economy [9].

Also, an FAO report has suggested that to meet the demands of the increasing global population, agriculture will need to produce almost 50% more food, feed, and biofuel in 2030 than it did in 2012, when the world population may reach 9.73 billion. This is particularly very important in Sub-Saharan Africa and South Asia where agricultural output will need to be doubled. This target can only be achieved with heavy and intensive application of insecticides and new technological innovations [82].

7. Search for new insecticidal agents and future research

In solving the challenges of insect resistance, health and environmental toxicity, climate change, and corrosion of household and agricultural tools/utensils associated with the use of existing synthetic insecticidal agents, current research initiatives must focus on natural products as sources of new insecticidal molecules. While natural products may be the source of structural guides to new molecules, structural modifications may be necessary to improve (i) stability, (ii) specificity, (iii) effectiveness, (iv) environmental tolerance, and (v) commercial viability. Evidence has shown that the mode of action of acaricidal agents is diverse with some agents modulating several pathways in exterminating the insect pest. However, structurally related substances employ similar pathways for their mode of action [83, 84]. Exploration of existing promising botanical insecticidal agents through chemical and/or biological modification to create new moieties that address existing challenges of resistance management, health and environmental toxicity, and depletion of the ozone layer is a necessary approach [85]. There must be renewed search for new natural chemical moieties from plants, animals, and microbes, and future research should cover the deployment of emerging technologies such as nanotechnology, co-crystallization techniques, computational chemistry, and genetic engineering such as mRNA technology to create nucleopolyhedroviruses [86], for the development of new insecticides to control environmental releases and effectiveness, safety in humans, and specificity. New researches must adopt lessons learned from human pharmaceutical research and development to unravel new insecticide targets and increase target diversity [87].

8. Conclusion

In conclusion, insecticidal agents play a critical role in pest control by providing efficient, targeted, and versatile solutions for managing pest populations. While challenges such as resistance, environmental impacts, and safety concerns must be addressed, the advantages of insecticidal agents in pest control are significant. One important approach to addressing the hazardous environmental impact of current commercially available insecticides is focusing on botanical pesticides to guide the development of new generation of insecticides for industrial, agricultural, and healthcare use. In addition, future researches must employ emerging technologies such as nanotechnology, co-crystallization, computational chemistry, genomic, and lessons learned from human pharmaceutical science to provide feasible solutions to existing and potential threats and challenges. By harnessing the strengths of diverse insecticidal agents and integrating them into sustainable pest management practices, we can mitigate the impact of pests on agriculture, promote environmental stewardship, and ensure food security and quality healthcare for future generations.

Acknowledgements

The authors wish to acknowledge the support of the management of their institutions during the development of the manuscript.

Conflict of interest

The authors declare no conflict of interest. There was no specific funding for the development of this work.

Author details

Simon Koma Okwute[1*] and Henry Omoregie Egharevba[2]

1 Faculty of Natural and Applied Sciences, Department of Pure and Applied Chemistry, Veritas University, Abuja, Nigeria

2 Department of Medicinal Plant Research and Traditional Medicine, National Institute for Pharmaceutical Research and Development, Abuja, Nigeria

*Address all correspondence to: profokwute@yahoo.com

References

[1] Maton SM, Dodo JD, Nesla RA, Ali AY. Environmental impact of pesticides usage in farmlands in Nigeria. International Journal of Innovative Research and Development. 2016;**5**(4):311-317

[2] Olufemi P. Introduction to Public Health Pest Management. 1st ed. Ibadan: Ibadan Book Wright Nigeria Publishers; 2009. pp. 1-180

[3] Oaya CS, Malgwi AM, Degri MM, Samaila AE. Impact of synthetic pesticides utilization on humans and the environment: An overview. Agricultural Science and Technology. 2019;**11**(4):279-286. DOI: 10.15547/ast.2019.04.047

[4] Tooley P. Crop protection. In: Food and Drugs: Chemistry in Industry. London: John Murray; 1971. pp. 92-160

[5] Kulkarni J, Kapse N, Kulkarni DK. Plant-based pesticides for control of *Helicoverpa amigera* on cucunmis. Asian Agricultural History. 2009;**13**(4):327-332

[6] Agrios GN. Transmission of plant disease by insects [Internet]. n.d. Available from: https://www.entnemdept.ufl.edu/capinera/eny5236/pest1/content/03/3_plant_diseases.pdf [Accessed: May 16, 2024]

[7] Laroche M, Bérenger JM, Delaunay P, Charrel R, Pradines B, Berger F, et al. Medical entomology: A reemerging field of research to better understand vector-borne infectious diseases. Clinical Infectious Diseases. 2017;**65**(1):30-38. DOI: 10.1093/cid/cix463

[8] Laroche M, Raoult D, Parola P. Insects and the transmission of bacterial agents. Microbiology Spectrum. 2018;**6**(5):1-6. DOI: 10.1128/microbiolspec.MTBP-0017-2016

[9] Ojo J. Pesticides use and health in Nigeria. Ife Journal of Science. 2016;**18**(4):981-991

[10] United States Environmental Protection Agency. Public Health Issues Caused by Pests. [Internet]. 2023. Available from: https://www.epa.gov/pesticides/public-health-issues-caused-pests [Accessed: May 16, 2024]

[11] Lopes A, Sylvestre M, Tolke DE, Tavares JF, Barbosa-Filho JM, Cebrian-Torrejon G. Plant-derived pesticides as an alternative to pest management and sustainable agricultural production: Prospects, application and challenges. Molecules. 2021;**26**(16):4835. DOI: 10.3390/molecules2616-4835

[12] The World Bank Group. Population, Total. [Internet]. Washington DC: The World Bank Group; 2024. Available from: https://data.worldbank.org/indicator/SP.POP.TOTL?end= [Accessed: May 16, 2024]

[13] Byrd JD, Crow WD, Layton B. Pesticides: Benefits and Risks. Under: Insects-Crop Pests. [Internet]. Mississippi: Mississippi State University Extension Services; 2023. Publication Number 1962. Available from: http://www.extension. msstate.edu/publications/pesticidesbenefits-and-risks [Accessed: May 16, 2024]

[14] Ganai M, Khan Z, Tabasum B. Challenges and constraints in chemical pesticide usage and their solution: A review. International Journal of Fauna and Biological Studies. 2018;**5**(3):31-37

[15] Sarkar S, Dias Bernardes Gil J, Keeley J, Möhring N, Jansen K. The Use of Pesticides in Developing Countries and their Impact on Health and the Right

to Food. Belgium: European Union; 2021. pp. 1-44. DOI: 10.2861/28995

[16] Sălceanu C, Paraschivu M, Cotuna O, Sărățeanu V, Prioteasa MA, Flondor LS. Global pesticide market: Size, trends, forecasts. Annals of the University of Craiova - Agriculture, Montanology, Cadastre Series. 2022;**52**(2):146-157

[17] Sharma A, Kumar V, Shahzad B, et al. Worldwide pesticide usage and its impacts on ecosystem. SN Applied Sciences. 2019;**1**:1446. DOI: 10.1007/s42452-019-1485-1

[18] Barathi S, Sabapathi N, Kandasamy S, Lee J. Present status of insecticide inputs and eco-friendly approaches for remediation - a review. Environmental Research. 2024;**240**(1):117432. DOI: 10.1016/j.enres.2003.117432

[19] United States Environmental Protection Agency. Insecticides. Causal Analysis/Diagnosis Decision Information System (CADDIS). [Internet]. Washington DC: United States Environmental Protection Agency; 2024. Available from: https://www.epa.gov/caddis/insecticides [Accessed: May 21, 2024]

[20] Ware GW, Whitacre DM. An introduction to insecticides. In: Radcliffe EB, Hutchison WD, Cancelado RE, editors. Radcliffe's IPM World Textbook. 4th ed. St. Paul, MN: University of Minnesota; 2004. Available from: https://www.ipmworld.umn.edu/ware-intro-insecticides [Accessed: May 16, 2024]

[21] World Health Organisation. Environmental Health Criteria 9. [Internet]: DDT and its Derivatives (EHC9, 1979). Geneva: World Health Organisation; 1979. 170 p. Available from: https://www.iris.who.int/bitstream/handle/10665/39562/9241540699-eng.pdf?sequence=1 [Accessed: May 16, 2024]

[22] United Nations Environment Programme (UNEP). Ridding the World of POPs: A Guide to the Stockholm Convention on Persistent Organic Pollutants. Nairobi: United Nations Environment Programme (UNEP); 2005. 17 p. Available from: http://www.chm.pops.int/portals/0/repository/chm-general/unep-pops-chm-guid-ridding.english.pdf [Accessed: May 6, 2024]

[23] Stephen DO. Natural pesticides from plants. In: Janick J, Simon JE, editors. Advances in New Crops. Portland: Timber Press; 1990. pp. 511-517

[24] Abraham EF, Chain E. An enzyme from bacteria able to destroy penicillin. Nature. 1940;**146**:837

[25] Srivastava A. Utility of some plant products as pesticides. Wutan Huatan Jisuan Jishu. 2021;**17**(6):221-227

[26] Stephenson GR, Ferris IG, Holland PT, Nordberg M. Glossary of terms relating to pesticides (IUPAC recommendations 2006). Pure and Applied Chemistry. 2006;**78**(11):2075-2154. DOI: 10.1351/pac200678112075, Available from: https://www.old.iupac.org/publications/pac/2006/pdf/7811x2075.pdf [Accessed: May 6, 2024]

[27] Murray BI. Botanical insecticides, deterrents, repellents in agriculture and increasingly regulated world. Annual Review of Entomology. 2013;**51**:45-66

[28] Gupta RC. Classification and uses of organophosphates and carbamates. In: Gupta RC, editor. Toxicology of Organophosphate & Carbamate Compounds. Waltham MA: Academic Press; 2006. pp. 5-24. DOI: 10.1016/B978-012088523-7/50003-X

[29] Mencke N, Jeschke P. Therapy and prevention of parasitic insects in veterinary medicine using. Current Topics in Medicinal Chemistry. 2002;**2**:701-715

[30] Umetsu N, Shirai Y. Development of novel pesticides in the 21st century. Journal of Pesticide Science. 2020;**45**(2):54-74. DOI: 10.1584/jpestics. D20-201

[31] Babajide AM. Microbial insecticides in Nigeria: State of the art. Nigerian Journal of Entomology. 1989;**10**:23-34. DOI: 10.36108/NJE/8891/90.0140

[32] Mertz F, Yao RC. Sacharopopyspora spinosa sp nov. isolated from soil collected in a sugar mill run still. International Journal of Systematic and Evolutionary Microbiology. 1990;**40**(1):34-39. DOI: 10.1099/00207713-40-1-34

[33] Bell CH. Pest control of stored food products: Insects and mites. In: Lelieveld HLM, Holah J, Napper D, editors. Hygiene in Food Processing. Principles and Practice, Woodhead Publishing Series in Food Science, Technology and Nutrition. 2nd ed. Cambridge: Woodhead Publishing (Elsevier); 2014. pp. 494-538

[34] Gregson M, Ollis WD, Sutherland IO, Gottlieb OR, Magalhães MT. Violastyrene and isoviolastyrene, cinnamylphenols from *Dalbergia miscolobium*. Phytochemistry. 1978;**17**:1375-1377

[35] Dhra G, Ahmad M, Kumar J, Patanjali PK. Mode of action of azadirachtin: A natural insecticide. International Research Journal. 2018;**7**:41-46

[36] Kilani-Morakchi S, Morakchi-Goudjil H, Sifi H. Azadirachtin-based insecticide: Overview, risk assessments, and future directions. Frontiers in Agronomy. 2021;**3**(676208):1-13. DOI: 10.3389/fagro.2021.676208

[37] Rytwinski T, Taylor JJ, Donaldson LA, Britton JR, Browne DR, Gresswell RE, et al. The effectiveness of non-native fish removal techniques in freshwater ecosystems: A systematic review. Environmental Reviews. 2018;**27**(1):71-94. DOI: 10.1139/er-2018-0049

[38] Tanner CM, Kamel F, Ross GW, Hoppin JA, Goldman SM, Korell M, et al. Rotenone, Paraquat and Parkinson's disease. Environmental Health Perspectives. 2011;**119**(6):866-872. DOI: 10.1289/ehp.1002839

[39] Gupta RC. Rotenone. In: Wexlerp P, editor. Encyclopedia of Toxicology. 3rd ed. Waltham MA: Academic Press; 2014. pp. 185-187. DOI: 10.1016/B978-0-12-386454-3.00194-9

[40] Tat J, Heskett K, Gerry R, Boss GR. Acute rotenone poisoning: A scoping review. Heliyon. 2024;**10**(7):1-9. DOI: 10.1016/j.heliyon.2024.e28334

[41] Bond C, Buhl K, Stone D. Pyrethrins General Fact Sheet. Oregon: National Pesticide Information Center, Oregon State University Extension Services; 2014. [Internet]. Available from: http://www.npic.orst.edu/factsheets/pyrethrins.html [Accessed: May 6, 2024]

[42] Subba B, Thapa S, Gurung B, Pokhrel P. Plant-based organic insecticides. In: Recent Trends in Entomology. Tamil Nadu: Cape Comorin Publisher; 2023. pp. 154-164

[43] Okwute SK. Plants as potential sources of pesticidal agents: A review. In: Soundarajan RP, editor.

Pesticides-Advances in Chemical and Botanical Pesticides. London: IntechOpen Limited; 2012. pp. 207-232. DOI: 10.5772/46225

[44] Rajapake RSH, Ratnaseka D. Pesticidal potential of some selected tropical plant extracts against *Callosobruchus maculatus* F. and *Callosobruchus chinenis* L. Tropical Agricultural Research and Extension. 2008;**11**:69-71

[45] Okwute SK. Plant-derived pesticidal and antimicrobial agents for use in agriculture. A review of phytochemical and biological studies on some Nigerian plants. Journal of Agriculture, Science and Technology. 1992;**2**(1):62-70

[46] Okwute SK, Onyia R, Anene C, Amodu OP. Protectant, insecticidal and antimicrobial potentials of *Dalbergia saxatilis* hook f. (Fabaceae). African Journal of Biotechnology. 2009;**8**(23):6556-6560

[47] Ivbijaro MF, Agbaje M. Insecticidal activities of *Piper guineense* Schum and Thonn and *capsicum* species on the cowpea bruchid *Callosobruchus maculatus* F. Insect Science and its Application. 1986;**7**(4):521-524

[48] Adesina JM, Ileke KD, Yallappa R, Ofuya TI. Insecticidal evaluation of *Bridelia micrantha* and *Dalbergia lacteal* aqueous extracts for the control of *Podagrica uniforma* (Jacoby) and *Nisotra dilecta* (Jacoby) (*Coleoptera: Chysomelidae*) infestation on okra. AGRIVITA Journal of Agricultural Science. 2016;**38**(3):269-274

[49] Asadollahi A, Khoobdel M, Zahraei-Ramazani A, et al. Effectiveness of plant-based repellents against different anopheles species: A systematic review. Malaria Journal. 2019;**18**(436):1-20. DOI: 10.1186/s12936-019-3064-8

[50] Vasudeva N, Vats M, Sharma SK, Sardana S. Chemistry and biological activities of the genus Dalbergia – A review. Pharmacognosy Reviews. 2009;**3**(6):307-319

[51] Park IK. Insecticidal activity of isobutylamides derived from *Piper nigrum* against adults of two mosquito species, *Culex pipiens pallens* and *Aedes aegypti*. Natural Product Research. 2012;**26**(22):2129-2131. DOI: 10.1080/14786419.2011.628178

[52] Scott IM, Jensen HR, Philogène BJR, et al. A review of piper spp. (Piperaceae) phytochemistry, insecticidal activity and mode of action. Phytochemistry Reviews. 2008;7:65-75. DOI: 10.1007/s11101-006-9058-5

[53] Batish DR, Singh HP, Kohli RK, Kaur S. Eucalyptus essential oil as a natural pesticide. Forest Ecology and Management. 2008;**256**(12):2166-2174. DOI: 10.1016/j.foreco.2008.08.008

[54] Frances SP, Rigby LM, Chow WK. Comparative laboratory and field evaluation of repellent formulations containing deet and lemon eucalyptus oil against mosquitoes in Queensland, Australia. Journal of the American Mosquito Control Association. 2014;**30**(1):65-67. DOI: 10.2987/13-6366.1

[55] Iqbal T, Ahmed N, Shahjeer K, et al. Botanical insecticides and their potential as anti-insect/pests: Are they successful against insects and pests? In: Global Decline of Insects. London: IntechOpen Limited; 2021. pp. 1-15. DOI: 10.5772/intechopen.100418

[56] Gulsevin A, Glazer AM, Shields T, Kroncke BM, Roden DM, Meiler J. Veratridine can bind to a site at the

mouth of the channel pore at human cardiac sodium channel NaV1.5. International Journal of Molecular Sciences. 2021;**23**(4):2225. DOI: 10.3390/ijms23042225

[57] Chemical Book. Ryanodine. [Internet]. 2023. Available from: https://www.chemicalbook.com/ChemicalProductProperty_EN_CB1480087.htm [Accessed: May 20, 2024]

[58] Deedat DY. Problems associated with the use of pesticides: An overview. International Journal of Tropical Insect Science. 2011;**15**(3):247-251. DOI: 10.1017/S174275840017537

[59] Matwo NS, Tanner M, Munhenga G, Mupa AS, Finda M, Utzinger J, et al. Use of insecticides to control mosquito via long-lasting insecticidal nets (LLINs) and indoor residual spraying (IRS). Malaria Journal. 2020;**19**:257

[60] Alraodion AI, Ngwogu KO, Ngwogu AC, Megwas AU, Ekenjoku JA, Awosanya OO. Household insecticides used in Nigeria induced oxidative stress in Wistar rats. Asian Journal of Immunology. 2020;**3**(2):39-45

[61] Usta C. Microorganisms in biological Pest control - a review (bacterial toxin application and effect of environmental factors). In: Current Progress in Biological Research. London: IntechOpen Limited; 2013. pp. 287-317. DOI: 10.5772/55786

[62] United States Environmental Protection Agency. Phaseout of Ozone-Depleting Substances: Methyl Bromide. [Internet]. Washington DC: United States Environmental Protection Agency; 2024. Available from: https://www.epa.gov/ods-phaseout/methyl-bromide [Accessed: May 22, 2024]

[63] Tostado L. Pesticides and agriculture: Dangerous substances. In: Pesticide Atlas 2022. Brussels: Heinrich Boll Stiftung Publishers; 2022

[64] Raj A, Dubey A, Malla AM, Kumar A. Pesticide pestilence: Global scenario and recent advances in detection and degradation methods. Journal of Environmental Management. 2023;**338**:117680. DOI: 10.1016/j.jenvman.2023.117680

[65] Souza MCO, Cruz JC, Cesila CA, Gonzalez N, Rocha BA, Adeyemi JA, et al. Recent trends in pesticides in crops: A critical review of the duality of risks-benefits and the Brazilian legislation issue. Environmental Research. 2023;**228**:115811

[66] Gogi MD, Nawaz A, Sufyan M, Sarfraz RM, Liburd OE. Biorational approaches in Pest management. In: Khan IA, Farooq M, editors. Sustainable Insect Pest Management. Pakistan: University of Agriculture, Faisalabad, Pakistan; 2017. pp. 231-283. Available from: https://www.dokumen.pub/sustainable-insect-pest-management.html [Accessed: May 21, 2024]

[67] Maia MF, Moore SJ. Plant-based insect repellents: A review of the efficacy, development and testing. Malaria Journal. 2011;**10**(1):11. DOI: 10.1186/1475-2875-10-S1-S11

[68] Labunmi L, Escoubas P, Mizutani J. Termite antifeedant activity in *Xylopia aethiopica* (Annonaceae). Phytochemistry. 1995;**40**(4):1105-1112. DOI: 10.1016/0031-9422(95)92653-P

[69] Dalziel JM. The Useful Plants of West Africa. (2nd Rep.) (1st Reprint, 1948; End Reprint, 1955). London: Crown Agents for Overseas Governments; 1937. 612 p

[70] Ayensu S. Medicinal Plants of West Africa. Algonac Michigan, USA: Reference Publications Inc.; 1978

[71] Siegler DS. Plant-derived insecticides. In: Plants and their Uses. Illinois: University of Illinois; 2005

[72] Sikam KG, Ntabo KV, Happi GM, Meikeu LZ, Wansi JD. Chemistry and pharmacological aspects of Aridanin, a lead compound from *Tetrapleura tetraptera* (Fabaceae). Natural Resources for Human Health. 2023;**3**(1):1-6. DOI: 10.53365/nrfhh/152273

[73] Babu M, Joseph KH, Sree A, Scariya S. In-vitro evaluation of anti-urolithiatic and larvicidal activity of Alternanthera Sessilis. Biomedical and Pharmacology Journal. 2021;**14**(02): 671-680. DOI: 10.13005/bpj/2169

[74] Rawani A, Ghosh A, Chandra G. Mosquito larvicidal potential of four common medicinal plants of India. The Indian Journal of Medical Research. 2014;**140**(1):102-108

[75] Singla RK, Dhir V, Madaan R, Kumar D, Singh Bola S, Bansal M, et al. The genus alternanthera: Phytochemical and ethnopharmacological perspectives. Frontiers in Pharmacology. 2022;**13**:769111. DOI: 10.3389/fphar.2022.769111

[76] Chiu S-F. Effectiveness of Chinese insecticidal plants with reference to the comparative toxicity of botanical and synthetic insecticides. Journal of the Science of Food and Agriculture. 1950;**1**(9):276-286

[77] Samarasekera J. Insecticidal Natural Products from Sri Lanka. [Thesis]. Walton Hall: The Open University; 1997. DOI: 10.21954/ou.ro.00004d49

[78] Idoko JE, Ileke KD. Comparative evaluation of insecticidal properties of essential oils of some selected botanicals as bio-pesticides against cowpea bruchid, *Callosobruchus maculatus* (Fabricius) [Coleoptera: Chrysomelidae]. Bulletin of the National Research Centre. 2020;**44**(119):1-7. DOI: 10.1186/s42269-020-00380-2

[79] Seiyaboh EI, Odubo TC, Izah SC. Larvicidal activity of *Tetrapleura tetraptera* (Schum and Thonn) Taubert (Mimosaceae) extracts against *Anopheles gambiae*. International Journal of Advanced Research in Microbiology and Immunology. 2020;**2**(1):20-25

[80] Aremu JF. Pesticide Usage: Benefits, Challenges and Recommendations for Safety. Abuja: Cedro Royal Multiventures Ltd; 2018

[81] Torto B. Innovative approaches to exploit host plant metabolites in malaria control. Pest Management Science. 2019;**75**(9):2341-2345. DOI: 10.1002/ps.5460

[82] Food and Agricultural Oganisation (FAO). The Future of Food and Agriculture: Trends and Challenges. Rome: FAO; 2017

[83] Ayilara MS, Adeleke BS, Akinola SA, Fayose CA, Adeyemi UT, Gbadegesin LA, et al. Biopesticides as a promising alternative to synthetic pesticides: A case for microbial pesticides, phytopesticides, and nanobiopesticides. Frontiers in Microbiology. 2023;**14**:1040901. DOI: 10.3389/fmicb.2023.1040901

[84] Insecticide Resistance Action Committee (IRAC) International MoA Working Group. Mode of action classification scheme version 11.1. [Internet]. 2024. Available from: https://www.irac-online.org/documents/moa-classification/ [Accessed: May 16, 2024]

[85] Sparks TC, Bryant RJ. Innovation in insecticide discovery: Approaches to the discovery of new classes of insecticides. Pest Management Science. 2022;**78**(8):3226-3247. DOI: 10.1002/ps.6942

[86] Williams T, López-Ferber M, Caballero P. Nucleopolyhedrovirus coocclusion technology: A new concept in the development of biological insecticides. Frontiers in Microbiology. 2022;**12**:810026. DOI: 10.3389/fmicb.2021.810026

[87] Swale DR. Perspectives on new strategies for the identification and development of insecticide targets. Pesticide Biochemistry and Physiology. 2019;**161**:23-32. DOI: 10.1016/j.pestbp.2019.07.001

Chapter 4

The Role of Insecticides within IPM Strategies

Paul Horne and Rebecca Addison

Abstract

Integrated pest management (IPM) involves the integration of biological, cultural and insecticidal controls and a key aspect for implementing an IPM strategy is knowing how to select insecticides that are compatible with biological control agents. This chapter discusses the issues involved in this process and how insecticides can be assessed in different ways. The aim is not to label insecticides as "good" or "bad" but to evaluate their compatibility with biological control agents within specific strategies. An insecticide may be safe for some beneficial species while highly disruptive to others. The types of testing used and the ways in which results can be presented are discussed. Interpreting and using the results of pesticides' side-effects are important aspects when developing and implementing IPM and examples of how to incorporate these are presented.

Keywords: IPM, pesticides, biological control, side-effects, sublethal effects

1. Introduction

There are very many definitions of integrated pest management (IPM) [1], but it involves using a combination of biological and cultural control methods along with insecticides. In general, pesticides are the support tools rather than the primary control options. Biological controls are usually the foundation of an IPM strategy [2–5] and these can be either predators, parasitoids or pathogens with some commercially produced pathogens being formulated as pesticides. Cultural controls are management practices that can either benefit biological control agents or hinder pest species. Where possible, pesticides should be selected not only for their impact on pests but also for their compatibility with important biological control agents in the crop. Loss of biological control can lead to secondary pest outbreaks or pest flares [6], caused not only by insecticides but also by miticides and fungicides [7]. Integrating the three types of control methods in a compatible way ensures no overreliance on any single management option.

2. Pesticides and biological control agents

2.1 Pesticide impacts on pests

Usually, the first factor to be considered when selecting a pesticide is the impact on the pest. Obviously, it needs to be effective so considerations such as insecticide resistance are very important. If the pest has developed resistance, then there may be no point in applying it. If there is no resistance, then the frequency of past applications and rotation with other pesticide groups must be considered.

The pesticide will need to be registered on the crop and growers will need to adhere to regulations regarding withholding periods (WHP) and maximum residue levels (MRL). The crop's growth stage influences these factors, as the time until harvest affects the feasibility of meeting WHP and MRL guidelines. The pesticide's knockdown time is also important, particularly as harvest time nears, to ensure effective pest control within the available time frame. The cost of the pesticide is always going to be a consideration but is often secondary to its efficacy, especially when dealing with insecticide-resistant pests.

The life-stage of the pest is often crucial, with some label recommendations advising to target the smaller instars. For example, from the insecticide flubendiamide label: *Apply at egg hatch in crops with nil thresholds for damage. Otherwise, apply at egg hatch or very soon after egg hatch to target young larvae. For potato moth (tomato leafminer), commence sprays when young larvae are first detected in leaves.* For pesticides such as insect growth regulators, the life-stage is critically important as they only affect the juvenile forms during moulting. If several pest species are present, then a seemingly easy answer is to choose a broad-spectrum insecticide.

It is important to know how long a pesticide remains effective after it is applied to ensure that the crop is adequately protected. For example, some products are volatile and may only give protection for 24 hours while others can persist for days or even weeks. For some insecticides, the residual effect can be severely reduced because of degradation by UV light or wash-off by rainfall or irrigation. This is especially important for products based on bacteria or viruses, where degradation can occur very rapidly at times of the year when UV levels are high.

Application factors must be carefully managed to ensure adequate coverage of the chosen insecticide for optimal efficacy. This includes ensuring the sprayer has been calibrated, using the appropriate water rate for different crop stages and incorporating wetters or stickers where useful or required. This is important as it is not just the active ingredient that is involved in efficacy but how well it is applied.

2.2 Side-effects on beneficials

The factors described above are important in a pesticide-based strategy but in an IPM strategy, there are also some other factors that are equally important. In a pesticide-based approach, pests are controlled solely by pesticides. In contrast, an IPM strategy relies partially or predominantly on biological control agents, or beneficial species, which naturally prey on or inhibit pests.

In many instances, biological and cultural methods can provide most, if not all, of the control required. Insecticides and miticides can be used as support tools when these methods are insufficient. Therefore, when selecting an insecticide or miticide, it is necessary to choose one that minimally impacts beneficial species while also effectively reducing pest populations. The aim of using pesticides in an IPM strategy is not to replace the control provided by biological control agents but rather to supplement or complement it.

Numerous studies have examined the impact of pesticides on beneficial species with a database maintained by the International Organisation for Biological Control (IOBC). Some insecticide groups such as organophosphates, carbamates, fiproles and synthetic pyrethroids are generally very disruptive in IPM strategies, as most products kill beneficial insects and mites, although there are some exceptions. (These exceptions include pirimicarb, which is not toxic to most predatory insects or mites and is highly volatile, and so is not residual. Tau-fluvalinate is toxic to most predatory insects and mites but is not toxic to Hymenoptera.) However, with other groups there is usually a range of toxicities to different species or groups of biological control agents.

2.2.1 Acute toxicity

The simplest measure of the side effects of pesticides is the acute toxicity of a product to individuals of a beneficial species. There are some standard tests (e.g. IOBC) that involve exposing individuals in a test chamber to a defined dose of pesticide (typically a single spray at the label rate, often a dried deposit) for a set period of time (i.e. 24, 48 or 96 hours). The number of individuals alive after the allocated time is recorded, and the result is assigned to a category indicating the relative toxicity of the product to the species. There are often 3 or 4 categories based on % mortality such as "Low" <30%; "Moderate" 30–79%; "High": 79–99%; "Very High" >99% [8] or the IOBC ratings of "Harmless", "Slightly Harmful", "Moderately Harmful" and "Harmful". This information is commonly presented as a colour-coded chart [8] (see https://sideeffects.koppert.com/side-effects) with either green, yellow or red boxes indicating low, medium or high toxicity. One problem with this is the hard division between categories occurring at a particular percentage (e.g. categorised as low at 29% and moderate at 30%). A clearer way of presenting the information is on a continuum (as shown in https://ausveg.com.au/biosecurity-agrichemical/crop-protection/#IPM) rather than in discrete boxes.

The information presented in these charts has certain limitations which must be understood. The test results are based on particular life-stages, which may not always represent the most susceptible phases to insecticides. They are conducted over a defined period of exposure and may involve treated surfaces or topical application, with exposure occurring to either wet or dried spray deposits. Additionally, tests are typically conducted under controlled laboratory conditions where factors such as UV degradation and exposure to wind or rainfall are absent. Consequently, these tests are considered "worst-case scenarios" as outdoor crops are subject to various environmental influences that could significantly affect pesticide activity and in some cases significantly reduce the impact on beneficial species. Typically, this information is designed to illustrate the relative impact of pesticides on beneficial, not an absolute measure of if they are safe or not safe. Moreover, even insecticides that are deemed highly disruptive to beneficial species can potentially still be used in IPM strategies. For example, they could be used as a border spray to prevent pest invasion from adjacent farms or used at the end of the crop life (at or after crop senescence) after transient beneficial have left the crop.

Some sources highlight another important aspect of these test results which is the residual time during which the impact is expected to last. This is especially important in protected cropping where the introduction of biological control agents is common. If, for example, two products are both highly toxic to a key beneficial species but have different residual times; one remaining toxic for a few days, while the other may persist for several months, then clearly, these two products are not equal. In protected cropping, residues of some products (such as most synthetic pyrethroids and organophosphates) can prevent the introduction of key beneficial species for 3 months (see https://sideeffects.koppert.com/side-effects).

The side-effects guides are developed using a single spray at the label rate, but multiple sprays can have a cumulative impact on beneficial species [8]. Thus, even if a product is listed as essentially harmless, multiple applications can still cause harm. The limitations of testing are considerable, and the methodologies can vary [8] so published results should only be viewed as a guide rather than an absolute measure of toxicity. Multiple sprays of the same pesticide can be more disruptive than a single application, and results can vary even for closely related species [8].

It is usually considered good practice to rotate through different pesticide groups to avoid or slow the development of resistance. This is true, but care needs to be taken as different products can have very different impacts on different beneficial species. For example, product 1 may be safe to ladybirds and lacewings but harmful to wasps, while product 2 might be safe to wasps and lacewings but not ladybirds and product 3 might be safe to ladybirds and wasps but not lacewings. So, if the 3 products (all with legitimate claims to have a fit in IPM strategies) are used in a sequence then potentially ladybirds, lacewings and wasps will all have been disrupted, and their populations reduced.

2.2.2 Secondary poisoning

Insecticides used in baits, even if they are broad-spectrum products, are a means of targeting pests without directly killing beneficial species. Various "attract-and-kill" methods have been developed for different pests [9]. However, generalist predators can still receive a lethal dose of secondary poisoning when they consume prey killed by the baits. The same effect can occur when plants are treated with systemic insecticides that are present within the vascular system of the plant and not on leaf surfaces. Predatory species can walk on the plant surface without any ill effects yet can subsequently be killed by ingesting treated pests such as aphids [10].

2.2.3 Sublethal effects

The acute effects of insecticides are quickly apparent (within hours or days usually) while sublethal effects are more subtle but can be highly disruptive. For example, pesticides that do not immediately kill beneficial species but rather reduce fecundity or longevity can have massive impacts on populations. If a product reduces egg production or sterilises individuals, then not a single individual may be killed immediately but the population could be destroyed. This has been observed with some growth regulators and ladybird beetles [11, 12] and some fungicides impacting predatory mites [7]. Similarly, if the lifespan of a parasitic wasp is halved it can be assumed that the wasp will parasitise fewer hosts, resulting in considerable sublethal effects [13]. In addition, it can also be assumed that the overall fitness of any biological control agent could be compromised if its lifespan is shortened due to pesticide application.

3. Examples of an IPM system incorporating pesticides of different types

3.1 IPM and insecticide use in potatoes

There is a range of pests to be dealt with in potato crops, and the species vary with location around the world. Some of the key pests in Australia are caterpillars including the potato tuber moth (PTM), *Phthorimaea operculella*, and loopers, (*Chrysodeixis spp.*) as well as aphids (including green peach aphid, *Myzus persicae*, and potato aphid *Macrosiphum euphorbiae*) and potato wireworm (*Hapatesus hirtus*). There are more pests in many regions, but these are sufficient to describe IPM and insecticide selection.

3.1.1 Biological controls

Each of these pests needs to be dealt with but not necessarily with the use of insecticides as there are biological control agents (usually predators and parasitoids) that will attack them. PTM caterpillars are attacked by three species of parasitoid wasps (*Orgilus lepidus, Apanteles subandinus* and *Copidosoma koehleri*) and predatory damsel bugs (*Nabis kinbergii*). Aphids are attacked by parasitoid wasps (*Aphidius spp.*) and predators including brown lacewings (*Micromus tasmaniae*), ladybird beetles (Coccinellidae) and hoverflies (Syrphidae). These biological control agents provide a significant degree of control [14] so care must be taken to avoid killing or disrupting them. There are predators of wireworms, but when these pests are present then usually biological control alone is not sufficient.

3.1.2 Insecticide support tools

There are many options for growers wanting to use insecticides to control PTM, loopers and aphids (registered products: https://www.apvma.gov.au/registrations-and-permits/search-registered-chemical-products-and-permits), and the range goes from extremely safe to highly toxic and thus disruptive to beneficial species. Products mentioned here are simply to illustrate how an IPM strategy can include different products.

PTM caterpillars are leaf-miners, living between the upper and lower surfaces of potato leaves, making them difficult to contact with many insecticides (including insecticides based on *Bacillus thuringiensis*). The ideal insecticide for PTM would have trans-laminar activity and be safe for the beneficial listed. In contrast, looper caterpillars are exposed on the leaf surface and ingest large amounts of leaf material making them easier to target with various insecticides. If both PTM and loopers are present, a product such as chlorantraniliprole should be effective against both species as it is translaminar (see https://ag.fmc.com/au/en/products/insecticides-miticides/coragen-insecticide) and is safe for most beneficial species of interest here. If only loopers are present, a bacterial-based insecticide such as *Bacillus thuringiensis kurstaki* or *B.t. aizawai* could be used effectively. Broad-spectrum insecticides that are not translaminar are likely to be ineffective against PTM inside leaf-mines and would disrupt biological control agents. However, broad-spectrum insecticides can be safely used in the IPM strategy when the crop is senescing (or is killed) and the populations of biological control agents have left. Such applications late in the life of the crop will help protect against PTM adults laying eggs on the soil surface or exposed tubers.

For insecticide management of aphids, the ideal choice is a product that effectively targets aphids but without disrupting the biological control agents for aphids or other pests in the crop. At present, there are several such products available including Afidopyropen, Pymetrozine, Spirotetramat and Flonicamid, which are all currently registered for use on potato crops in Australia (see: https://www.apvma.gov.au/registrations-and-permits/search-registered-chemical-products-and-permits). Pirimicarb is safe to use in IPM programmes, but green peach aphid has developed resistance to it in some areas [15] thus the species of aphid present will determine if it should be used or not.

If wireworms are present, then any insecticide applications need to be made prior to planting. These could include broad-spectrum insecticides that are soil incorporated pre-planting or applied to seed-pieces. Neither of these options will disrupt the biological control agents described above but seed-piece dressings will be less disruptive to a range of soil biota.

What this example illustrates is that with a range of pests and insecticides available, it is possible to have both biological control agents and insecticides working together. This means that total control does not need to be achieved by either method alone and less insecticide is required (case study; see https://nla.gov.au/nla.obj-1036654286/view, p. 18)). It is important to note that cultural control strategies have not been discussed here but can also play a significant role in managing potato pests to reduce the reliance on insecticide use. A hypothetical IPM strategy for control of a set of pests in a potato crop is shown in **Table 1**. This illustrates a set of control options that are complimentary and so the overall control is not dependent on any one method.

3.2 IPM and insecticide use in strawberries

Strawberries are grown in both outdoor and protected cropping systems in Australia, each facing a similar range of pests, such as two-spotted mite (*Tetranychus urticae*), western flower thrips (*Frankliniella occidentalis*), several aphid species, greenhouse whitefly (*Trialeurodes vaporariorum*), and several species of Miridae bugs. The key pests (thrips, mites and whitefly) are largely resistant to most available insecticides and miticides, so control is primarily achieved through the release of commercially produced biological control agents. When western flower thrips became established in Australia and were treated with insecticides, it practically ended the IPM strategies available at that time [16]. However, when this pest became resistant to all insecticides available at the time (broad-spectrum or selective) then a new approach was required. An improved IPM strategy was developed in Victoria, Australia, which eliminated the need for insecticides and miticides for the key pests of two-spotted mites and western flower thrips [17] and this approach has become the standard used in southern Australia. A very similar sequence happened in England when western flower thrips arrived and developed resistance to IPM compatible insecticides, prompting development of IPM based on biological control and cultural methods [18].

Pest	Beneficial	Cultural	Pesticide
Potato tuber moth	Parasitoid wasps	Soil Management	Chlorantraniliprole
	Damsel bugs	Irrigation	
Looper caterpillars	Parasitoid wasps	Nectar source for beneficials	*Bacillus thuringiensis*
	Damsel bugs		
Aphids	Parasitoid wasps	Border plantings as trap crop	Afidopyropen
	Hoverflies, Lacewings, Ladybird beetles	Location	Pymetrozine
		Seed source	(See text above for more)
Wireworms	Carabid beetles	Location	Bait/seed piece dressing

Table 1.
A hypothetical IPM strategy for pests of potatoes using a compatible set of options including biological and cultural controls as well as pesticides (examples only).

Therefore, the requirement for insecticides is often low and sometimes highly undesirable, as it could disrupt control of the key pests and lead to severe damage. In protected cropping systems, it could mean that pesticide residues could result in preventing the re-establishment of biological control agents for several months. Therefore, deciding to use a disruptive, broad-spectrum insecticide against pests such as mirids, needs to be carefully considered and well-timed. If the potential damage by mirids, however, is expected to be severe, then insecticides will need to be applied. Minimising disruption can be achieved by using border sprays or spot-sprays instead of treating the entire crop, or by applying lower rates of insecticide. When no effective biological control agents or IPM-compatible insecticides are available, more effort is typically directed towards cultural controls.

In any crop, when a new insecticide or miticide effective against resistant pests enters the market, there is a risk of overuse and subsequent resistance development. This is especially true in pesticide-based strategies with few effective products available for rotation. However, if the new product is IPM compatible then there is not so much reliance on this pesticide slowing resistance development. Of course, if such a product is developed then it is extremely valuable for use within IPM strategies and care must be taken not to overuse it.

These two very different examples illustrate how pesticides and biological control agents can be used in a compatible way within IPM strategies. Selecting pesticides involves more than just effectiveness against the target pest; it requires careful consideration of compatibility with biological control agents and thus its wider impact on the crops' ecological system, aiming to keep it in balance. Decisions on the use of insecticides in IPM strategies rely on monitoring both pest and beneficial species and assessing if the biological control is sufficient or if support is needed by insecticide application.

4. Conclusions

Farmers who incorporate pesticides within an IPM strategy benefit from effective pest control while minimising the use of pesticides leading often to lower costs, improved ecological outputs and reduced risk of pesticide contamination. This integrated approach, combining biological, cultural and pesticides ensures greater sustainability in pest management than relying on pesticides alone. The selection of pesticides within an IPM strategy involves far more than knowing efficacy on the pest, and knowledge of the side-effects on beneficial species is a key aspect. Producers of pesticides also benefit as their products will have greater longevity since resistance is avoided much more effectively.

Acknowledgements

We thank Janet Horne for comments on the manuscript. We also thank Jessica Page whose work on pesticide testing and IPM adoption over many years contributed greatly to the ideas presented in this chapter.

Conflict of interest

The authors declare no conflict of interest.

Author details

Paul Horne* and Rebecca Addison
IPM Technologies Pty Ltd, Hurstbridge, Australia

*Address all correspondence to: paul@ipmtechnologies.com.au

References

[1] Bajwa WI, Kogan M. Compendium of IPM definitions (CID) – What is IPM and how is it defined in the worldwide literature? In: IPPC Publication No. 998, Integrated Plant Protection Center (IPPC). Corvallis, OR: Oregon State University; 2002

[2] Kogan M. Integrated Pest management: Historical perspectives and contemporary developments. Annual Review of Entomology. 1998;**43**:243-270

[3] Page J, Horne P. Controlling Invertebrate Pests in Agriculture. Victoria, Australia: CSIRO Publishing; 2012. 116 p

[4] Horne P, Page J. IPM. In: Alyokhin A, Rondon S, Gao Y, editors. Insect Pests of Potato: Global Perspectives on Biology and Management. Second ed. Oxford, UK: Academic Press; 2022. pp. 483-490

[5] Stenberg JA. A conceptual framework for integrated pest management. Trends in Plant Science. 2017;**22**:759-769

[6] Demirozer O, Tyler-Julian K, Funderburk J, Leppla N, Reitz S. *Frankliniella occidentalis* (Pergande) integrated pest management programs for fruiting vegetables in Florida. Pest Management Science. 2012;**68**:1537-1545. DOI: 10.1002/ps.3389

[7] Pozzebon A, Borgo M, Duso C. The effects of fungicides on non-target mites can be mediated by plant pathogens. Chemosphere. 2010;**79**:8-17

[8] Mata L, Knapp RA, McDougall R, et al. Acute toxicity effects of pesticides on beneficial organisms – Dispelling myths for a more sustainable use of chemicals in agricultural environments. Science of the Total Environment. 2024;**930**:172521. DOI: 10.1016/j.scitotenv.2024.172521

[9] Kroschel J, Zegarra O. Attract-and-kill: A new management strategy for the management of the potato tubermoths *Phthorimaea operculella* (Zeller) and *Symmetrischema tangolias* (Gyen) in potato: Laboratory experiments towards optimising pheromone and insecticide concentration. Pest Management Science. 2010;**66**:490-496

[10] Cole PG, Horne PA. The impact of aphicide drenches on *Micromus tasmaniae* (Walker) (Neuroptera: Hemerobiidae) and the implications for control in lettuce crops. Australian Journal of Entomology. 2006;**45**:244-248

[11] Hattingh V, Tate B. Effects of field-weathered residues of insect growth regulators on some Coccinellidae (Coleoptera) of economic importance as biocontrol agents. Bulletin of Entomological Research. 1995;**85**:489-493

[12] Noelia FM, Clara SA, Graciela M, Ines AM. Toxicity assessment of two IGR insecticides on eggs and larvae of the ladybird *Eriopis connexa*. Pest Management Science. 2023;**79**:1316-1323

[13] McDougall R, Mata L, Ward S, Hoffmann AA, Umina PA. Assessing the sub-lethal impacts of insecticides on aphid parasitoids through laboratory-based studies. Austral Entomology. 2024;**63**:119-129

[14] Horne P, Page J. Australia and New Zealand. In: Alyokhin A, Rondon S, Gao Y, editors. Insect Pests of Potato: Global Perspectives on Biology and Management. Second ed. Oxford, UK: Academic Press; 2022. pp. 401-405

[15] Umina P, Edwards O, Carson P, van Rooyen A, Anderson A. High levels of resistance to carbamate and Pyrethroid

chemicals widespread in Australian *Myzus persicae* (Hemiptera: Aphididae) populations. Journal of Economic Entomology. 2014;**107**:1626-1638. DOI: 10.1603/EC14063

[16] Ullio L. Australia's national strategy for the management of western flower thrips (WFT), *Frankliniella occidentalis* (Pergande). Acta Horticulturae. 2002;**567**:687-689. DOI: 10.17660/ActaHortic.2002.567.150

[17] Horne P, Page J. Develop an Effective IPM Strategy to Deal with Pests in the Victorian Strawberry Industry. Sydney, Australia: Report to Horticulture Australia Limited; 2011. Project BS08011. ISBN 0734126441

[18] Mouden S, Facun Sarminto K, Klinkhamer PGL, Leiss KA. Integrated pest management in western flower thrips: Past, present and future. Pest Management Science. 2017. DOI: 10.1002/ps.4531. Available from: wileyonlinelibrary.com

Chapter 5

Insecticides: Treatment and Resistance in Insects

Akanksha Singh, Himanshu Pande and Gaurava Kumar

Abstract

Insects are the most successful group of creatures, thriving in any habitat. Insect pests can have far-reaching consequences, such as destroying crops that are essential food supplies for humans and livestock. Insecticides are essential in modern-day agriculture; besides being effective as a crop safety measure, they can also prove beneficial in preventing the spread of hazardous pests that can cause human diseases. Because of economic and medicinal reasons, developing potent and efficient pest control chemicals is a critical challenge in agrochemical research and the industrial sector as well. In this chapter, we are concerned about the exorbitant use of insecticides and their negative impacts on agriculture and so on the environment as well. Excessive pesticide treatment can contribute to the development of resistance in target insects, requiring larger concentrations or stronger chemicals, raising production costs, and disrupting natural pest control systems. Consequently, non-target animals, such as beneficial insects and aquatic life, suffer the inadvertent outcomes of insecticide use, resulting in ecosystem disparities and possible food chain pollution. This section offers a valuable understanding of merging present-day advances in science and practical applications, insect pest management approaches, and the promise of greater resilience and sustainability in agriculture.

Keywords: agriculture, insect pests, insecticides, insect pest management, insecticidal resistance

1. Introduction

The agricultural sector, a foundation stone of the Indian economy, employs over 70% of the population and substantially contributes to the nation's GDP. India's agricultural prowess is evident in its diverse crop production, encompassing grains, vegetables, and cash crops [1]. The country's horticultural exports, including fruits, vegetables, and flowers, significantly generate foreign exchange.

India's agricultural sector, a vital pillar of the economy, is under variable constant threats like insect pests, diseases, and other biological and abiotic factors that are waging a fierce war on agricultural operations worldwide. In particular, insect pests pose the most significant threat to vegetable crops, striking at various stages of growth and inflicting a significant decrease in production and quality [2]. These pests are not confined to open fields but infiltrate protected structures like net houses, polyhouses, and glasshouses [3]. Insect pests are not just a local menace but a global threat, causing substantial economic losses and environmental issues. Arthropods and weeds account for approximately 35% of yield losses in crucial crops worldwide. In developing regions with limited pest management options, losses can exceed 50% [4]. Pests, especially arthropods, can inflict significant crop losses or even trigger total crop failure [5–7]. The situation is dire, and the need for effective crop protection strategies is not just critical; it is a matter of utmost urgency for maintaining yields and ensuring global food security.

The world is on the brink of a daunting task-feeding a huge population in the next two decades. The constraints of limited land and renewable resources further compound this challenge. Over half of this population growth will be concentrated in Asia and Africa. Population growth outpaces the expansion of the agricultural sector by a significant margin. The production of high-quality food will become increasingly arduous as consumer concerns about food quality escalate. This future scenario underscores the pressing need for innovative and sustainable solutions in crop protection. The consequences of not addressing these challenges could be catastrophic, with potential food shortages and economic instability looming large if we fail to develop and implement effective crop protection strategies.

There is a significant concern about caring for our environment, human health, food safety, and protecting it. This has led to a renewed interest in the use of pesticides in agriculture. Pesticides are essential in agricultural production as they help farmers to control weeds and insects. They have also been responsible for the remarkable increase in agricultural products. The increase in the world's population in the twentieth century would not have been possible without a parallel increase in food production.

Any living organism resulting in economic losses or harming human health is considered a pest. Pests can ruin crops, spread diseases in plants or humans, and create other forms of damage. Pesticides are substances that can eradicate these pests. These chemicals inhibit, eliminate, repel, or alleviate pests. They can also function as plant regulators, defoliants, or desiccants [8].

Insecticides have long been used for pest control. Early methods involved using inorganic agents such as sulfur, arsenic, hydrogen cyanide, or cryolite. However, these agents were found to be highly toxic not only to pests but also to non-targeted species, including vertebrates. In the middle of the twentieth century, efficient insecticides such as organochlorine, organophosphorus, and carbamate were introduced. These insecticides were a significant breakthrough in the agrochemical sector because they allowed for the minimization of crop losses caused by insect activity. DDT was one of the first organochlorine insecticides introduced during World War II. It had a wide range of uses and a prolonged residual activity. However, it was later discovered that organochlorine insecticides, including DDT, caused severe environmental damage in terrestrial and aquatic ecosystems. These insecticides remained in the environment longer and accumulated in animals through the food chain. As a result, most of these insecticides were banned in many countries, but DDT is still in use in some countries where malaria is endemic.

In addition to the adverse environmental effects, insect populations have developed resistance to organochlorine, organophosphorus, and carbamate insecticides [9]. To combat this resistance, new pesticides with different mechanisms of action are essential [10]. Therefore, the primary objective of pesticide study is to create selective and highly effective substances that do not harm human health or the environment [11].

Insecticides are essential for modern agriculture as they protect crops and ensure optimal yields. However, their excessive use raises concerns about their adverse effects on agriculture and the environment. When insecticides are overused, target insects can develop resistance, which leads to the need for higher concentrations or more potent chemicals. This results in increased production costs and disturbance of natural pest control mechanisms. Insecticide use also harms non-target organisms like beneficial insects and aquatic life, causing unintended consequences and potentially leading to ecosystem imbalances and food chain contamination [12].

2. Pest outbreak and need for insecticidal treatment

Insect pests pose significant challenges to global agriculture, causing substantial economic losses and environmental concerns. These can cause extensive damage, such as destroying crops that are essential food sources for humans and livestock. In addition, invasive insect species can upset the balance of local ecosystems, leading to the decline of native species and increased environmental stress. Insecticides are crucial for modern agriculture, protecting crops and preventing the spread of harmful pests that cause human diseases. These pests can also threaten farmers' livelihoods and impact the overall economic stability of regions that rely heavily on agriculture.

Insects in various developmental phases, such as larvae and adults, attack diverse sections of plants, including roots, stems, and leaves. When hazardous insects attack plants, they are more likely to perish [13].

Ceutorhynchus napi is a gray-colored beetle measuring 3.2–4 mm. The female beetles puncture plant stems to lay their eggs inside. After 10–20 days, the larvae hatch and begin to grow. As they grow, abnormal bulbous growths appear on the stem, eventually rupturing and causing deformities. The larvae also tunnel through the stems, causing damage that can create pathways for pathogens to enter the plants. *Ceutorhynchus napi* beetles produce one generation per year [14].

Ceutorhynchus quadridens are brown beetles, approximately 2.5–3 mm long, with characteristic whitish dots on their wing covers. In spring, the female beetles lay eggs, which take up to 3 months to develop into white larvae that can grow up to 6 mm in size and have brown heads. The beetle's chrysalis is yellowish and has a long snout. The larvae feed inside the plant stems without distorting them, creating tunnels that act as sites for infection. This damage hinders the growth of the plant [13].

Grubs are beetle larvae. They have a distinctively curved, white body with a darker head and three pairs of strong legs. These insects live in the soil and can injure plants' root systems, causing them to wither and eventually die [13].

Phorbia brassicae flies have a body length of 5–6 mm. They can be distinguished by the red dots on their silver-white foreheads. Females are larger than males. The larvae of these flies attack the roots, causing plants to die [13].

Baris spp. are little beetles about 3–4 mm long with a blue or dark green metallic shine. Females lay eggs in plant roots, and larvae tunnel through them, causing damage and poisoning. Chapman [13] states that this damage slows plant growth and facilitates pathogen infection.

Zabrus tenebrioides is a black beetle that is 12–15 mm in length. It has a distinctive light brown abdomen, red-brown antennae, and legs. During the autumn, the larvae of this beetle might harm germinating seedlings. They feed on leaf bases and blades, which can cause wilting and death [14].

Thripidae are tiny, thin insects with black-brown bodies and narrow wings. They can grow up to 2 mm long and are typically found among flowers, behind tree bark, or in plant litter. The larvae and adults can damage seeds, allowing infections to infect the plant. This often results in the demise of the diseased plant [14].

These are only a few pests that damage plant parts and eventually cause the plant's death. Several such insect pests can cause significant damage to crops, leading to withering or total failure.

The following are different types of pests that can affect crops based on their occurrence:

i. *Regular pests*: These are the ones that appear regularly and are directly related to the crop. Ex: Rice Stem Borer and Brinjal Fruit Borer.

ii. *Occasional pests*: These occur infrequently and are not closely associated with the crop. Examples are caseworms on rice and Mango Stem Borer.

iii. *Seasonal pests*: These occur annually during a specific season. Examples are the red-haired caterpillar on groundnuts and mango hoppers.

iv. *Persistent pests*: These appear on the crop throughout the year and are challenging to manage. Examples are chili thrips and mealy bugs on guava.

v. *Sporadic pests*: These occur in isolated locations for a long time. One example is the coconut slug caterpillar [15].

Pests are categorized into two groups based on their infestation level: epidemic pests and endemic pests. An epidemic is defined as the unexpected emergence of a serious pest in a specific territory at a certain period. Examples include the Brown Plant Hopper (BPH) in Tanjore, the Red Hairy Caterpillar (RHC) in Madurai, and Pollachi. On the other hand, endemic pests exist in small pockets and are typically limited to a specific area—for example, rice gall midges in Madurai and mango hoppers in Periyakulam [15].

2.1 Causes of pest outbreak

Pest outbreaks are primarily caused by certain human activities that undermine ecological balance. Here are various human interventions that contribute to this problem: [15].

a. Deforestation and land cultivation

   i. Pests that feed on forest trees are compelled to eat crops.

   ii. Forests have higher biomass per unit area than agricultural land.

   iii. Weather changes can impact insect growth.

b. Destruction of natural enemies

i. Excessive use of pesticides can destroy natural enemies, disrupt natural control mechanisms, and lead to pest outbreaks.

ii. Synthetic pyrethroid insecticides are one example of this.

c. Intensive and extensive cultivation

i. Intensive cultivation multiplies pests.

ii. Extensive cultivation of sensitive varieties in large areas eliminates competition for food.

iii. This multiplies pests. For example, stem borers are found in rice and sugarcane.

d. Introduction of new types and crops

i. Succulent, dwarf rice cultivars encourage leaf folders, whereas Cambodian cotton attracts stem weevils and spotted bollworms.

ii. Hybrid sorghum (CSH1) and cumbia (HB1) attract shoot flies and gall midges.

e. Improved agricultural methods

i. Increased nitrogen fertilizer use has led to a rise in leaf folders on rice.

ii. Closer planting has led to an increase in brown planthoppers (BPH) and leaf folders. Additionally, granular pesticides have been linked to phytotoxicity in rice.

f. Because of the lack of natural enemies, introducing a new pest into a new environment might cause it to multiply rapidly. For example, the apple wooly aphid, Eriosoma lanigerum, proliferated without its natural predator, Aphelinus mali.

g. Pests like the diamondback moth (Plutella xylostella) can be found on cauliflower, the potato tuber moth (Phthorimaea operculella) can be present, and the cottony cushion scale (*Icerya purchasi*) can infest wattle trees. The wooly aphid (Eriosoma lanigerum) can affect apple trees, while the psyllid (*Heteropsylla cubana*) can be found on subabul. Additionally, the spiraling whitefly (Adeyrodichus dispersus) can be seen on various horticultural crops, and these pests might unintentionally be brought in from other countries through air or sea ports [16].

3. Broad groups of insecticides

3.1 Group 1: acetylcholinesterase (AChE) inhibitors

Organophosphate (OP) and carbamate insecticides inhibit acetylcholinesterase (AChE), an essential enzyme in the functioning of the central nervous system of insects. These are a broad range of insecticides that can be used in crop and non-crop settings. This category includes important insecticides applied to the soil and foliage for row crops and vegetables. Additionally, certain members of these chemical groups also have the ability to control nematodes and mites [17].

3.1.1 Group 1A: carbamates

Carbamates are primarily broad-spectrum insecticides used on cotton, fruit, vegetable, row, and fodder crops. Few carbamates are systemic and can be used for soil or seed treatment (**Table 1**) [18].

3.1.2 Group 1B: organophosphates

Notable members of the organophosphates group include dimethoate, BASF's Perfekthion insecticide, temephos, and Abate mosquito larvicide. Perfekthion is utilized to manage aphids and other pests in wheat, rye, triticale, sugar beet, other beet crops, seed crops, and ornamental plants. Abate mosquito larvicide is essential in the battle against malaria and targets mosquitoes that spread human diseases (**Table 2**) [18].

Common name	Trade name
Carbaryl	Sevin
Carbosulfan	Marshal, Aatank
Methomyl	Lannate, Methavin
Aldicarb	Temik, Cardinal
Thiodicarb	Larvin
Oxamyl	Vydate, Oxamyl
Triazamate	Aztec
Carbofuran	Furadan

Table 1.
Carbamates.

Common name	Trade name
Terbufos	Counter, Aragran
Dimethoate	Perfekthion, Rogor
Chlorpyrifos	Lorsban, Warhawk
Methamidophos	Monitor, Filitox
Acephate	Orthene, Asataf
Profenofos	Curacron, Benofos
Chlorethoxyfos	Fortress
Phorate	Thimet, Agrimet
Temephos	Abate

Table 2.
Organophosphates.

3.2 Group 2: gABA-gated Chloride Channel antagonists

3.2.1 Group 2A: cyclodienes

The group became well-liked globally because of cyclodienes' stability in soil and the environment, extensive range of effectiveness, high level of performance, and affordability. Nevertheless, the utilization of cyclodienes has swiftly decreased because of widespread insect resistance, long-lasting presence, widespread dispersal, and accumulation in wildlife food chains (**Table 3**) [18].

3.2.2 Group 2B: phenylpyrazoles

The first phenylpyrazole to be discovered was fipronil. It is a very effective insecticide at minimal quantities against various pests commonly found on crops like sugarcane, rice, and maize (**Table 4**) [18].

3.3 Group 3: sodium Channel modulators

3.3.1 Group 3A: pyrethroids

Alpha-cypermethrin, a third-generation pyrethroid, is effective in managing various crop pests and plays an important role in preventive measures against tropical disease vectors such as malaria and dengue fever [17]. These chemicals are commonly used to combat crop pests like caterpillars, beetles, aphids, and mites. They are also applied to control mosquitoes, termites, and cockroaches in non-crop areas. Additionally, certain members of this class are used to manage ectoparasites in both pets and humans. Critical applications include foliar sprays in vegetable crops, oilseed rape, and cotton, as well as soil and foliar applications in maize (**Table 5**) [18].

3.3.2 Group 3B: DDT, Methoxychlor

DDT was commonly used as a crop insecticide because it had low toxicity for mammals. However, it was found to be very persistent, leading to over one billion pounds accumulating in the environment over time, causing disruptions in ecological food chains. The Environmental Protection Agency (EPA) prohibited the use of

Common name	Trade name
Thiodan	Discontinued
HCH or Lindane	Discontinued

Table 3.
Cyclodienes.

Common name	Trade name
Fipronil	Regent, Standak, Cosmos, Termidor, Impede
Ethiprole	Curbix, Kirappu

Table 4.
Phenylpyrazoles.

Common name	Trade name
Alpha-cypermethrin	Fastac, Fendona, Mageos, Interceptor, Carifend
Esfenvalerate	Asana XL
Cyfluthrin	Baythroid
Permethrin	Ambush, Pounce
Bifenthrin	Capture, Brigade
Deltamethrin	Decis
Tefluthrin	Force
Zeta-cypermethrin	Fury, Mustang Max
Lambda-cyhalothrin	Karate, Warrior
Fluvalinate	Mavrik
Etofenprox	Trebon

Table 5.
Pyrethroids.

DDT in 1972 due to environmental worries, and it has since been largely phased out globally. However, it has recently been reintroduced in Africa to control disease-carrying mosquitoes by treating interior walls and home ceilings. Another chemical in the same category, methoxychlor, was suggested as a substitute for DDT but has been banned in the United States and Europe since 2003 due to concerns about its ability to build up in the environment and its impact on the endocrine system (**Table 6**) [18].

3.4 Group 4: nicotinic acetylcholine receptor (nAChR) agonists

3.4.1 Group 4A: neonicotinoids

Neonicotinoids effectively control both acute and residual infestations of sucking insects like aphids, leafhoppers, planthoppers, and whiteflies, as well as certain chewing insects such as the Colorado potato beetle, rice water weevil, and codling moth. Additionally, two neonicotinoids, thiacloprid and acetamiprid, have demonstrated effectiveness in managing a broad spectrum of lepidopteran pests. Imidacloprid is approved for various non-crop uses, including spot-on flea treatment, grass treatment for white grubs, and termiticide applications (**Table** 7) [18].

3.4.2 Group 4B: nicotine

Nicotine, produced by plants like tobacco, is a natural insecticide that protects against insects. In 2001, the EPA banned insecticides containing nicotine due to their high acute toxicity [18].

Common name	Trade name
DDT	Gesarol/Anofex
Methoxychlor	Chemform/Methoxo

Table 6.
DDT, Methoxychlor.

Common name	Trade name
Imidacloprid	Admire, Gaucho, Merit, Provado
Acetamiprid	Assail, Intruder
Clothianidin	Poncho
Dinotefuran	Safari, Starkle, Venom
Nitenpyram	Bestguard
Thiacloprid	Calypso
Thiamethoxam	Cruiser, Actara

Table 7.
Neonicotinoids.

3.4.3 Group 4C: sulfoxaflor

Sulfoxaflor, a sulfoximine nAChR agonist, has exceptional broad-spectrum effectiveness against significant sap-feeding pests. It also exhibits excellent residual activity and manages a variety of imidacloprid-resistant insects (**Table 8**) [18].

3.5 Group 5: nicotinic acetylcholine receptor (nAChR)

3.5.1 Allosteric modulators: spinosyns

Spinosyns are microbial insecticides with a unique mechanism of action that controls certain persistent pests in vegetables, such as the Colorado potato beetle, diamondback moth, European maize borer, hornworms, thrips, and leaf miners. They are primarily used to manage pests in fruits, vegetables, cotton, maize, and vineyards. Because they are natural products, they are well-suited for organic farming (**Table 9**) [18].

3.6 Group 6: chloride channel activators

3.6.1 Avermectins and milbemycins

Avermectins and Milbemycins are derived from macrocyclic lactones produced by actinobacteria in the soil. All of them have insecticidal, acaricidal, and nematicidal

Common name	Trade name
Sulfoxaflor	Transform, Closer

Table 8.
Sulfoxaflor.

Common name	Trade name
Spinosad	Spintor, Success, Tracer
Spinetoram	Delegate, Radiant

Table 9.
Allosteric modulators—spinosyns.

Common name	Trade name
Abamectin	Agri-Mek, Zephyr
Emamectin benzoate	Proclaim
Milbemectin	Milbeknock, Ultiflora

Table 10.
Avermectins and milbemycins.

properties, although the products in this group differ in their range and animal safety advantages. Abamectin was derived from the fermentation of *Streptomyces avermitilis*. It controls pests like leaf miners and mites and is mainly used in the production of vegetable and fruit crops (**Table 10**) [18].

3.7 Group 9: selective homopteran feeding blockers

Insecticides, known as selective homopteran feeding blocks, are utilized to manage homopterous insects like aphids, cicadas, whiteflies, and leafhoppers on various crops such as vegetables, potatoes, rice, stone fruits, and ornamentals. These insecticides induce homopterans and other insects to cease feeding. Selective homopteran feeding blockers, an important complement to neonicotinoids in rotation, function by coming into direct contact with the pest, moving systemically within the plant, or locally moving through the plant, spreading through leaf tissues and creating a store of active chemicals within the leaf [18].

3.7.1 Group 9B: pymetrozine

In 1999, Pymetrozine was launched by Novartis, which is now known as Syngenta (**Table 11**).

3.7.2 Group 9C: flonicamid

ISK Biosciences Corporation and its parent company, Ishihara Sangyo Kaisha, developed and introduced the Flonicamid globally in the late 1990s. Nevertheless, FMC distributes it in America and Europe (**Table 12**) [18].

Common name	Trade name
Pymetrozine	Chess, Plenum, Fufill

Table 11.
Pymetrozine.

Common name	Trade name
Flonicamid	Aria, Beleaf, Carbine, Teppeki

Table 12.
Flonicamid.

3.8 Group 14: nicotinic acetylcholine receptor channel blockers

These substances are known as thiocarbamate insecticides or nereistoxin analogs because they are all converted into the active form, nereistoxin, which is toxic to insects. Cartap controls micro-Lepidoptera, such as the diamondback moth (*Plutella xylostella*) and the leaf-mining moth (*Tuta absoluta*). The diamondback moth is a significant pest of cruciferous crops worldwide, primarily feeding on plants containing glucosinolates. *Tuta absoluta*, or the tomato leaf miner, is a major threat to tomato crops in Europe and South America, as well as other plants in the nightshade family, including potatoes, eggplant, and tobacco (**Table 13**) [18].

3.9 Group 19: octopamine receptor agonists

Amitraz is mainly used as an insecticide/acaricide to control ticks, lice, and mange mites in cattle, swine, and dogs. It can also control psyllids, whiteflies, and mites on certain crops (**Table 14**) [18].

3.10 Group 22: voltage-dependent sodium channel blockers

3.10.1 Group 22A: indoxacarb

In 2000, DuPont unveiled indoxacarb as a prodrug that undergoes bioactivation by amidases and esterases in insects but is broken down in mammals. Indoxacarb effectively combats a wide range of insects, including most Lepidoptera and other orders such as Coleoptera, Hemiptera, Diptera, Orthoptera, and Hymenoptera (**Table 15**) [18].

3.10.2 Group 22B: metaflumizone

Metaflumizone effectively controls most Lepidoptera and certain members of other insect orders, including Coleoptera, Hemiptera, Diptera, Orthoptera, and

Common name	Trade name
Bensultap	Bancol
Cartap	Padan
Thiocyclam	Evisect
Thiosultap	Dimehypo

Table 13.
Nicotinic acetylcholine receptor channel blockers.

Common name	Trade name
Amitraz	Mitaben, Preventic

Table 14.
Octopamine receptor agonists.

Common name	Trade name
Indoxacarb	Avaunt, Provaunt, Steward

Table 15.
Indoxacarb.

Common name	Trade name
Metaflumizone	Alverde, Siesta

Table 16.
Metaflumizone.

Common name	Trade name
Flubendiamide	Belt, Phoenix
Chlorantraniliprole	Altacor, Coragen
Cyantranilipore	Exirel, Benevia

Table 17.
Diamides.

Chemical group	Examples of insecticides
Nuclear polyhedrosis Virus (NPV)	*Helicoverpa* viruses
Metarhizium	Entomopathogenic viruses
Petroleum spray oil	Paraffinic oil

Table 18.
Insecticides which are not a member of any group.

Hymenoptera. Metaflumizone effectively manages pests in different agricultural areas, such as cotton, potatoes, fruits, and leafy vegetables. Moreover, it is used to manage ants and other nuisance pests (**Table 16**) [18].

3.11 Group 28: ryanodine receptor modulators—Diamides

These insecticides can effectively eliminate Lepidoptera pests, and they can be used on various types of crops. They successfully manage pest populations that have become resistant to other insecticides. Cyantraniliprole was introduced by DuPont in 2012 as a second-generation ryanodine receptor modulator, delivering comprehensive foliar and systemic effects against chewing and sucking insects (**Table 17**) [18].

There are insecticides which are not a member of any group. These are as follows (**Table 18**):

4. Insecticides for major insect pests in agriculture and horticulture crops

Insect pest infestations and damage can limit the production of fruits and vegetables in farming systems. They attack a wide range of agriculture and horticulture

DOI: http://dx.doi.org/10.5772/intechopen.1006498

S. No.	Trade name	Chem. composition	Active ingr.	Recomdn.	Against
1.	Acephate/ Starthene	O-S Dimethyl acetylphosphoramidothioate	75% WP	0.75–1 g/l	Aphids, Jassids, Bollworm
2.	Actara	Thiamethoxam	25%	0.2 g/l	Sucking pest, Aphids, Whiteflies, Root grub
3.	Coragen	Chlorantraniliprole	18.50%	0.5 ml/l	Borer, Caterpillar
4.	Decis	Deltamethrin	2.8%EC	1 ml/l	Leafminor, White flies
5.	Dursban	Chlorpyrifos	20%EC	1.5–2.5 ml/l	Leafhopper, BPH, Termites
6.	Furadon/Tata Furan	Carbofuran	3%G	5 kg/acre	Stem borer, Cutworm, White grub, Termites, Shootfly, Stemborer, Aphids, Thrips, Jassids
7.	Karate	Lambda-cyhalothrin	5%EC	0.5 ml/l	Borer, Weevils, Aphids, Jassids, Thrips, Whiteflies
8.	Colonel-S	Dicofol	18.5%	2.7 ml/l	Mites
9.	Cythion	Malathion	50%EC	1 ml/l	Caterpillars, Fruitborers, Gallmidge fly, Leaf- eating caterpillars, Stemborer
10.	Matador	Lambda-cyhalothrin	4.9%EC	0.5–0.7 ml	Borer in paddy and cotton
11.	Starphor	Phorate	10%	25–30 kg/ ha	Termites, Nematodes
12.	Vertimec	Abamectin	1.90%	0.25–0.5ml	Spoders & Mites in Ornamentals
13.	Avaunt	Indoxacarb	15.8%EC	0.5– 0.8 ml/l	DBM, Bollworm
14.	Wettasul-80	Sulfur	80%WP	2–5 g/l	Mites
15.	Spark	Deltamethrin+Triazophos	1% + 35%	1.5 ml	Sucking pests, Bollworms, Leaf miner
16.	Nuvan	Dichlorvos	76%EC	0.5–1 ml/l	Cutworms, Leafroller, Aphids, Sucking pests
17.	Proclaim	Emamectin benzoate	5%SG	00.5 g/l	Bollworm, Fruit and shoot borer
18.	Marshal	Carbosulfan	25%EC	1–1.5 ml/l	Aphid, Borer,

Table 19.
Insecticides used against major insect pests.

crops and cause extensive damage, resulting in less production and sometimes even complete crop failure. This leads to economic instability and loss. Many significant fruits and vegetables like mango, apple, banana, papaya, strawberries, okra, tomatoes, chili, cucumbers, brinjal, carrots, and onions are attacked by insect pests, leading to their withering and deterioration. The major insect pests of various crops, which include white grubs, caterpillars, aphids, white flies, jassids, bollworms, sucking pests, borer, root grub, gall midge fly, and mites, are mentioned here with the specific insecticide used to kill or prevent them (**Table 19**) [19].

5. Insecticidal resistance in insects

Insecticide resistance refers to the decreased sensitivity of an insect population to insecticides. This becomes evident when insecticides do not effectively manage insects as anticipated according to the product label instructions. The swift development of resistance in various insects and mites is attributed to genetics and the widespread usage of insecticides. When insects come into contact with insecticides, those carrying resistance genes are more likely to survive and transmit this characteristic to their offspring [20]. As a result, the population of resistant insects gradually supersedes the susceptible ones, rendering the insecticide ineffective in the long term [21].

The development of insecticide resistance occurs at varying rates, influenced by factors such as the rapidity of insect reproduction, the extent of resistance in the insects, the migration patterns and range of hosts for the insects, the persistence and specificity of the insecticide, as well as the frequency, timing, and quantity of insecticide applications. Resistance accelerates more quickly in environments like greenhouses, where insects or mites reproduce swiftly, there is minimal or no influx of susceptible insects, and growers may frequently use insecticides from the same chemical category [21].

In 1945, the use of synthetic chemical pesticides like DDT, BHC, and dieldrin marked a new chapter in chemical pest control. These substances displayed great potential for managing pests due to their ease of use, rapid action, and ability to eliminate most targeted pests effectively. Introducing these new chemical pesticides generated widespread excitement, and they were quickly embraced in the United States and other regions. However, challenges arose when certain insect pests developed resistance to DDT and other pesticides within two years of their initial use. It was found that houseflies and other insect pests had developed resistance to the chemical. This resistance meant that higher doses of pesticides had to be used to control the pests. Additionally, not all pests were susceptible to chlorinated pesticides, leading to reductions in certain bird and fish populations [21].

The issue of resistance aside, insecticides have destroyed various natural enemies, permitting non-pest species to flourish and develop into significant pests. Despite ongoing production and use of pesticides, many newer compounds are highly toxic at current application rates per acre. Consequently, even though small amounts are applied per hectare, the toxicity of modern pesticides is much greater than that of previous pesticides. Under certain conditions, the dosage of modern pesticides is one-thousandth that of DDT and similar chlorinated insecticides.

Insects can build resistance to agricultural chemicals using different approaches and exhibit several resistance mechanisms simultaneously [21].

i. *Behavioral resistance*: Insects that are resistant may be able to sense or identify the danger and avoid the poison. This resistance mechanism has been documented for various types of insecticides, such as organochlorines, organophosphates, carbamates, and pyrethroids. If insects encounter certain insecticides, they may cease feeding or relocate from where the spraying occurred [20].

ii. *Penetration resistance*: Insects that are resistant may absorb the toxin slower than susceptible insects. When insects develop barriers in their outer cuticle, it can slow down the absorption of chemicals into their bodies, resulting in penetration resistance. This resistance can provide protection from a wide variety of insecticides. Penetration resistance often coexists with other forms of resistance, and reduced penetration can enhance the effects of those other mechanisms [21].

iii. *Metabolic resistance*: Resistant insects may metabolize or eliminate the toxin more rapidly than susceptible insects or swiftly expel the toxic molecules from their bodies. Metabolic resistance is the most prevalent mechanism and often poses the most significant difficulty. Insects utilize their internal enzyme systems to break down insecticides. Resistant varieties may have elevated levels or more effective types of these enzymes. Furthermore, these enzyme systems may be more effective and have a wide range of activity [21].

iv. *Altered target-site resistance*: In insects, the site where the toxin typically binds is altered to diminish the impact of the insecticide. This modification represents the second most prevalent mechanism of resistance [21].

Insecticides are essential for managing pests but can have negative consequences, such as leaving harmful residues in food, water, air, and soil. They have the potential to result in the resurgence and resistance of insect pests and can have adverse effects on non-target organisms. It is worrisome that over 645 species of insects and mites have developed resistance to pesticides, and at least one chemical has led to resistance in 542 species of arthropods [22].

6. Ways to prevent insecticidal resistance

The excessive use of insecticides has caused insects to develop resistance and has negatively impacted the environment. As a result, a new generation of modern chemicals with innovative action methods has emerged, offering better effectiveness in controlling insects and being safer for the environment. This new category of insecticides includes neonicotinoids, spinosyns, avermectins, oxadiazines, IGRs, fiproles, pyrroles, pyridine azomethine, ketoenols, and benzene dicarboxamides. Most of these types of pesticides play a crucial role in effectively managing numerous arthropod pests. They have strong bio-efficacy, high selectivity, and low toxicity to mammals, making them an appealing substitute for synthetic organic pesticides. These new categories of pesticides are likely to have a significant role in future Integrated Pest Management (IPM) programs [23].

Small-scale farmers in underprivileged nations frequently use homemade botanical insecticides because they have restricted access to or cannot afford commercial pesticides. Agricultural extension services and certain development organizations suggest the use of homemade botanical insecticides, but there may be insufficient scientific evidence to support the effectiveness and safety of these preparations. While there has been extensive research on botanicals with insecticidal properties, specific studies have not focused on homemade preparations used in real-world fields or storage conditions [23].

The Global Agricultural Plantwise program includes various plants recommended by national extension partners in 20 different countries. Some of these plants include garlic (*Allium sativum*), neem (*Azadirachta indica*), chili pepper (*Capsicum* spp.), Siam weed (*Chromolaena odorata*), mother of cocoa (*Gliricidia sepium*), chinaberry (*Melia azedarach*), moringa (*Moringa oleifera*), tobacco (*Nicotiana tabacum*), clove basil (*Ocimum gratissimum*), tephrosia (*Tephrosia vogelii*), tree marigold (*Tithonia diversifolia*), and bitter leaf (*Vernonia amygdalina*) [23].

The active ingredients found in these specific plants have insecticidal, antifeedant, or repellent properties. Homemade insecticides made from these plants have successfully controlled pests or prevented damage. However, their effectiveness varied and was often lower than that of synthetic pesticides. The efficacy of homemade botanical insecticides is influenced by factors such as variations in the content of active ingredients, the concentration of plant material, and the preparation method [23]. Globally, over 2500 different plant species belonging to 235 families have been identified as having biological activities that can effectively combat pests [24].

The most effective strategy is to prevent insecticide resistance. Pest control experts increasingly recommend insecticide resistance management programs as part of a comprehensive, integrated pest management (IPM) strategy [25, 26].

i. *Monitor pests*: Scouting is crucial for managing insecticide resistance. Monitor insect populations and natural enemies to decide on control measures. After treatment, continue monitoring pest populations and control to evaluate the situation [21].

ii. *Focus on economic thresholds*: Insecticides should only be applied when the number of insects causing financial losses exceeds the cost of the insecticide and its application. Seek advice from local crop experts for specific economic thresholds [21].

iii. *Take an integrated approach to managing pests*: To effectively manage pests, incorporate a range of control strategies in your integrated pest management (IPM) programs. These include synthetic and biological insecticides, beneficial arthropods, cultural practices, transgenic plant varieties, crop rotation, pest-resistant crop varieties, and chemical attractants or repellents. Exercise caution when selecting insecticides to minimize their impact on the environment and future pest populations, prioritizing targeted options over broad-spectrum ones whenever feasible [21].

iv. *Time applications correctly*: Remember to apply insecticides when the pests are at their most susceptible, often right after they have emerged. Follow the application rates and schedules provided by the manufacturer or seek advice from a local pest management professional [21].

v. *Mix and apply carefully*: Precise insecticide application becomes crucial as resistance potential increases. Adjust water pH to product specifications for tank mixes—Mark swath widths in the aerial application. Check nozzles for blockage and wear and ensure they can handle sufficient pressure. Regularly calibrate and check spray equipment. Thorough pruning of tree fruits improves canopy penetration [20].

vi. *Alternate different insecticide classes*: Avoid using the same insecticide multiple times or using insecticides from the same chemical group repeatedly, as this may lead to resistance or cross-resistance. Consider the impact of pesticides on beneficial insects and apply products at the rates and intervals specified on the label [21].

vii. *Protect beneficial arthropods*: Remember to choose insecticides that cause the least harm to beneficial arthropod populations. For instance, applying insecticides in the furrow during planting or in a band over the row instead of broadcasting can help preserve specific natural enemies [21].

viii. *Preserve susceptible genes*: Protect vulnerable individuals in the target population by creating safe spaces for susceptible insects, like untreated areas in treated fields, neighboring refuge fields, or habitat features that encourage migration. These vulnerable individuals may compete with and mate with resistant individuals, reducing the concentration of resistant genes and, thus, the effect of resistance [21].

ix. *Consider crop residue options*: Disposing of remaining harvests during the winter can result in a scarcity of food and shelter for insects. This technique can eradicate pests that have developed resistance to insecticides and those that have not and prevent them from producing resistant offspring next season. However, it's essential to assess your soil conservation requirements before disposing of remaining crops [21].

Dhanuka Agritech, an agrochemical company, has recently introduced a new pesticide called 'LaNevo' and a bio-fertilizer known as 'MYCORe Super'. These products are anticipated to revolutionize crop protection and increase agricultural yields. The launch of 'MYCORe Super' has initially taken place in Tirupati (Andhra Pradesh), Bengaluru (Karnataka), and Nashik (Maharashtra), with plans for expansion to other regions of the country in the near future. The company claims that 'LaNevo' operates in a distinctive manner, providing dual benefits for enhanced crop protection against sucking and chewing pests. They are engineered to reduce the development of resistance to insect pests and promote the growth of robust crops with increased yields. These essential components for agriculture are expected to offer innovative solutions for sustainable farming methods. The company declares that this insecticide is potent enough to efficiently manage numerous pests such as jassid, thrips, whiteflies, seedlings, fruit borers, and leafhoppers.

7. Conclusion

For many years, the unrestricted use of highly toxic insecticides has led to harmful environmental effects, poisoning of non-targeted species, and the development of resistance in insects. Millions of people encounter pesticides yearly, mainly through agricultural activities.

Overusing insecticides may develop resistance in the insects being targeted, requiring more potent chemicals or increased concentrations. Consequently, this raises production expenses and disturbs the natural mechanisms for controlling pests. This, in turn, leads to unintended harm to non-target organisms, such as helpful insects and aquatic life, resulting in disruptions to ecosystems and potential contamination of the food chain. In addition, insects have developed widespread immunity to many traditional insecticides, and the industry may not have sufficient resources to constantly create and provide the market with new insecticide products to substitute older ones. Farmers experiencing resistance issues cannot afford to wait for new chemical solutions. It is crucial for farmers to preserve the effectiveness of available insecticides by implementing practices for managing insecticide resistance. We are constantly confronted with the challenge of reducing the prevalence of insect pests and vectors while also ensuring a safe environment for future generations.

Researchers and scientists are committed to continually advancing the development of compounds that offer improved stability. These compounds need to be able to undergo photo-degradation, microbial degradation, and chemical degradation while leaving behind minimal residue in the environment. This emphasis on enhanced stability has driven the replacement of numerous conventional pesticides with more sophisticated, targeted insecticides. These modern insecticides shield crops and protect the natural predators of pests, thereby playing a crucial role in ensuring overall environmental safety.

Author details

Akanksha Singh[1*], Himanshu Pande[1] and Gaurava Kumar[2]

1 Department of Zoology, D.S.B. Campus, Kumaun University, Nainital, Uttarakhand, India

2 Department of Entomology, G. B. Pant University of Agriculture and Technology, Pantnagar, Uttarakhand, India

*Address all correspondence to: akankshasingh.pth@gmail.com

References

[1] Vanitha SM, Chaurasia SNS, Singh PM, Naik PS. Vegetable Statistics. Technical Bulletin No. 51. Varanasi: IIVR; 2013. 250

[2] Sharma S, Kooner R, Arora R. Insect pests and crop losses. In: Breeding Insect Resistant Crops for Sustainable Agriculture. 2017. pp. 45-66

[3] Rai AB, Halder J, Kodandaram MH. Emerging insect pest problems in vegetable crops and their management in India: An appraisal. Pest Management in Horticultural Ecosystems. 2014;**20**(2):113-122

[4] Oerke EC. Crop losses to pests. The Journal of Agricultural Science. 2006;**144**(01):31

[5] Abate T. Maize stalk borers of Ethiopia: Quantitative data on ecology and management. In: Proceedings of the 3rd National Maize Workshop of Ethiopia. Mexico: CIMMYT; 2011. pp. 174-184

[6] Grzywacz D, Stevenson PC, Mushobozi WL, Belmain S, Wilson K. The use of indigenous ecological resources for pest control in Africa. Food Security. 2014;**6**(1):71

[7] Lingappa S, Basavanagoud K, Kulkarni KA, Patil RS, Kambrekar DN. Threat to vegetable production by diamondback moth and its management strategies. In: Fruit and Vegetable Diseases. Dordrecht: Kluwer Academic; 2004. pp. 357-396

[8] USEPA. About Pesticides. United States of Environmental Protection Agency; 2014

[9] Hemingway J, Ranson H. Annual Review of Entomology. 2000;**45**:371

[10] Grapov AF. Novel insecticides and acaricides. Russian Chemical. 1999;**68**(8):773-784

[11] Richardson ML. Chemistry, Agriculture and Environment. Cambridge: The Royal Society of Chemistry; 1991

[12] Barathi S, Sabapathi N, Kandasamy S, Lee J. Present status of insecticide impacts and eco-friendly approaches for remediation-a review. Environmental Research. 2023;**240**:117432

[13] Chapman RF. The Insects. Cambridge University Press, United Kingdom by the MPG Books Group; 2013

[14] Taiz L, Zeiger E. Plant Physiology. 5th ed. Sunderland: Sinauer Associates Inc.; 2010. 782 p

[15] Kumar S, Biswas SK, Shukla P, Jaisval GK, Kumar S. What Is Pest, Categories of Pests and Causes of Pest Outbreak. Modern Aspects of Entomology, Integrated Publications; 2023. p. 116

[16] Idrees A, Afzal A, Li J. Laboratory efficacy of selected synthetic insecticides against second instar invasive fall armyworm, *Spodoptera frugiperda* (Lepidoptera: Noctuidae) larvae. PLoS One. 2022;**17**(5):e0265265

[17] List of Insecticide. Directorate of Agriculture, Govt. of Assam

[18] Salgado VL. BASF Insecticide Mode of Action Technical Training Manual. BASF Corporation; 2013

[19] Yogui GT, Taniguchi S, Silva JD, Miranda DDA, Montone RC. The legacy of man-made organic compounds in

surface sediments of Pina sound and Suape estuary, northeastern Brazil. Brazilian Journal of Oceanography. 2018;**66**(1):58-72

[20] Can Insects Adapt to Pesticides?—Pet Brilliant.com

[21] Insecticide Resistance: Causes and Action Mode of Action. A joint effort between the Regional Integrated Pest Management Centers and the Insecticide Resistance Action Committee

[22] Gouws MC. Practical and scalable synthesis of N-alkyl-N, N'Diacylhydrazines [Ph.D. thesis]. Nelson Mandela Metropolitan University; 2008

[23] Dougoud J, Toepfer S, Bateman M, et al. Efficacy of homemade botanical insecticides based on traditional knowledge. A review. Agronomy for Sustainable Development. 2019;**39**(4)

[24] Pimentel D. Rational pesticide use. Agriculture, Ecosystems & Environment. 1988;**21**(3):310-312

[25] Stevenson PC, Isman MB, Belmain SR. Pesticidal plants in Africa: A global vision of new biological control products from local uses. Industrial Crops and Products. 2017;**110**:2

[26] Casida JE, Quistad GB. Annual Review of Entomology. 1998;**43**:1

Chapter 6

Pesticide Effects on Human Health and Pest Management

Rajendra Sangaraju, Rajesh Kumar K, Tien Huynh and Sukesh Narayan Sinha

Abstract

This chapter covers public health, agriculture, pesticide regulation, and control. Pesticide resistance leads to higher costs. Agricultural and industrial xenobiotics are harmful to human health and food security. Insecticides and pesticides harm ecosystems. Pesticides pollute agriculture, the environment, air, and water. It does not affect harmless species such as fish, birds, parasites, pollinators, bacteria, or plants. Toxic chemicals can accumulate in tissues as well as in the environment. To reduce insect resistance, use synthetic and biological pesticides while boosting the population of natural predators and parasites. Integrated pest management is used in agriculture to control diseases, weeds, and insects. Integrated pest management is both affordable and safe. IPM uses chemicals to manage pests and protect crops. Pesticide efficacy is reduced by pollution, insect resistance, and non-target organisms. Because of the risks to human health and insect resistance, these pesticides are ineffective. IPM addresses these difficulties using biological, cultural, and mechanical means rather than insecticides. This complete method improves pest management, decreases chemical use, and preserves human and environmental health. Pesticides with different modes of action, proper administration, and insect population monitoring can help you make better decisions. Integrating pesticides with other strategies enhances agricultural sustainability and pest control.

Keywords: integrated pest management, bioaccumulation, insecticides, ecological contamination, biopesticides

1. Introduction

Global Plant Protection Pesticides are primarily organophosphates. Organophosphorus pesticides account for half of US pesticide consumption, at 75 million pounds each year. Industry and agriculture have contaminated soil and groundwater with dangerous pollutants. Manmade compounds known as xenobiotics contaminate the world's ecosystems. Pollution, primarily from industry and agriculture, impacts human health and food quality. Pesticides (fungicides, herbicides, and insecticides) have become more widely used and unregulated over the world, resulting in increased soil and water pollution. Pesticides and fungicides were applied recklessly, resulting in less soil contamination and environmental damage.

1.1 A brief outline of the significance of insecticides in the management of pests

Insecticides are substances or organisms used to manage populations of insects. Control can be attained by eliminating or pre-empting the bug from causing harm. Natural or synthetic insecticides are employed to specifically target pests via sprays, baits, and slow-release dispersion methods. Biotechnology has enabled the introduction of insecticidal microbial genes into several food plants. These proteins lethally affect insects unknowingly when they consume these plants [1]. There is a correlation between crop productivity and pesticide use. This document provides a detailed description of the environmental toxicity pathway and the necessary steps to ensure pesticide safety. Specific pesticides possess inherent characteristics that can potentially influence the health of the environment. Pesticides can be emitted by many means such as spraying, drifting, vaporisation, surface runoff, criminal activity, spills and drainage discharges, leaching, and soil dust. Pesticides, including endosulfan, BHC, endrin, heptachlor, heptachlor epoxide, DDT, and pyrethroids, affect both crops and non-target animals. Pesticide prevention techniques encompass many strategies such as crop rotation, utilisation of protective garments, integrated pest control methods, safeguarding practices, and adherence to pesticide application regulations. Improved agricultural systems require the use of chemical alternatives. Implementing a pest management monitoring method would enhance food production while also safeguarding human and environmental well-being [2]. In the twenty-first century, the problems of pest management involve reducing the use of synthetic pesticides, particularly broad-spectrum insecticides, for protecting plants and increasing the use of "birational" approaches [3].

1.2 An overview of the multifaceted strategies, challenges, and impacts that are linked to the use of insecticides

Invasive pests pose a menace to the entire world. People accidentally or purposefully import non-native invasive pests. Simple pesticides work. These compounds kill insects but are harmful to humans, animals, and plants. Pesticides terrify animals. Because of misuse, many invasive species are resistant to man-made insecticides. Resistance makes it difficult to handle invasive organisms. Controlling pesticide resistance requires understanding how insects develop it. Insects regulate hazardous substances using behavioural, pharmacological, physiological, genetic, and metabolic routes. Resistance is caused by the overexpression of detoxifying enzymes. Pesticide metabolism enzymes reduce parasite-fighting abilities. Esterase [4], glutathione S-transferases, cytochrome P450 monooxygenases, and hydrolyses are all required for metabolic resistance. Digestive microbial communities benefit insects. The majority of pesticide resistance results via detoxification. Discover how invasive pest species withstand hurdles and build efficient management strategies. IPM helps to reduce chemical and environmental pollutants, as well as health dangers. Additionally, it may provide the most effective insect control [5].

Multiple ecosystem roles make insects vital to worldwide agri-food systems. Pollution, habitat loss, and intense agriculture may lower insect numbers, threatening essential services. We often underestimate insects' long-term societal effects, even though they help achieve many sustainable development goals. Insect-mediated biological management decreases invasive pests, increases crop yields, and benefits society and ecosystem. Trans-disciplinary research and "systems thinking" are needed to enhance biodiversity-driven solutions and insects' benefits to agriculture, environment, and society by examining the pan-tropical cassava mealybug invasion and its suppression by the monophagous parasitoid *Anagyrus lopezi* [6]. As plant pathology advances, a more precise and interdisciplinary approach is needed to properly understand the multiple links in each illness, disease complex, or syndrome. Plant pathologists must participate in transdisciplinary research to overcome these issues. Disease management, precision agriculture, plant growth and development, decision analysis, and disease risk are covered in plant pathology. New plant protection agents and host genetic diversity strategies like host resistance are also explored. Plant pathology also requires novel biological control perspectives [7].

1.3 Importance of sustainable pest control through the implementation of integrated pest management strategies

Bugs and plants are controlled by pesticides. Excess pesticide and fertiliser use contaminates food with trace substances. Global environmental issue: agriculture pesticide-related diseases. Pesticides can accumulate in farms, field sizes, and landscape loss from overuse. Poor biodegradability makes these chemicals chronic hazards. Organochlorine herbicides cause developmental and neurological system problems in humans and animals. Increasing lipophilic substances in food and pesticide stability in biological systems affect humans and animals. Pesticide deposition and magnification cause most plant, microbial, and animal biodiversity loss. Microbial-based bioremediation is recommended to safely and sustainably remove harmful pollutants from damaged ecosystems. Environmental contamination and health issues caused by pesticides [8].

Modern agriculture needs insecticides to protect crops and maximise output. However, widespread usage of these medications raises worries about their environmental and agricultural impacts. Insecticides also harm beneficial insects and aquatic life. This affects ecosystem and may taint food. IPM techniques, which combine insecticide application with biological control and cultural therapies, may reduce pesticide use to address these challenges. Targeted pesticides with minimal environmental persistence are essential. To reduce insecticide use, farmers need training, equipment, and precision agricultural technologies [9].

Neonicotinoid insecticides are used worldwide in agriculture and cities. Neonicotinoid insecticides were once considered safe alternatives to organophosphates due to their low environmental and health impacts. This research critically investigates neonicotinoids and their metabolites' ecosystem harm and exposure risks. The environment, water, and food contain neonicotinoids below the daily dose. Animal studies reveal low doses may cause genotoxicity, cytotoxicity, immune system damage, and growth and reproductive problems. Oxidative genetic damage, birth abnormalities, and acute respiratory, cardiovascular, and neurological problems are linked due to insufficient ecological or cross-sectional epidemiological study. Long-term exposure to widely used neonicotinoids has health risks, and thus they must be adequately evaluated [10].

2. Effects of insecticides

2.1 Assessment of the environmental consequences

Pesticides as environmental pollutants are well known. Pesticides, acaricides, and insecticides can affect phytophagous insect predators, competitors, and the intended target species. Pesticide emissions can have unanticipated and frightening ecological effects. Consumers may be harmed by agricultural and animal waste residues. Many pesticides have been used on humans, animals, plants, pastures, crops, water, the atmosphere, stored grains, and cereals. These treatments aim to minimise arthropod pests and vectors. Unfortunately, pesticide active component biodegradation rates vary by environment. They also accumulate there and become organic material in plant and animal excrement. Our knowledge of pesticides in the environment, regardless of source, use, or elimination, is growing. It is commonly accepted that pesticides are in the environment and have effects beyond toxicity [11].

The global usage of pesticides has led to the widespread dissemination and persistence of toxins across the planet. The mitigation of pesticides in contaminated soil and water is a significant issue due to its adverse effects on human health and the environment. Certain indigenous bacteria possess the ability to metabolise or detoxify persistent contaminants, rendering them well-suited for the process of bioremediation. Therefore, these microorganisms that degrade xenobiotics are increasingly being utilised as an effective tool for environmental conservation [12].

2.1.1 Impact on non-target organisms

Non-target organism pesticide toxicity depends on the mode of action. However, the wide range of vulnerability across animal taxa suggests that species-specific biochemical features explain their sensitivity. All pesticides affect water arthropods because they are physiologically similar to target insects. Neurotoxic and respiratory inhibitor chemicals harm various animals, including fish and amphibians. They are less harmed by selective pesticides like IGRs and stomach poisons. Except for insecticides produced from plant or fungal toxins like pyrethroids, neonicotinoids, avermectins, and spinosad, neurotoxicants and respiratory inhibitors damage terrestrial animals. This insecticide appears safe for birds and mammals [13].

Mechanism determines non-target animal pesticide toxicity. Significant differences in vulnerability among animal species show that biochemical features of a group dictate sensitivity. Every pesticide affects water arthropods because they are like insects. Broad-spectrum neurotoxic and respiratory inhibitor pesticides are more dangerous to fish and amphibians than IGRs and stomach poisons. Except for pyrethroids, neonicotinoids, avermectins, and spinosad, neurotoxicants and respiratory inhibitors impact terrestrial vertebrates. Some insecticides are safe for mammals and birds.

Pyrethrum, nicotine, and synthetic pesticides are safer for land animals than OC, OP, and carbamate. Most natural pesticides bind agonistically.

Products manufactured by humans are popular. Food seeds and fruits come from birds and mammals in toxic plant ecosystems. Their neurophysiology adapts to avoid poisoning. As expected, synthetic insecticides are more harmful. The majority of species lack physiological traits that minimise their effects.

Effective and affordable pest management has enhanced agricultural productivity over the past 50 years. But social and legal pressure to reduce pesticide ecotoxicity is growing. Key is novel pesticide target organism specificity [14].

The broader environmental consequences of neonicotinoid insecticides in agroecosystems are worrying. Because of their highwater solubility, they can contaminate soil, water, and unintended plant targets. Many recent studies have looked at the effects of neonicotinoid exposure on solitary, bumble, and honeybee populations, as well as wild pollinator declines. Nothing is known about how exposure affects helpful nontarget species. The relationship between agricultural changes, such as neonicotinoids, and bird species counts in agricultural areas was investigated using UK time series data.

Overall, the findings support the impacts of neonicotinoids on non-target animal development, food consumption, and death. A soil clothianidin decreased *Lumbricus terrestris* survival. Real-life conditions also have a short-term impact on field food consumption. Clothianidin occasionally kills *Polyommatus icarus* larvae in a sublethal dose. Clourianidin and thiamethoxam did affect *Ostracoda* and *Diptera*, but in different ways. Thiamethoxam doses mimicked in the field had no effect on immature hoverflies *Eristalis tenax*. Low chemical concentrations had no effect on adult hoverfly behaviour or survival. No larvae survived. Reassessing the relationship between agricultural development and bird population shifts reveals that farmland birds remain impacted. Neonicotinoid exposure is associated with a decrease in bird populations.

Endocrine disruptive chemicals are chemicals that resemble hormones and bind to receptors that are specifically intended for them. They establish an attachment, which initiates a chain reaction that irrevocably initiates the signalling cycle and results in uncontrolled growth. Pesticides have the potential to cause cancer, congenital birth abnormalities, and reproductive ailments in unintended animals by affecting hormones. Non-target organisms are poisoned by insecticides. Pesticides have numerous adverse effects. Little research has been conducted on the impacts of pesticides on the endocrine system. The objective of this literature review is to comprehend the endocrine disrupting effects of pesticides. ROS-induced endocrine, neurological, genotoxic, and pesticide toxicity are also discussed. The molecular mechanisms by which pesticides affect non-target animals have also been investigated. The deleterious effects of chlorpyrifos on non-target organisms have been widely investigated, including species identification [15].

2.1.2 The pollution of water and its impact on aquatic ecosystems

In the last decade, aquatic environments have been more poisoned with insecticides. Pesticides have an impact on aquatic life even before they enter the food chain. As a result, aquatic creatures must be assessed during prolonged ambient pesticide exposure. These compounds minimise insect, rodent, and weed damage. Microorganisms in humans, animals, plants, and ecosystems can all affect food production, storage, and consumption. Pesticides are classified based on their target species. Pesticides encompass insecticides, fungicides, herbicides, rodenticides, acaricides, and nematocides. Healthy eating, aquatic foods' nutritional value, and high protein content have piqued interest in aquatic ecosystems.

The dynamic ecology of aquatic species enables them to respond swiftly to stress caused by rapid changes. This adaptation depletes energy during development and reproduction. Short-term adaptation allows creatures to live. Chronic stress inhibits physiological adaptation, leaving the organism vulnerable to future stressors. This kills the organism. Industrial and domestic pollution can have a negative impact on water ecosystem species and structures. Consequently, species diversity may decline. Each aquatic animal study illustrates how chemical diversity influences ecosystems.

Marine life provides us with food. Pesticides have been used to meet food demand throughout the last few decades. Pesticide residues on surfaces influence aquatic life. Pesticide toxicity rises with population. Pesticides poison aquatic food webs mostly because they contaminate water and reduce dissolved oxygen. Pollution damages algae, plankton, and benthos, resulting in decreased fish variety and water quality.

Poor water quality harms aquatic life. We assess pesticides' impact on water quality and aquatic ecosystem diversity. Pesticides harm fish. Pesticides damage fish livers, kidneys, gills, and brains. Pollution-exposed fish harm humans.

Aquatic animals provide critical food. Pesticide use has skyrocketed in recent decades to meet food demands. Pesticide residues in surface water can affect aquatic plants and animals. Pesticide toxicity increases with population. Pesticides reduce dissolved oxygen levels, affecting aquatic food systems. Pollutants damage fish and water via disrupting algae, plankton, and benthos. Pollution lowers marine biodiversity. This study found that pesticides harm water biodiversity and quality. Pesticides hurt fish. Typically, fish contaminants harm the liver, kidneys, gills, and nervous system. Eating fish is not healthy.

Pesticides, heavy metals, and petroleum hydrocarbons all have an impact on aquatic ecosystems. Environmental changes that are fatal or sublethal, such as competition releases or trophic cascades, may kill sensitive species. This could unsettle tolerant animals. Contaminants affect ecosystems, nutrient and oxygen dynamics, as well as foundation and keystone facilitator species. Toxins have both direct and indirect effects on ecosystems. Toxicity diminishes directly injured organisms while indirectly affecting abundance. This study examines 150 academic publications on aquatic chemical collateral effects. Long-term contamination, controlled studies, and accidental pollution all damaged benthic and pelagic creatures. Contaminants influence species diversity and community composition via behaviour, competition, and predation/grazing. Changes can exacerbate, obscure, or confuse pollution. In 60% of the trials, trophic cascades enhanced primary producers, whereas pollution reduced grazers. Few studies use indirect effect pathways to distinguish competitive release from trophic cascades. Large trophic cascades in freshwater pelagic zones can result in severe indirect contamination. Pollution hurts nature while revealing population competition and trophic mechanisms [16].

Farmers' widespread use of systemic neonicotinoid insecticides pollutes crop soil and aquatic habitats. These insecticides harm aquatic insects and other arthropods, but their effects on aquatic habitats and the environment are unknown. These insecticides are polluting creeks, rivers, and lakes at low micrograms per litre, according to new monitoring studies in numerous nations. Understanding is limited by the difficulty of obtaining empirical evidence linking species impacts to population and environmental effects.

2.1.3 The impact of soil health on microbial communities

As fertile land becomes scarce, food yields must rise to feed a growing population, straining agricultural systems. With increased farming, herbicides and insecticides, known as "biocide," affect the soil microbiome. Thus, scientists must understand how insecticides and pesticides affect bacteria. Recent research on biocides and soil microbial diversity and types is reviewed here. This bibliometric evaluation statistically analyses recent advances using Scopus document searches. Pesticides and insecticides disrupt microorganism metabolism and cellular activity. This is controversial: biocides disrupt plant-microbial symbiosis. Most microbe receptors bind biocides. Biocides chemically activate biomarker receptors. Redox-active chemicals, free radicals, and electrophilic/nucleophilic species metabolise biocides. Most studies examined biocides' effects on soil microorganisms. Technology for insecticides and herbicides that do not harm soil microbes is needed [17].

Food security is dependent on agricultural production due to the world's growing population. There are growing concerns that agricultural intensification harms ecosystems and yields by affecting soil fertility. Overuse of fertiliser and pesticides to boost crop yields degrades soil, emits greenhouse gases, accumulates pesticides, and depletes water. Soil pesticide behaviour influences microbial-plant interactions that support ecosystem services. Synthetic fertiliser reduces soil biodiversity by suppressing nitrogen-fixing microorganisms while encouraging nitrogen-dependent species. Feeders accelerate humus and organic waste breakdown. The composition of soil changes as organic matter decreases. Several soil physiological processes are affected. However, research on the adverse effects of pesticides and synthetic fertilisers is inconsistent. This chapter discusses how pesticides and synthetic fertilisers affect soil microbiomes and processes [18].

Farmers use a wide range of hazardous pesticides to control harmful microbes during food production, harvesting, and storage. These potent chemicals endanger crops, fertility, and human health. These compounds are accumulating in the food chain and will harm humans. Microbial resources can degrade these compounds, lowering their toxicity. Toxic exposure increases the risk of health problems for field workers and farmers. Farmers may be unaware that they have diabetes or hypertension until tested due to exposure. *Actinobacteria, Ascomycota, Bacteroidetes, Basidiomycota, Chlorophyta, Cyanobacteria, Firmicutes, and Proteobacteria* are the most efficient degraders of carbamate, organochlorine, organophosphate, and pyrethroids. Pesticide degradation was tested on *Arthrobacter, Aspergillus, Bacillus, Burkholderia, Chlamydomonas, Methylobacterium, Nocardioides, Nostoc, Phanerochaete, Pseudomonas,* and *Sphingobacterium* [19].

2.1.4 The quality of the air and the movement of airborne particles

Agricultural practices release pesticides into the atmosphere, focussing on volatilisation and dispersion and how pesticides harm health and cause air pollution. To reduce pesticide application dispersion and volatilisation risks, multidisciplinary research in science, technology, agriculture, and public policy is essential. More research is needed to improve risk assessment and predictive model development, develop new technologies for real-world applications, implement sustainable pest management plans, and understand pesticide airborne dispersal. Pesticide safety and sustainability for the environment and public health require strong policies and regulations supported by extensive research, comprehensive education, rigorous training, and ongoing development [20]. Following the submission of an application, wet leaf surfaces and soil can allow pesticides to seep into residential areas, exposing residents [21]. Italy's Veneto Region is well known for its thriving viticulture and agriculture industries, which have grown significantly over the last 10 years and have strong output indices. Understanding the effects on the environment and public health requires assessing the amount of pesticide pollution in the air in rural areas and how it spreads to high altitudes [22].

Brazil is one of the countries that uses the most pesticides. However, there are serious consequences for human health and the ecosystem when there is a lack of scientific information about airborne pollution. To counteract lower atmospheric pollution, green barriers that function as air filters could be implemented in areas where pesticides are sprayed. The quantity of pesticide that evaporates into the air is reduced when the pulverisation process is moved from the morning to the afternoon. Other approaches to reduce the use of pesticides are being thought about. Studies on volatilisation show that banning goods with high vapour pressure is a practical way to reduce air pollution. Determining the best strategies for reducing pesticide contamination in the air is crucial for both local and international stakeholders. Pollution is becoming a serious threat to human health and the environment [23].

2.1.5 The impact of ecological disturbance on biodiversity loss

Humans value aquatic organisms as food. Pesticide use has skyrocketed in recent decades to meet food demands. Surface runoff transports pesticide residues into aquatic ecosystems, endangering plants and animals. Pesticide toxicity increases with population. Pesticides pollute water and reduce dissolved oxygen levels, which harms the aquatic food chain. Pollutants harm algal, plankton, and benthos, which reduces water quality and fish diversity. Contaminated water also depletes aquatic life. This assessment focusses on pesticides' impact on water quality and aquatic ecosystem biodiversity. Pesticides can harm aquatic organisms, including fish. Fish pesticides primarily harm the liver, kidneys, gills, and nervous system. Fish consumed by humans can develop health issues after being exposed to chemicals [24]. Pesticides, like habitat loss and climate change, reduce biodiversity. These substances can cause immediate harm to species and long-term disruption of the food chain. Pesticide drift, secondary poisoning, discharge into nearby waterways, and groundwater contamination all have the potential to harm both aquatic and terrestrial biodiversity. Pesticides can contaminate soil, water, turf, and plants. Pesticides kill insects and weeds but are toxic to birds, fish, beneficial insects, and non-target plants. Insecticides are the most toxic, but herbicides can also harm animals unintentionally. Pesticides' multifaceted impact on biodiversity reduction necessitates immediate and targeted action. Pesticide damage to biodiversity can be reduced through scientific advancements, regulatory changes, and environmentally friendly farming practices. The fragile life-sustaining web will be safeguarded. Future biodiversity preservation is dependent on our ability to strike a balance between agricultural needs and life on the Earth. The goal of research and development should be to create pesticides that target specific organisms while minimising non-target harm. Green chemistry can produce environmentally friendly alternatives. Farmers, legislators, and the general public must be aware of pesticides' environmental effects. With education, people can make more informed decisions and practise sustainable agriculture. Governments must regulate sustainable agriculture and the use of pesticides. Farmer incentives for environmentally friendly practices can help drive change [25].

2.2 Health consequences

Pesticides boost food production and public health by lowering food-borne and vector-borne diseases. They may be harmful depending on the agent and the level of exposure. Use agricultural pesticides to prevent foodborne illness and crop loss. For years, the toxic debate has raged around the world over pesticide concerns and environmental impacts. They may endanger human and animal health. Pesticides kill larger organisms, posing toxicological challenges. Long-term pesticide exposure poses a cancer risk. Pesticides are widely used, raising health concerns. Pesticide use in developing countries is increasing on an annual basis. Pesticides boost food yields and efficiency, lowering costs while improving quality. Diseases, weeds, and pests destroy 40% of the world's food each year. Insecticide-free agriculture would result in quadrupling crop losses. Residential pesticides can kill termites, ants, roaches, and other insects that endanger public infrastructure and human health. Pesticides benefit public health, research, and food production by controlling pests. Pesticides, unlike many household and commercial chemicals, are generally considered harmful to humans and the environment unless proven otherwise. Pesticides are major environmental pollutants. These poisons eradicate pests and disease-carrying organisms. These are among the more than 1000 active ingredients found in fungicides, insecticides, and herbicides. Researchers and producers must create new and effective pesticides to maintain hygiene, combat insect resistance, and feed a growing population. Pesticides increase crop yield and prevent disease, but they are harmful to humans when used in the workplace or environment. Chronic pesticide exposure impairs organ function and can be fatal. These include nervous, endocrine, immune, kidney, and reproductive issues. Coronary artery disease, asthma, COPD [26], and chronic nephropathies are all included. Pesticide use is increasingly linked to diabetes, cancer, ageing, chronic kidney disease, neurological conditions such as ALS, PD, AD, and MS, and cardiovascular disease. Pesticides cause birth defects, rheumatoid arthritis, chronic fatigue syndrome, and SLE. Chronic diseases are defined by disruptions in cellular homeostasis. Pesticides have a direct effect on enzymes, ion channels, and receptors. Non-primary channels can also achieve this. Pesticide use is strongly associated with chronic diseases in this chapter. Pesticides can harm DNA, epigenetics, hormone function, oxidative stress [26], endoplasmic reticulum stress, mitochondria, the ubiquitin-proteasome system, UPR, and autophagy.

Classification of pesticides is mentioned in **Tables 1** and **2**.

2.2.1 Hazards to human health (farmers, pesticide applicators, consumers) and health issues

According to literature data, human cancer can be caused by exposure to pesticides.

According to the WHO, cancer encompasses a broad spectrum of neoplastic diseases that impact every bodily component. In 2012, cancer was the biggest cause of death globally. That year, an additional 14 million cases of cancer were predicted, with a 70% increase over the next two decades. The most common malignancies that cause death are those of the liver, lungs, stomach, breast, colorectal, and oesophageal regions. According to WHO (2015), environmental factors include physical, biological, and chemical exposures as well as inherited factors can cause cancer. Numerous substances, such as fungicides, herbicides, and insecticides, have been connected to cancer.

2.2.2 The enduring health consequences of prolonged exposure

Diverse pesticides have been employed to manage insect populations in different habitats. Human beings have been concerned about pest control for a significant period of time. The literature indicates that this determination was previously conducted using both inorganic and organic chemicals. Nevertheless, the utilisation of novel and more powerful carbon-based chemical compounds has resulted in a significant rise in pesticide application during the last century. Pesticides have significantly enhanced human efforts in enhancing agricultural productivity, safeguarding crops, controlling household pests, and eradicating contagious diseases. The malaria eradication campaign demonstrates the effectiveness of insecticides. Humans can be exposed to pesticides through various means, such as occupational involvement in pesticide production, transportation, delivery, and application, residing in areas contaminated with pesticides, and the accumulation and spread of pesticides in the food chain. Pesticides were developed as noxious substances intended to inflict

Target pest	Types of pesticide
Algae	Algicides or algaecides
Birds	Avicides
Bacteria	Bactericides
Fungi and oomycetes	Fungicides
Plant	Herbicides
Insects	Insecticides
Lampreys	Lampricides
Mites	Miticides or acaricides
Snails	Molluscicides
Nematodes	Nematicides
Rodents	Rodenticides
Slime moulds	Slimicides
Viruses	Virucides

Table 1.
Pesticides can be classified by the target organism.

Chemical group	Chemical name
Carbamates	Formetanate hydrochloride Aldicard Carbaryl
Organochlorine	DDT Chlordane Dieldrin Mirex
Organophosphates	Dimethoate Phosmet Dimethoate Malathion Parathion Sarin Diazinon Tribufos (DEF)
Pyrethroids	Deltamethrin Cypermethrin Fenpropathrin Allethrin Cyfluthrin Permethrin
Neonicotinoids	Thiamethoxam Acetamiprid Imidacloprid Thiacloprid Clothianidin novaluron
Neonicotinoids- sulfoximines	Sulfoxaflor+Spinetoram

Chemical group	Chemical name
Neonicotinoids-Butenolides	Flupyradifurone
Spinosyns	Spinetoram Spinosad
Avermectins	Abamectin
Pyridinecarboxamides	Flonicamid
Pyropenes	Afidopyropen
Mite growth inhibitors	Clofentezine
BT microbials	*Bacillus thuringiensis*
Benzoylureas	Novaluron Acetamprid + novaluron
Diacylhydrazines	Methoxyfenozine
Acequinocyl	Acequinocyl
METI acaricides	Pyridaben
Tetronic acids	Spirodiclofen Spirotetramat
Beta-ketonitrile derivatives	Cyflumetofen
Diamides	Chlorantraniliprole Cyantraniliprole Tetraniliprole
Flonicamid	Flonic acid
Others	Bifenazide Calcium polysulfide Fatty acids Kaolin clay Mineral Oil Granulosis virus

Table 2.
List of insecticides.

harm on living organisms, thereby making human and animal poisoning an unavoidable consequence. The correlation between pesticide exposure and numerous health problems is underscored, highlighting this fact.

Exposure to pesticides can result in enduring health consequences. Pesticides are used to manage and regulate pests in various environments such as homes, farms, and other locations. Although it is efficient, extended exposure has been associated with a range of health problems [27].

I. Various chronic health consequences are followed:

i. Neurotoxicity:

Prolonged exposure to some pesticides, particularly organophosphates and carbamates, can result in neurotoxic consequences, such as cognitive decline, memory issues, and impaired motor function. Prolonged exposure has been associated with neurodegenerative conditions like Parkinson's disease and Alzheimer's disease [28].

ii. Cancer:

The quality or property of being able to cause cancer. Extended exposure to pesticides has been linked to a heightened risk of various forms of cancer. Exposure to organochlorines such as DDT and glyphosate has been associated with the development of breast cancer, non-Hodgkin lymphoma, and leukaemia. Individuals who have occupational exposure or live near agricultural areas may have a higher risk **Table 3**.

iii. Health related to reproduction and the process of growth and maturation

Reproduction:

Prolonged exposure to pesticides can have a detrimental impact on reproductive health [29], resulting in decreased fertility, disruption of hormonal balance, and problems during pregnancy. For instance, certain research indicates that there may be a correlation between pesticide exposure and a heightened likelihood of birth abnormalities and miscarriage.

Impact on development:

Exposure to pesticides during pregnancy or early childhood can lead to persistent health problems in children, such as delays in development, behavioural disorders, and heightened vulnerability to chronic diseases in the future.

Some of the epidemiological studies on pesticide-related cancers in human	
Type of cancer	**Pesticide**
Brain cancer	Chlorpyrifos
Bladder cancer	Imazethapyr
Colon cancer	Aldicarb Dicamba EPTC Imazethapyr Trifluralin
Leukaemia	Diazinon Chlorpyrifos Chlordane/Heptachlor EPTC Fonofos
Lung cancer	Pendimethalin Dieldrin Diazinon Chlorpyrifos Dicamba
Multiple myeloma	Permethrin
Melanoma	Carbaryl, Parathion, Maneb/Mancozeb Toxaphene Carbaryl

Table 3.
Some of the epidemiological studies on pesticide-related cancers in humans.

iv. Endocrine Disruption:

Certain pesticides function as endocrine disruptors, which means they interfere with the hormone systems in the body. This can result in a range of health issues, such as thyroid malfunction, disrupted reproductive health, and metabolic

v. Respiratory Problems:

Extended contact with pesticide residues, particularly in work environments, can result in persistent respiratory ailments such as asthma, chronic bronchitis, and other pulmonary disorders.

vi. Effects on the Immune System:

Prolonged exposure to pesticides has the potential to impact the immune system, rendering individuals more vulnerable to infections and potentially resulting in autoimmune disorders.

vii. Skin-related Problems:

Prolonged exposure to pesticides can result in skin issues such as rashes, dermatitis, and, in more severe instances, long-lasting skin diseases.

viii. Overall Health Impact:

Extended exposure to pesticides might potentially contribute to overall health problems, such as fatigue, migraines, nausea, and other nonspecific symptoms. These factors can have a substantial impact on one's quality of life.

II. Measures to reduce and prevent

In order to mitigate the potential hazards linked to pesticide exposure, it is advisable to take into account the following approaches.

i. Utilise protection gear:

When dealing with insecticides, ensure the usage of suitable personal protection equipment (PPE) such as gloves, masks, and goggles.

ii. Adhere to safety rules:

Comply with the prescribed safety practices and rules for the application, storage, and disposal of pesticides.

iii. Minimise Exposure:

Whenever feasible, choose integrated pest management strategies that employ fewer chemical pesticides or investigate alternate, less deleterious approaches to pest control.

iv. Routine Health Surveillance:

Regular health check-ups and monitoring are crucial for individuals who have work exposure in order to detect early indicators of health problems.

Through comprehending and controlling the hazards linked to pesticide exposure, it is feasible to alleviate certain probable enduring health repercussions.

2.3 The socio-economic impact

2.3.1 The health and income of farmers

Growing vegetables is profitable because of its high ROI. Recent analysis showed that insecticides are applied randomly to specific plants. Misuse of insecticides leads to pest resistance and the elimination of native predators, promoting pest growth. This disrupts the agricultural ecosystem, pollutes the environment, and endangers human health. Most farmers used no biological or cultural methods. A minority of farmers practise basic sanitation. The media should report insecticide dosage and side effects in vegetable agriculture [30].

2.3.2 Healthcare expenses

Farmers frequently used insecticides to protect crops and increase yields. Pesticides have caused farm workers to experience high fever, dry cough, shortness of breath, skin redness, and nervous system abnormalities. The government should impose harsh penalties on pesticide manufacturers. This can be accomplished through extensive training, short courses, and farmer pesticide safety advice. These safeguards protect farmers and other agricultural workers from pesticides. A new rural health strategy should include monthly farmworker physicals. Mobile health units and rural clinics can assist. Technical, methodical, and community-based initiatives are required to develop and implement affordable and effective pesticide alternatives in both controlled and natural environments [31].

2.3.3 Enhancing food security and improving agricultural productivity

The global population is expected to grow by 70 million per year, or 30%, to 9.2 billion by 2050. Food production demand will rise 70% due to population density. A shift in eating habits in developing countries towards high-quality food, including more meat and dairy, is the main cause. Furthermore, cereals are increasingly fed to cattle. Limited agricultural land is available. The expansion will mostly invade forests and natural environments that support wildlife, crop relatives, and crop pest predators. Instead of growing food, more agricultural land will be used to grow biofuels and fibre. Thus, we must grow crops on less land while using less water, energy, fertiliser, and pesticides than current agricultural practices. Given these constraints, large-scale sustainable production must be implemented quickly. Reducing pest-related crop losses hinders agricultural efficiency. The public influences agricultural pesticide use. As long as pesticides are needed to manage pests and ensure food security, it is important to recognise and address their external effects on human and environmental health [32].

Global population is expected to grow 30%, or 70 million, to 9.2 billion by 2050. Food demand rises 70% with population density. Developing nations eat more high-quality meat and dairy for this reason. Animals eat grains. Arable land is scarce.

Instead of food and feed, more land will be used for bioenergy and fibre. We must conserve water, energy, fertilisers, and pesticides while growing crops on less land. Due to these constraints, eco-friendly industrial methods must be implemented quickly. Minimising pest-related crop losses lowers productivity. Pesticides are necessary for pest control and food security, but they harm humans and the environment [33].

Agrochemicals were developed decades ago to boost crop yield and protect against pests. Farmers must use more insecticides to protect their crops as pests become more resistant. Unexpected events increase food manufacturing costs. Innovative methods like pest-resistant GMOs may reduce agrochemical use in agriculture. Although biological chemical-free agriculture is gaining popularity, it cannot meet large-scale food production needs. Pesticides are still common in tropical and poor countries. DDT, HCH, and lindane, which have long-term environmental effects, are banned in developed countries but widely used in developing ones. Pesticide residues harm food and the environment. Collaboration is needed to prioritise food production, quality, safety, and pesticide residue management [34].

3. Difficulties in the application of insecticides

Pesticide resistance: Pests can develop pesticide resistance. These substances become ineffective when resistant, making pest management more difficult and requiring stronger pesticides. Insecticides can harm pollinators like bees and beneficial predators like ladybirds. This injury could disrupt ecosystems and bug populations. Environmental impacts: Insecticides pollute soil, water, and air, harming non-target organisms and disrupting the ecosystem. This can cause lasting ecological damage and ecosystem disruptions. Pesticide spraying requires precise timing. If pests are unresponsive, applying a pesticide too early or late may reduce its efficacy or worsen the problem. Application: Sprays, granules, and systemic drugs are difficult to choose. Pest type, environment, and life cycle stage must be considered.

Safety: Poor pesticide handling can harm humans and animals. To reduce exposure and injuries, PPE and safety regulations are essential. Premium pesticides are expensive, especially for farmers. Pesticide, application equipment, and labour are included in the price. Pesticide regulations vary by location. Following these standards can be complicated and require permission or guidelines. Perception and acceptance: Pesticides may cause health and environmental concerns. Managing public opinion and addressing these issues requires education and transparency.

Weather can significantly impact pesticide efficacy. Rain, strong winds, and extreme temperatures can prevent pesticides from sticking to or penetrating, causing runoff and pollution.

3.1 Emergence of resistance

3.1.1 Resistance mechanisms

Several mechanisms give insects resistance. Metabolic resistance lets insects produce pesticide-degrading enzymes. Target site resistance alters insect receptors or ion channels, reducing pesticide efficacy. Insects can avoid pesticides and treated areas. Reasons for resistance: several factors cause resistance.

Pesticide overuse or misuse accelerates resistance. Unvarying insecticide use can cause resistance. Unsuitable application: Poor dispersion or dosage can cause insect control failure. It lets pest populations resistant to certain measures reproduce and spread traits. Due to their rapid reproduction, insect populations change quickly. Selects traits that resist certain factors. Genetic diversity: Insects may be resistant. Pesticide-resistant people live longer and reproduce better. Application implementation effect challenges: Coverage issues arise when bugs persist due to device malfunction, environment, or uneven distribution. Increases pest resistance. Weather conditions like wind, rain, and temperature affect pesticide effectiveness. Rain and extreme temperatures can make pesticides ineffective, creating pest-friendly environments. Complex vegetation and varying plant height make pesticide dispersion difficult. Treatment gaps may let pests survive. Supervisory and conflict resolution: The following are effective management conflict resolution strategies: Integrated pest management uses chemicals, biological control, crop rotation, and habitat management to control pests. This method cuts pesticide uses and slows resistance. Pesticides with multiple modes of action reduce resistance. Combining pesticides with other methods improves control. Pest management can be adjusted quickly by monitoring pest populations for resistance. Education and Training: Educating farmers and applicators about pesticide dosage and administration may reduce resistance [35].

3.1.2 Effects on pest control

Reduced effectiveness: Pesticide coverage may vary due to equipment or method failure. Pests can reproduce unaffected. Poor dosage: Inadequate pesticide treatment may not control pests, whereas excessive application may harm nontarget species and the environment. Both could jeopardise pest control. Weather conditions such as wind, rain, and extreme temperatures influence insecticide application. Air currents and rain can disperse pesticides, resulting in uneven distribution and significant damage to nearby areas. Growing survivor resistance: Poor implementation may lead to pest resistance. Pest survivors reproduce, leading to increased resistance. Pest resistance increases as a result of selection in suboptimal application conditions. These traits spread because only weak bugs can withstand treatment.

Economic and Environmental Impacts: Increased costs: Poor application may necessitate additional treatments, increasing pest control costs. Insecticides and other management strategies may become less effective, increasing the cost to farmers. Poor pest control leads to increased crop damage. Crop output and quality may decline, affecting farmer profits and food availability. Environmental risks: Pesticide overuse or accidental release pollutes ecosystems. It could harm beneficial insects, wildlife, and water.

Pest Behaviour and Management: Pests avoid treated areas, making control difficult. Untreated, they have the potential to multiply. Pest control becomes more complicated as the difficulty increases. More monitoring, method changes, or controls may be required to achieve the desired results. Getting past application problems: Improved training and equipment to apply insecticides consistently, applicators require better training and more reliable equipment. IPM reduces application-related issues. Insecticides, biological, cultural, and mechanical controls all contribute to better pest management. Analysing pest control technologies and adapting to application complexity is an example of adaptive management. Pesticides, applications, and technology may evolve. GPS-enabled equipment increases the accuracy and efficacy of insecticide applications.

3.1.3 Approaches for mitigating resistance

3.1.3.1 Application enhancements

Precision Application: Insecticides are evenly distributed using GPS, drones, or automated sprayers. Technologies may help to reduce application errors. To prevent errors and ensure accurate dosage, application equipment must be calibrated and maintained on a regular basis. Regular inspections and precise adjustments help to maintain application efficiency. Avoid spraying pesticides in high winds or rain. To reduce drift and wash-off, schedule applications during good weather.

Multiple control methods in IPM: Add traps, barriers, and predators to insecticides. This reduces reliance on a single method while improving insect population management. Monitor pest populations to determine when and where pesticides should be applied. Determine therapy requirements using action thresholds to reduce resistance and unnecessary treatments.

Controlling Insecticide Resistance: Rotating and combining insecticide mechanisms reduces resistance. Insecticides with multiple mechanisms can reduce resistance by targeting pest routes. Insecticides with multiple mechanisms of action are more effective at delaying resistance. Use combinations to increase efficacy while avoiding cross-resistance. Follow the insecticide instructions and avoid overuse. Overuse generates selective pressure, which increases resistance. Minimal doses lower insect populations and resistance.

Biological Control and Options: Encourage insect-eating native predators and natural enemies. Biocontrol reduces insecticide use. Culture reduces insecticide use and pest populations. This includes using pest-resistant crops, adjusting planting dates, and controlling irrigation.

3.1.3.2 Plan academic and instructional

Trained Applicators: Instruct pest control and agricultural workers on insecticide measurement, timing, and procedure. Learning minimises application errors and enhances pest control.

Public Awareness: Teach farmers the best practices for managing resistance. Understanding the effects of resistance can help you manage more effectively.

3.1.3.3 Research and development

Fund new insecticide research to combat pest resistance. Continuous innovation improves pest control. Continue researching pest resistance. Data can be used to guide resistance management and pest control efforts.

3.1.3.4 Laws and regulations

Advice and Guidelines: Comply with pesticide and resistance regulations. We can reduce pest resistance by implementing comprehensive pest control measures and limiting insecticide use.

3.2 Environmental issues

3.2.1 Challenges and adherence to regulations

Agriculture requires strict compliance with rules for safe, efficient, and environmentally responsible pesticide use. Pesticide regulations reduce dangers, but they can be tough for farmers. Understanding the challenges and laws is essential for efficient and legal pesticide use.

3.2.1.1 Pesticide use issues

Resistance emergence, issue: Excessive or incorrect pesticide application can cause insect populations to become resistant to chemicals, limiting chemical control efficacy. Resistance may increase application rates, expenses, and the necessity for more expensive control approaches. Environment-health impact: Pesticides can pollute land and water and offer health concerns through direct contact or food residue. Impact: Addressing these issues frequently demands additional steps and costs, such as optimising implementation and safety. Economic cost is the financial cost of an activity or circumstance. Pesticide purchase, application, resistance management, and compliance can be expensive. Consequence: High expenses can strain small-scale growers, lowering agricultural profitability. Operations and technology challenges: Effective pesticide application involves proper equipment, technology, and knowledge of proper methods. Insufficient use can impair effectiveness, increase resistance, and harm the environment, necessitating more training and equipment [36].

Regulation compliance: The US Environmental Protection Agency (EPA), European Food Safety Authority (EFSA), and other national and international bodies set pesticide usage regulations. Maximum Residue Limits (MRLs), labelling, records, and application and handling safety must be followed to comply with regulations. Communicating and Recording: Producers must record pesticide application rates, dates, and locations. They may also have to report incidents or unfavourable consequences. Accurate documentation helps monitor compliance, evaluate pest management plans, and resolve regulatory enquiries or inspections. Certification and education: Many countries require pesticide training and certification for safety and efficiency. Training programs teach safe handling, application, and regulatory compliance. Risk-management methods: Risk management strategies like buffer zones to protect water bodies and pesticide limits are often required by regulations. Encompasses: Following these recommendations reduces ecological and medical dangers, ensuring pesticide efficiency and durability.

Adherence and administration method: Integrated Pest Management, Strategy: Integrated pest management can reduce pesticide use, manage resistance, and save costs. Advantages: Integrated pest management combines biological, cultural, and mechanical pest control with pesticide administration to create sustainable, compliant practices. Technology innovation, approach: Precision agriculture technology improve pesticide application, reduce waste, and meet regulations. Advantages: GPS-guided sprayers and remote sensors improve pesticide application precision. Continuing education, strategy: Producers and applicators receive ongoing training on regulatory changes, new technology, and best practises. Benefits: Regular updates and refresher training maintain safety and compliance.

Pesticide laws must be strictly followed to reduce health and environmental concerns and improve pest management. Agricultural producers encounter criticism, costs, and technological issues. These challenges can be overcome with strategic measures like IPM, modern technologies, and continual education. Efficient management and compliance protect health, the environment, and agricultural viability and profitability.

3.2.2 The way the general public views and the level of interest shown by consumers

Pesticides kill agricultural, animal, and human pests. These include fungicides, insecticides, and herbicides. Pesticide interest and public opinion differ greatly. Health and safety concerns: Pesticides in food and the environment worry many. Pesticide concerns and organic farming drive pesticide-free product demand. Pesticides' destructive effects on soil, water, and biodiversity are becoming known. This popularises integrated pest management and sustainable agriculture. Pesticide usage and regulation transparency are desired by consumers. Consumers may like pesticide-free products and food cultivation info. Demand for organic products without synthetic pesticides is rising. This trend shows consumer concern for sustainability, the environment, and health. Public education and advocacy have improved pesticide safety. Consumer tastes have changed, favouring safer pesticides. Modern agriculture uses pesticides, but people want safer, greener alternatives.

3.2.3 The effect on species and ecosystems that were not intended as targets

Pesticides harm nontargets. Food and water pesticide residues can acutely poison birds, mammals, and amphibians. Pesticides affect all food chain animals. Pesticides that kill insects may indirectly affect birds and bats, which eat them. Pesticides, especially neonicotinoids, kill pollinators. Loss of pollinating insect's harms plant reproduction and biodiversity. Effects on nature: Pesticides kill beneficial bacteria in soil microbiota, which fixes nitrogen and degrades organic matter. The disturbance may deplete soil fertility and plant health. Farm runoff pesticides pollute rivers, lakes, and aquifers. Pollution harms fish and invertebrates in aquatic ecosystems. Continuous pesticide use reduces biodiversity. Pesticide-intensive ecosystems may change or simplify ecosystems by threatening non-target species. Pesticides can harm ecosystem services by extincted key species. This could harm insect control, soil fertility, and water purification. Reduced insectivorous birds may increase pest populations with insecticides.

Management and relief: These effects can be mitigated by IPM. Integrated pest management reduces pesticide use and uses mechanical, cultural, and biological methods. Ecosystems and non-target species can benefit most from organic farming and safer pesticides.

3.3 Potential hazards to health

3.3.1 Occupational exposure

Occupational exposure refers to the potential contact or interaction with hazardous substances or conditions in the workplace.

3.3.1.1 Exposure routes

Inhalation: Pesticides are inhaled when they are sprayed, mixed, or present in the air. Potent pesticides are frequently handled, mixed, applied, dispersed, or sprinkled. Dermis interaction: Pesticide skin contact can result from contaminated equipment, spills, or splashes. Using pesticides without proper protection; touching affected surfaces or plants. Hand movements: Poor hygiene, such as eating or drinking without washing your hands after coming into contact with pesticides, can result in pesticide ingestion. Cross-contamination with pesticide residues is common, as is contaminated food or water. Eye exposure: Pesticides that splash or spill can irritate or damage the eyes. Insecticides are commonly applied using spray equipment without eye protection.

3.3.1.2 Pesticide's impact on health

Direct health impacts: Coughing, wheezing, shortness of breath, and respiratory inflammation are all short-term respiratory symptoms. Burns, dermatitis, and hypersensitivity are typical skin reactions. Tremors, headaches, nausea, and dizziness are all potential side effects of certain pesticides. Pesticide-induced ocular irritation can cause redness, discomfort, and vision problems. Long-term health effect: Long-term pesticide exposure raises the risk of lymphoma and leukaemia. Long-term exposure can result in Parkinson's disease, peripheral neuropathy, and cognitive impairment. This substance may cause developmental delays or atypical birth defects, which can have an impact on reproduction and growth. Pesticides can disrupt hormonal function, resulting in long-term health problems [37].

3.3.1.3 Prospects & standards

OSHA: OSHA sets PELs and PPE for pesticides. Workers must complete OSHA pesticide handling and use training. EPA: According to EPA regulations, pesticide labels must include first aid, personal protective equipment, and safe handling instructions. Pesticide handlers and agricultural workers are protected under the Protective Worker Standard. NIOSH: NIOSH recommends safety protocols and pesticide exposure limits.

3.3.1.4 Risk assessment and management

Risk evaluation: Identify occupational pesticide exposure sources and pathways. Evaluate health risks, incidence, and exposure. Control measures include PPE, engineering, and administrative methods for reducing exposure. Regulators: Engineering controls entail constructing barriers or enclosed spaces to mix and spray pesticides. Ventilation systems help to reduce airborne pesticides. Administrative controls include safe handling policies, employee training, and safety precautions. PPE: Ensure that employees wear gloves, masks, goggles, and protective clothing when handling pesticides. Directions: Training programs cover pesticide application and handling, as well as emergency response. Staff should be trained on safety, personal protective equipment (PPE), and pesticide symptoms. Evaluate and monitor: Pesticide concentrations in the air and on surfaces are measured, and exposure levels are determined.

Regular pesticide-related health screenings. Assess and improve: Conduct regular reviews of exposure data and safety protocols to identify areas for improvement.

3.3.2 Emergency response

3.3.2.1 Urgent medical care

Immediate action: Learn how to clean your skin and eyes, and seek medical attention for pesticide exposure. Emergency contacts: A list of contacts is required for medical emergencies involving poisoning or pesticides. Incident Report: Report all pesticide exposures to ensure proper monitoring and avoid future incidents.

3.3.3 Consumer exposure to food residues

Pesticide residues in food can be consumed by consumers. Pesticides protect crops from pests, diseases, and weeds, but they can remain on foods after harvesting and processing.

3.3.3.1 Residue origins

Use in agriculture: Crops are grown with pesticides. Food residues may remain on the surface or inside. Insecticides are applied to harvested foods after harvesting. This prevents product deterioration and pest infestation during storage and transport. Potential health risks: High pesticide residues can cause nausea, migraines, and other health issues. Cancer, endocrine disruption, and reproductive issues can result from prolonged pesticide residue exposure. Regulations: To prevent pesticide residues on food from exceeding safe levels, regulatory agencies like the EPA in the US and the EFSA in Europe set Maximum Residue Limits (MRLs). Surveillance and implementation: Regular food product surveillance ensures MRL compliance and consumer safety. Reduce or prevent problem impact: Rinsing fruits and vegetables with running water and removing their skins may reduce residue. Organic options: Since organic farming uses less pesticides, organic products have lower residue levels. Consumer Awareness: Knowing and following food safety guidelines can reduce pesticide residue exposure. Food safety is threatened by pesticide residues. Pesticide residue risks in food can be managed and reduced through regulation, food handling standards, and consumer education.

3.3.4 Health consequences for susceptible populations

Pesticides have a greater effect on children, pregnant women, the elderly, and people with pre-existing conditions. Physiological and developmental differences may exacerbate health in certain populations.

Health hazards on various age group:

3.3.4.1 Children

Children are more likely to be exposed to pesticides due to their developing bodies and minds. Exposure can impact behaviour, slow down development, and impede cognitive function. Because of their higher surface area-to-volume ratio and faster metabolism, children absorb more pesticides than adults.

3.3.4.2 Pregnancies among women

Premature birth, low birth weight, and birth defects can all result from pesticide use during pregnancy. Pesticides may have an impact on endocrine function, which may impact reproductive health. Pregnancies exposed to pesticides may result in children's long-term cognitive and developmental problems.

Senior citizens: Because their immune systems are weakened and they are less able to flush out poisons, pesticides can be harmful to elderly people. In older adults, pesticides can exacerbate long-term conditions.

3.3.4.3 Patients in medicine

Insecticides can exacerbate symptoms of cardiovascular disease, diabetes, and asthma. Pesticides can affect people with weak immune systems or general health.

Strategies for mitigation and avoidance: Strict adherence to safety regulations lowers exposure. Vulnerable populations are safeguarded by surveillance and food chemical residue limitations.

Education and Awareness: Reducing risks may be achieved by teaching vulnerable populations how to handle food and select organic goods. PPE and safety procedures can lower exposure and safeguard those who are more susceptible.

Pesticides put people who are vulnerable at risk. To safeguard their health from these risks, appropriate laws, safeguards against harm, and focused education are needed.

3.4 Economic factors

3.4.1 The cost associated with the development and registration of insecticides

Pesticide development and registration are costly. Costs can affect pesticide availability and pricing from inquiry to approval.

R&D expenses: Discovering and testing insect-killing chemicals or biological agents costs money. This is where efficacy research, laboratory testing, and high-throughput screening begin. To evaluate the new insecticide's safety and efficacy, preclinical testing involves extensive laboratory and field research. Formulation development, environmental impact assessments, and toxicity testing.

Rules, registration fees: Health, safety, efficacy, and environmental data are collected for regulatory filings. This usually requires extensive data analysis and long-term research. For application and evaluation, the European Chemicals Agency (ECHA) and EPA charge regulatory fees. Region-specific costs vary greatly. Creating regulatory-compliant reports and documents. Regulatory compliance requires legal and scientific knowledge.

Field tests and experiments: Pilot implementation involves extensive field studies of pesticide performance in agriculture. Establishing the scenario, deploying controls, monitoring, and collecting data cost money. Trial results are analysed for pesticide efficacy and safety. Often, this requires extensive modelling and statistics.

Product Launch and Marketing Costs: Distribution networks, marketing strategies, and consumer preferences are covered in market research. Costs include marketing, sales, and branding. End-user training and technical support ensure deployment and use.

Maintaining Compliance: Tracking insecticide effects after approval. This includes regulatory enquiries and record protection. Updates to formulas, labelling, and safety data due to scientific or legal changes are expensive. Insecticide development, field testing, regulatory compliance, and launch are expensive. Innovating pesticides ensures safety, efficacy, and environmental compatibility. Costs will affect pesticide prices and availability.

3.4.2 Cost-effectiveness for agricultural producers

Agricultural producers must use pesticides efficiently and cheaply due to many issues. Environmental impacts, resistance management, and regulations are issues. Pest control and crop protection must be compared to chemical costs to determine cost-effectiveness. Price: High-quality insecticides are expensive. Pesticide application costs include labour, equipment, and operations. Rotating insecticides or combining products to manage pesticide resistance costs money. Productivity: Pest control increases agricultural production and quality, increasing profits immediately. Avoids insect damage that lowers crop yield, sales, and profit. Better Market Access: Quality and safety lower prices and increase profits [36].

The Three Challenges and Key Points:

1. More restrictions may raise registration, monitoring, and reporting costs.

2. Pesticide resistance reduces efficacy, requiring more expensive alternatives or IPM.

3. Health and environmental concerns may require safer application methods or protective gear.

These actions may affect cost-effectiveness.

Maximising scale economy: Among other methods, IPM uses insecticides, cultural and biological controls. Aim for less chemicals and money. Precision agriculture boosts yield, cuts waste, and cuts costs by targeting pesticides. Alternative solutions: Investigating eco-friendly, cost-effective pest control methods can pay off.

Economic evaluation: Estimate pesticide costs using multiple cost-benefit analyses. Damage reduction and agricultural output are weighed against costs. This simplifies evaluating current methods' profitability. Planning should consider pest management's long-term economic effects, like ecological damage and resistance. To use pesticides efficiently, farmers must weigh pros and cons. Technology, pest management strategies, and guidelines can help farmers reduce pest control costs and issues.

3.4.3 Comparative analysis of the economic consequences of pest damage and the implementation of management methods

Infestation: Costs now pests lower crop yields and quality. Sales revenue drops. Because illness degrades food quality, damaged products may be discounted or discarded, lowering market value and consumer approval. Pest-affected crops may require more labour and resources. Costs of overhead: Reduced Market Access: Intense pest damage may restrict market entry into profitable sectors. Farmers may need more expensive pest-control supplies or technologies. Pests reduce biodiversity and soil health, lowering production and sustainability.

Economic impact of insecticide pest control:

Costs include pesticide, application equipment, and labour. Used frequently or severely can be expensive. Pesticide overuse can cause resistance, requiring more expensive alternatives or combinations. Advantages: Rapid control: Strong insecticides reduce pests and preserve crop quality quickly. Less insect damage increases agricultural productivity and profitability [37].

3.4.3.1 Integrated pest management

IPM implementation may cost money for monitoring, biological controls, and other methods. Training and education cost money. Installing IPM requires time and regular monitoring and adjustment. Advantages: IPM reduces pest resistance and environmental damage with many sustainable pest control methods. In the long run, IPM saves money by reducing chemical inputs and improving pest management efficiency.

IPM improves soil and ecosystem health, boosting productivity. Biological and cultural control:

Expenses: Certain cultural and biological controls require initial investments in new practices or species. Cultural norms and biological constraints may require extra effort and monitoring. Cost and resistance are reduced by using fewer chemical pesticides. Long-term sustainability: They boost productivity, ecology, and insect control.

3.4.3.2 Quickly monitor and identify

Expenses: Monitoring systems: Equipment, technology, and labour costs for regular inspections. Monitoring analysis and management costs. Advantages: Single-issue interventions. Early detection means spotting problems early. Improvements in accuracy and speed lower pest management costs. Prevention: Effective surveillance reduces severe infestations, injuries, and costs.

Summary of comparison. IPM and biological controls cost earlier but save money and promote sustainability. Pesticides control quickly but are expensive and can cause resistance, increasing long-term costs. Damage costs directly and indirectly and reduces profits. Effective management reduces these negative effects, but is expensive.

Compare long-term and short-term benefits, pest management may require more upfront investment. Avoiding harm, improving sustainability, and reducing resistance usually boosts long-term profits. Pest damage can lead to long-term financial losses and higher costs if ignored. Sustainable IPM and biocontrol's. Long-term economic benefits from ecosystem improvements and chemical reductions are more consistent.

4. Strategies for integrated pest management

4.1 Biological control

Biological control refers to the use of living organisms to control pests or diseases in agriculture or other ecosystems

Biological control employs living organisms to combat pests and diseases in agriculture and other ecosystems. This method reduces pesticide use by utilising natural predators, parasites, or pathogens to control invasive species. Biological control promotes ecological balance and biodiversity, making it a viable and environmentally friendly pest management strategy. Ladybirds can be used to control aphids, and fungi to treat plant diseases.

4.1.1 Preserving natural predators

The preservation of indigenous predators is critical to biological management because they naturally control pests and diseases. Protecting beneficial species such as birds, arthropods, and arachnids allows agriculturalists and land stewards to strengthen their ecosystems while reducing chemical use. Conservation practices include establishing ecosystems that support indigenous predators, reducing pesticide use, and avoiding disturbing predator populations. This method reduces pests while improving environmental health [38].

4.1.2 Augmentation and inoculative

Augmentation and inoculative releases refer to the introduction of additional organisms or the deliberate inoculation of organisms into a specific environment

Integrated pest management enhances biological pest control by introducing and establishing beneficial organisms in habitats.

Augmentation involves the deliberate increase of natural enemies or beneficial species within a specific region. Implement a consistent practice of introducing these organisms to enhance their population and effectively manage pests. Enhance the insect-controlling efficacy of these native adversaries to augment pest management.

Inoculative releases introduce advantageous organisms into an ecosystem to bolster a specific population. Unlike augmentation, which enhances the population of natural enemies, inoculative releases establish long-lasting beneficial species. This method reintroduces species that are either extinct or have experienced a significant decline in population.

IPM utilises a combination of approaches to minimise the use of chemical pesticides, encourage ecological equilibrium, and facilitate sustainable pest control.

4.1.3 Utilisation of microbial biopesticides

In integrated pest management, microbial biopesticides control pests while protecting the environment. These creatures contain beneficial bacteria, fungus, viruses, and protozoa. Their purpose is to fight specific pests and diseases [35].

4.1.3.1 Principal uses

Target Specificity: Microbial biopesticides are highly pest-specific, minimising harm to beneficial organisms and non-target species. This precision helps maintain ecological balance and reduce negative effects. Introducing or enhancing organisms that outcompete, parasitise, or kill pests strengthens natural biological control systems. Environmental safety: Microbial biopesticides are usually safer than chemical pesticides. They degrade quickly and leave little environmental impact. These pesticides' different modes of action reduce pest resistance compared to conventional chemical pesticides.

Application Flexibility: These biopesticides can be applied as foliar sprays, soil applications, or seed treatments, making them versatile pest control tools. Microbial biopesticides help Integrated pest management by providing precise and environmentally friendly pest control. These biopesticides work with other IPM strategies to promote sustainable agriculture and reduce pesticide use.

4.2 Cultural practices

Integrated pest management strategies use cultural practices to change farming and environmental conditions to prevent pests. These methods include the following:

Crop rotation changes the crops planted in an area each season. This approach disrupts pest life cycles and reduces pests and diseases.

Dates and timing of planting to avoid high insect populations or pest-friendly weather.

Selecting pest- and disease-resistant plant varieties.

Field hygiene removes agricultural trash and weeds that attract pests and diseases.

To promote plant vitality and resilience, soil management includes irrigation, organic matter, and balanced fertilisation.

IPM uses cultural practices to create an adverse habitat for pests and diseases, reducing chemical use and boosting sustainable agriculture.

4.2.1 Utilising crop rotation and diversification

Crop rotation is the process of altering the type of crop cultivated in a particular field on a seasonal or yearly basis. Through the practice of crop rotation, farmers can disrupt the life cycle of pests and pathogens that have adapted to certain crops, resulting in a decrease in their numbers and a reduction in soil-borne diseases.

Diversification is the cultivation of several types of crops either within the same region or across separate fields. This practice aims to establish a more intricate and less foreseeable ecosystem for pests. Implementing this measure can decrease the probability of pests and illnesses proliferating rapidly, as they typically flourish in monocultures (agricultural systems consisting of a single crop). Diversification additionally promotes the growth of advantageous organisms and improves the overall ability of the ecosystem to withstand and recover from disturbances.

4.2.2 Crop cultivars with resistance

Integrated pest management uses genetic remedies to reduce pest and disease damage by cultivating resistant crop varieties. Cultivars with innate pest and disease resistance are bred. Farmers can benefit from selecting and growing disease- and pest-resistant crop varieties. Minimise pest damage: Growing pest-resistant varieties improves crop health and yields. Resistance reduces pesticide use, saving money and the environment. Improve sustainability: Resilient varieties promote ecological balance and reduce pest resistance, advancing sustainable agriculture. Integrated pest management with resilient plant varieties protects crops and boosts agricultural productivity and environmental protection [36].

4.2.3 Planting and harvesting schedule

4.2.3.1 Planting schedule

Timing: Changing planting timing can reduce pest populations. Begin planting before or after the usual season to reduce the risk of pests that are more prevalent at certain times. A crop rotation schedule disrupts pest life cycles and reduces population growth. This method prevents pests from spreading.

4.2.3.2 Timetable for harvesting

Pest control must be timely to avoid crop damage. This can be achieved by harvesting crops at the right time. Quickly harvesting crops reduces pest damage.

Sanitation: Post-harvest management removes plant waste and debris that may attract pests. Effective cleaning can reduce insect populations, benefiting future planting seasons.

Integrated insect management (IPM) practitioners can control insect populations, reduce pest losses, and improve crop health and production by coordinating planting and harvesting schedules.

4.3 Mechanical and physical control

Mechanical and physical control refers to the use of physical methods and devices to manage and manipulate objects or systems.

Mechanical and physical control methods in integrated pest management use physical tools to reduce pest populations and crop damage. These methods are designed for fast and effective pest control.

Mechanical pest control uses traps, barriers, and screens. Sticky traps catch airborne insects, while crop barriers keep pests out. Hand-picking insects and larvae from plants is manual pest control. Physical control involves manipulating or directing physical objects. Environmental changes: Pests' surroundings can be made uninviting. Bug breeding grounds can be reduced by irrigation schedules or reflective mulches.

4.3.1 Traps and barriers

Traps and barriers refer to obstacles or preventive measures that are put in place to hinder or prevent access to a certain area or to protect against potential dangers or risks.

IPM uses traps and barriers to control insect populations by blocking entry or trapping them. Traps capture pests for monitoring or control. Pests can be lured with bait or pheromones or physically ensnared. Pest management traps include sticky traps, which catch flying insects, and pheromone traps, which attract certain pests. Traps help assess and reduce pest populations. Physical barriers are intentionally placed to keep pests out of specific areas or crops. They build a defence to keep pests out of plants and places. Row covers protect plants from insects, fences keep larger pests out, and mesh screens keep pests out. Traps and barriers reduce pest damage and chemical use in integrated pest management. This encourages sustainable pest control.

4.3.2 Mechanical cultivation

Mechanical cultivation refers to the process of using machinery and equipment to cultivate or prepare the soil for planting or growing crops.

Mechanical cultivation involves the utilisation of machines and equipment to prepare the soil for the purpose of planting or cultivating crops. This procedure encompasses a range of activities, including ploughing, tilling, and harrowing, which serve to loosen and aerate the soil, control weeds, and integrate organic material. Mechanical cultivation enhances soil structure, stimulates superior root growth, and aids seedbed preparation, resulting in more efficient planting and improved crop development. Through the utilisation of contemporary machinery, farmers can attain accurate soil management, enhance production, and diminish labour expenses in the field of agriculture.

4.3.3 Thermal therapies

Thermal manipulation: Employing methods such as applying heat or cold treatments can exterminate pests or eradicate their eggs. Soil solarisation is a technique that utilises sunshine to raise the temperature of the soil, thereby eliminating pests and pathogens.

Both mechanical and physical controls are essential components of Integrated pest management as they provide non-chemical approaches to pest management, decrease dependence on pesticides, and support sustainable pest control strategies.

4.4 Utilising chemical control as a final option

4.4.1 Choice of targeted pesticides

When implementing integrated pest management, the selection of pesticides should prioritise non-chemical methods and only resort to chemical control as a final option. The emphasis should be on utilising specific and efficient products that have minimal negative effects on the environment and do not harm beneficial creatures. When non-chemical approaches are not enough, specific pesticides are used based on the following:

Specificity refers to the process of choosing pesticides that specifically target the pest species in question, hence minimising harm to non-target organisms.

Timing: Utilising pesticides at the most opportune moment in the pest's life cycle to optimise efficacy and reduce the necessity for multiple treatments.

Application Methods: Employing accurate application techniques to minimise the dispersion of pesticide and decrease the quantity of pesticide utilised.

IPM, or Integrated Pest Management, seeks to efficiently control pest populations by carefully choosing and using specific pesticides. This approach promotes sustainable farming methods and reduces the need for chemical treatments.

4.4.2 Formulations with lower risk

Formulations with reduced risk objective: These formulations are specifically developed to minimise the adverse impacts on human health, beneficial creatures, and the environment. The options encompass biopesticides, naturally occurring compounds, and chemicals that have been modified to have lower toxicity.

Advantages: They offer efficient pest management while minimising any long-lasting effects, preventing harm to unintended species, and guaranteeing the safety of those who apply them and those who use them. Their utilisation promotes a more sustainable and ecologically conscious approach.

4.4.3 Specific application techniques

4.4.3.1 Precise application methods

Purpose: These strategies are specifically developed to optimise the effectiveness of pest control while minimising any unexpected consequences. They ensure the exact application of chemicals or other treatments to the specific areas where they are needed.

4.4.3.2 Examples

Targeted Application: Employing spot treatments or localised applications instead of wide-ranging spraying to concentrate on specific areas with high insect activity.

Precision Equipment: Utilising calibrated sprayers or drippers to apply materials with accuracy and efficiency.

Timing refers to the strategic use of treatments at precise moments, such as during the life phases of pests when they are particularly susceptible, in order to maximise their effectiveness.

4.5 Environmental alteration

Integrated pest management involves changing environmental factors to reduce pest populations and increase pest management. This strategy seeks to make pests miserable and beneficial creatures happy. Important parts include the following: Modifying Microclimates: Irrigation and shading can lower temperature and humidity, making pests and diseases less likely. Habitat management entails adding hedgerows, flowering plants, and water supplies to sustain predators and parasitoids. Soil and water management: Healthy soil and water management reduce pests and improve plant resilience. IPM manipulates environmental factors to disrupt pest life cycles, increase beneficial organism growth, and reduce chemical use.

4.5.1 Habitat manipulation

Habitat manipulation refers to the deliberate alteration of an environment to create specific conditions that are more favourable for certain organisms or ecological processes.

Habitat modification in Integrated pest management entails purposefully modifying a habitat to establish conditions that promote advantageous organisms and biological processes, while deterring pests. This method encompasses the following: Creating habitats or microenvironments that encourage natural enemies of pests, such as predatory insects or birds, in order to control pest numbers. Modifying planting patterns: Creating crop arrangements or intercropping systems to promote biodiversity and disturb pest habitats. Altering Environmental settings: Employing techniques such as manipulating irrigation, incorporating organic matter, or controlling agricultural wastes to establish less conducive settings for pests. The goal of IPM is to sustainably maintain ecological balance, promote beneficial creatures, and decrease pest burdens by altering habitats [39].

4.5.2 Utilising mulching and cover cropping

Utilising mulching and cover cropping are successful practices in Integrated pest management that improve soil health and decrease pest pressures. Mulching is the process of placing a layer of either organic or inorganic particles on top of the soil. Mulching aids in preventing Weed Growth: By obstructing sunlight, it inhibits the growth of weeds, which can serve as a habitat for pests. Sustaining soil moisture minimises water evaporation, hence ensuring stable soil conditions. Enhance Soil Health: Organic mulches degrade, contributing nutrients and improving soil structure.

Cover cropping is the practice of cultivating particular plants during periods when the main crops are not being grown in order to protect and nourish the soil. Cover cropping offers:

Pest Suppression: Specific cover crops possess the ability to discourage or accommodate natural predators of pests. Soil improvement involves enhancing soil fertility, structure, and preventing erosion. Weed control: They effectively outcompete weeds, hence lowering the availability of possible habitats for pests.

Both techniques such as utilising mulching and cover cropping which contribute to the promotion of a more robust ecosystem, diminish the dependence on chemical treatments, and uphold sustainable agriculture practices.

4.5.3 Buffer zones

Buffer zones play a crucial role in environmental modification methods as part of Integrated pest management. They entail the establishment of demarcated zones around crops or vulnerable ecosystems in order to the following:

Prevent Pesticide Drift: Buffer zones limit the potential for pesticides to disperse into unintended areas, such as nearby agricultural lands, bodies of water, and natural habitats, hence minimising their ecological repercussions.

Buffer zones contribute to the ecological balance and promote pest control by creating a habitat for natural predators and pollinators, therefore supporting beneficial organisms.

Prevent pest migration: Physical barriers can effectively restrict the movement of pests between fields, hence minimising the spread of pest populations.

Buffer zones play a crucial role in maintaining sustainable pest management by reducing environmental hazards and fostering a more robust ecosystem.

4.6 Education and outreach

Integrated pest management relies on education and outreach to improve stakeholders' pest management knowledge and skills. Education goal: To teach farmers, gardeners, and pest control professionals about IPM principles, methods, and techniques. This includes understanding pest biology, monitoring, and using various control methods. Approaches: Pest management workshops, seminars, extension services, and training programs provide practical knowledge. Manuals, comprehensive instructions, and online platforms are other educational tools. Outreach goal: To actively engage and include the community in Integrated pest management practises to promote sustainable pest control and IPM adoption. Methods: Outreach may include public awareness campaigns, community gatherings, local partnerships, and media coverage. These activities promote Integrated pest management and increase awareness of its benefits. Integrated pest management education and outreach aim to increase the adoption of sustainable pest control methods by educating individuals and communities [35].

4.6.1 Farmer and agronomic training programs

Farmers and agronomists are trained in integrated pest management strategies to improve their knowledge and skills.

4.6.1.1 Farmer education programs

Goal: Teach farmers pest management skills. This includes understanding pest identification, monitoring, and using various pest management technologies. The training covers crop rotation, biological controls, low-risk insecticides, and the best record-keeping systems. Delivery: Field demonstrations, interactive seminars, and local extension agencies spread these initiatives. They focus on practical applications to help farmers adopt Integrated pest management on their farms.

4.6.1.2 Agronomy education

Goal: To train agronomists and agricultural professionals in pest management, crop health, and sustainability. This training helps them guide farmers and implement IPM approaches more broadly. The topic includes pest behaviour, soil health, pest-nutrient integration, and the latest pest control research and technologies. Delivery: Professional development programs may include complete courses, seminars, and workshops. They often prioritise new scientific discoveries and managerial methods.

Both training programs provide personalised education to farmers to improve pest management, crop yields, and sustainability.

4.6.2 Initiatives to raise public awareness

Public awareness campaigns in Integrated pest management focus on sustainable pest management. These efforts include: Educational programs: Integrated pest management principles, practices, and benefits are taught to farmers, gardeners, and the public in workshops, seminars, and training. Engagement with a wider audience: Use media campaigns, booklets, and online resources to promote Integrated pest management and its best practices. Community Engagement: Working with local organisations, schools, and agricultural extension services to improve IPM and address community pest management issues. These projects aim to increase public awareness, understanding, implementation, and sustainable and efficient pest management.

4.6.3 Engaging in cooperation with stakeholders and policymakers

IPM implementation requires stakeholder and policymaker collaboration. Integrated pest management prioritises eco-friendly methods and minimises pesticide use.

4.6.3.1 Essential engagement elements

Identification of stakeholders: Identify and involve farmers, agricultural companies, researchers, environmental groups, and local communities. Every company has different pest management needs. Work closely with policymakers to ensure that integrated pest management initiatives comply with regulations and receive support. This partnership can help develop sustainable pest management policies and offer incentives to use them. Knowledge sharing: Share information and best practices among stakeholders. Workshops, training, and instructional materials can improve

IPM knowledge and understanding. Feedback mechanisms: Use stakeholder feedback to improve IPM tactics. This iterative method helps solve problems and adapt practices to local conditions. Resource allocation: Encourage and distribute resources for research, technology development, and extension to efficiently implement integrated pest management. Maintain systems to evaluate Integrated pest management strategies and their impact. This process maintains transparency and trust by involving stakeholders. Promoting stakeholder and policymaker collaboration can help agricultural practices integrate IPM tactics, improving pest management, sustainability, and environmental impact.

5. Conclusions

To summarise, the chapter on pesticide pest management emphasises the importance of chemical control methods in promoting public health, maintaining households, and sustaining agriculture. However, it acknowledges the limitations of these methods and emphasises the importance of implementing more sustainable practices. Pesticides, which include fungicides, herbicides, and insecticides, have been widely used for pest control and crop protection. Nonetheless, the unregulated use of these substances raises a slew of ecological and medical concerns, including the contamination of the atmosphere, water, and land, harm to unintended organisms, and the emergence of pest strains that are pesticide-resistant.

The chapter emphasises the importance of Integrated pest management as an environmentally friendly and long-term solution. Integrated pest management combines chemical and non-chemical approaches, such as biological control, cultural practices, and mechanical methods, to reduce reliance on pesticides and mitigate their negative effects. Integrated pest management aims to optimise pest control while protecting the environment and human health by strategically incorporating a variety of techniques. These strategies include using biological and synthetic pesticides sparingly and selectively, encouraging the presence of beneficial organisms, and implementing practices like crop rotation and the use of transgenic plants.

In general, the chapter recommends for a balanced approach to pest control that includes both chemical and non-chemical methods. This approach not only addresses current pest problems, but it also encourages long-term agricultural sustainability and environmental preservation.

Acknowledgements

Sangaraju Rajendra thanks the Indian Council of Medical Research (ICMR) for providing ICMR-Research Associate fellowship file no. (3/1/2/300/2021-Nut). All authors special thanks to the Director General of the ICMR in New Delhi, India, for providing the infrastructural and research facilities, and also to the Director of the ICMR-National Institute of Nutrition for supporting our research work by providing the necessary facilities, motivation, and funding.

Conflict of interest

The authors declare no conflict of interest.

Abbreviations

IPM	integrated pest management
PD	Parkinson's disease
MRLs	maximum residue limits
PPE	personal protective equipment
US EPA	United States Environmental Protection Agency
EFSA	European Food Safety Authority
GMOs	genetic modified organism
AD	Alzheimer's disease
AML	acute myelogenous leukaemia
SLE	systemic lupus erythematosus
COPD	chronic obstructive pulmonary disease
ALS	amyotrophic lateral sclerosis
MS	multiple sclerosis
UPR	unfolded protein response
WHO	world health organization
DDT	dichloro-diphenyl-trichloromethane
HCH	hexachlorocyclohexane
GMO	genetically modified organism
GPS	global positioning system
EPA	environmental protection agency
OSHA	occupational safety and health administration
PPE	personnel protective equipment

Author details

Rajendra Sangaraju[1], Rajesh Kumar K[1], Tien Huynh[2] and Sukesh Narayan Sinha[1*]

1 Division of Food Safety, Indian Council of Medical Research-National Institute of Nutrition, Jamai-Osmania, Tarnaka, Hyderabad, Telangana, India

2 School of Science, RMIT University, Bundoora, Melbourne, Australia

*Address all correspondence to: sukeshnr_sinha@yahoo.com

References

[1] Ware GW, Whitacre DM. An introduction to insecticides (4th edition). In: The Pesticide Book. 6th ed. Willoughby, Ohio: MeisterPro Information Resources, A Division of Meister Media Worldwide; 2004

[2] Sharma AK, Sharma D, Chopra A. An overview of pesticides in the development of agriculture crops. Journal of Applied and Natural Science. 2020;**12**(2):101-109

[3] Horowitz AR, Ishaaya I. Advanced technologies for managing insect pests: An overview. Advanced Technologies for Managing Insect Pests. 2013:1-11

[4] Sinha SN, Ungarala R, Kumar D, Sangaraju R, Kumar S. A novel RP-HPLC method for quantification of cholinesterase activity in human blood: An application for assessing organophosphate and carbamate insecticide exposure. PLoS One. 2022;**17**(12):e0279287

[5] Siddiqui JA, Fan R, Naz H, Bamisile BS, Hafeez M, Ghani MI, et al. Insights into insecticide-resistance mechanisms in invasive species: Challenges and control strategies. Frontiers in Physiology. 2023;**13**: 1112278

[6] Wyckhuys KA, Sasiprapa W, Taekul C, Kondo T. Unsung heroes: Fixing multifaceted sustainability challenges through insect biological control. Current Opinion in Insect Science. 2020;**40**:77-84

[7] Jeger M, Beresford R, Bock C, Brown N, Fox A, Newton A, et al. Global challenges facing plant pathology: Multidisciplinary approaches to meet the food security and environmental challenges in the mid-twenty-first century. CABI Agriculture and Bioscience. 2021;**2**(1):1-18

[8] Boudh S, Singh JS. Pesticide contamination: Environmental problems and remediation strategies. In: Emerging and Eco-friendly Approaches for Waste Management. Springer; 2019. pp. 245-269. ISBN: 978-981-10-8668-7

[9] Barathi S, Sabapathi N, Kandasamy S, Lee J. Present status of insecticide impacts and eco-friendly approaches for remediation-a review. Environmental Research. 2023;**240**(Pt 1):117432

[10] Thompson DA, Lehmler H-J, Kolpin DW, Hladik ML, Vargo JD, Schilling KE, et al. A critical review on the potential impacts of neonicotinoid insecticide use: Current knowledge of environmental fate, toxicity, and implications for human health. Environmental Science: Processes & Impacts. 2020;**22**(6):1315-1346

[11] Ikpeze O, Olofintoye O. Current review of the environmental impacts of insecticides. International Journal of Bioscience. 2009;**4**:1-11

[12] Vivekanandhan N, Duraisamy A. Ecological impact of pesticides principally organochlorine insecticide endosulfan: A review. Universal Journal of Environmental Research & Technology. 2012;**2**(5):369-376

[13] Sánchez-Bayo F. Insecticides mode of action in relation to their toxicity to non-target organisms. Journal of Environmental & Analytical Toxicology. 2012;**4**:S4-S002

[14] Duke SO, Menn JJ, Plimmer JR. Challenges of Pest Control with Enhanced

Toxicological and Environmental Safety: An Overview. Washington, DC: ACS Symposium Series; American Chemical Society; 1993

[15] Kumar V, Sharma N, Sharma P, Pasrija R, Kaur K, Umesh M, et al. Toxicity analysis of endocrine disrupting pesticides on non-target organisms: A critical analysis on toxicity mechanisms. Toxicology and Applied Pharmacology. 2023;**474**:116623

[16] Fleeger JW, Carman KR, Nisbet RM. Indirect effects of contaminants in aquatic ecosystems. Science of the Total Environment. 2003;**317**(1-3):207-233

[17] Jeyaseelan A, Murugesan K, Thayanithi S, Palanisamy SB. A review of the impact of herbicides and insecticides on the microbial communities. Environmental Research. 25 Dec 2023;**245**:118020. DOI: 10.1016/j.envres.2023.118020. Epub 2023 Dec 25

[18] Tripathi S, Srivastava P, Devi RS, Bhadouria R. Influence of synthetic fertilizers and pesticides on soil health and soil microbiology. In: Agrochemicals Detection, Treatment and Remediation. Butterworth-Heinemann: Elsevier; 2020. pp. 25-54. ISBN: 978-0-08-103017-2

[19] Kumar M, Yadav AN, Saxena R, Paul D, Tomar RS. Biodiversity of pesticides degrading microbial communities and their environmental impact. Biocatalysis and Agricultural Biotechnology. 2021;**31**:101883

[20] Zivan O, Segal-Rosenheimer M, Dubowski Y. Airborne organophosphate pesticides drift in Mediterranean climate: The importance of secondary drift. Atmospheric Environment. 2016;**127**:155-162

[21] Ramaprasad J, Tsai M-Y, Elgethun K, Hebert VR, Felsot A, Yost MG, et al. The Washington aerial spray drift study: Assessment of off-target organophosphorus insecticide atmospheric movement by plant surface volatilization. Atmospheric Environment. 2004;**38**(33):5703-5713

[22] Feltracco M, Barbaro E, Maule F, Bortolini M, Gabrieli J, De Blasi F, et al. Airborne polar pesticides in rural and mountain sites of north-eastern Italy: An emerging air quality issue. Environmental Pollution. 2022;**308**:119657

[23] Langenbach T, Caldas LQ. Strategies for reducing airborne pesticides under tropical conditions. Ambio. 2018;**47**:574-584

[24] Kumar P, Kumar R, Thakur K, Mahajan D, Brar B, Sharma D, et al. Impact of pesticides application on aquatic ecosystem and biodiversity: A review. The Biological Bulletin. 2023;**50**(6):1362-1375

[25] Kamlesh R, Santosh S, Sudhakar P, Sunita A. Role of pesticides in biodiversity loss. IJBB. 2024;**6**(1):01-03

[26] Nalban N, Alavala S, Sangaraju R, Mir SM, Sistla R. Therapeutic targeting of oxidative stress and inflammation in asthma and COPD and pharmacological interventions with phytochemicals. Oxidative Stress in Lung Diseases. 2019;**1**:429-449

[27] Kumar D, Sinha SN, Rajendra S, Sharma K. Assessing farmer's exposure to pesticides and the risk for non-communicable diseases: A biomonitoring study. Science of the Total Environment. 2023;**891**:164429

[28] Sinha SN, Kumpati RK, Ramavath PN, Sangaraju R, Gouda B, Chougule P. Investigation of acute organophosphate poisoning in humans

based on sociodemographic and role of neurotransmitters with survival study in South India. Scientific Reports. 2022;**12**(1):16513

[29] Banda VR, Perugu S, Kumar R, Sinha SN. In-silico studies on human reproductive hormones and interactions with organophosphate. European Journal of Biomedical. 2018;**5**(4):592-597

[30] Mohiuddin M, Hossain M, Rahman A, Palash M. Socio-economic study of insecticide use on vegetable cultivation at farm level in Chittagong region. Journal of the Bangladesh Agricultural University. 2009;**7**(2):343-350

[31] Athukorala W, Lee BL, Wilson C, Fujii H, Managi S. Measuring the impact of pesticide exposure on farmers' health and farm productivity. Economic Analysis and Policy. 2023;**77**:851-862

[32] Popp J, Pető K, Nagy J. Pesticide productivity and food security. A review. Agronomy for Sustainable Development. 2013;**33**:243-255

[33] Popp J, Pető K, Nagy J. Impact of pesticide productivity on food security. Sustainable Agriculture Reviews. 2014;**13**:19-33

[34] Carvalho FP. Agriculture, pesticides, food security and food safety. Environmental Science & Policy. 2006;**9**(7-8):685-692

[35] Greaves J. Biopesticides, regulatory innovation and the regulatory state. Public Policy and Administration. 2009;**24**(3):245-264

[36] Tago D, Andersson H, Treich N. Pesticides and health: A review of evidence on health effects, valuation of risks, and benefit-cost analysis. Advances in Health Economics and Health Services Research (Toulouse, France: Toulouse School of Economics). 2014;**24**:203-295. PMID: 25244910

[37] Ansari MS, Moraiet MA, Ahmad S. Insecticides: Impact on the environment and human health. In: Environmental Deterioration and Human Health: Natural and Anthropogenic Determinants. Dordrecht: Springer; 2014. pp. 99-123. DOI: 10.1007/978-94-007-7890-0_6. Published 8 December 2013. Print ISBN 978-94-007-7889-4, Online ISBN 978-94-007-7890-0. eBook Packages Earth and Environmental Science, Earth and Environmental Science (R0)

[38] Thompson GD, Head G. Implications of regulating insect resistance management. American Entomologist. 2001;**47**(1):6-10

[39] Kumar KR, Sinha SN. A review of the clinical management of current organophosphate poisoning treatments in humans. International Journal of Health Sciences. 2022;**6**(S9):2394-2415. DOI: 10.53730/ijhs.v6nS9.12950

Chapter 7

Adverse Effects of Pesticides: Regulatory Failures, Impacts on Public Health and Environmental Wellbeing

A.D. Dileep Kumar and Narasimha Reddy Donthi

Abstract

Synthetic pesticides are inherently toxic and are designed to kill or harm life of all organisms. As considerable risk is involved in pesticide use, many risk mitigation measures have been suggested. However, feasibility and suitability of these measures in relation to the local conditions of use have not been properly assessed. Biosafety and precautionary principles have been ignored in registrations and approvals. It is now widely recognized that synthetic pesticides have been contributing to deterioration of ecosystem health as well as diminishing human health, and hence there is a global demand for reduction in use of pesticides in agriculture, replaced by non-chemical methods. This paper explores intrinsic toxicity of pesticides, adverse health and environmental effects caused by them, factors contributing to the adverse effects, use conditions and suitability of risk mitigation measures, regulatory requirements and failures, and pesticide risk reduction strategy. Recognizing a wide array of adverse effects of synthetic pesticides, this paper emphasizes the urgent need to de-chemicalise agriculture and other sectors by using agroecological practices. International regulatory bodies, National and State governments have to put in concerted efforts to enable a paradigm shift in agriculture, in view of safety of agriculture workers from pesticides, agriculture sustainability, food and nutrition security, biosafety, enhancing livelihood options, climate change impacts, environmental safety and public health.

Keywords: pesticides, adverse effects, health, environment, assumed safety, regulation

1. Introduction

Pesticides are substances meant for killing pests, which includes insects, bacteria, fungus, virus, and wild animals that harm human food and comfortable life. According to the International Code of Conduct on Pesticides Management (ICCPM) the term pesticide refers to "*any substance, or mixture of substances of chemical or biological ingredients intended for repelling, destroying or controlling any pest, or regulating plant growth*" [1]. To elaborate, insecticides are used to kill various insects, herbicides are used to kill vegetation, fungicides and bactericides are used to kill or control fungus and bacteria that cause diseases to crops and plant growth regulators are used to modify growth stages of crops. Pesticides are also used for controlling vector organisms of human or animal disease, and other animals causing harm to human comfort. Pesticide application is commonly employed in the various processes of agricultural commodities such as production, processing, storage and transport. These toxic substances are used to prevent pest attacks on food, animal feed and wood based products. Additionally, these substances are also used as plant defoliant, desiccants, preventing the premature fall of fruit, and fruiting agents. Pesticide is a generic term to its variants such as insecticides, fungicides, herbicides/weedicides and plant growth regulators.

Pesticides are widely used in agriculture, as well as, in non-agriculture sectors such as construction, gardening & landscaping and public health programs for vector control. Extensive use of these toxic substances has been recognized as a major public health problem long back, due to their toxic effects on people and different forms of life [2].

Over the past decades, it has been widely acknowledged that pesticides have the potential to cause adverse health and environmental effects. Operationalizing that acknowledgement, pesticides are to be dealt carefully at all levels of regulatory intervention right from registration, approval for use, field use and post registration monitoring. Because considerable risk is involved in pesticide use, many risk mitigation measures have been suggested. However, feasibility and suitability of these mitigation measures in relation to who is using them for what has not been properly assessed. Over a period, regulation became lax due to a number of reasons. Due to inadequate monitoring of actual usage, regulatory failures are becoming obvious. There is no congruence between registered use and actual use. Consequently, pesticide poisonings are on the rise. Excessive and unnecessary usage has become the norm. Ease of doing business became the principle of regulation. Biosafety and precautionary principles are being ignored when pesticides are going through the regulatory scrutiny. Registrations and approvals are being given without applying precautionary principles, giving primacy to business. Recognizing the harm potential of pesticides to ecology and humans, agencies and other organizations have been echoing calls for reduction in pesticide usage, replaced by non-chemical methods of pest control. This paper explores unintended consequences of pesticides, their management and regulatory lacunae.

2. History of pesticides

Agriculture dates back to 10,000 years ago and since then mankind has been battling to control pest infestations to save crops. History of pest control dates back to Sumarians period when elemental sulphur compounds were used to control insects. About 3500 years ago, the Chinese used heavy metals such as mercury and arsenic compounds to control body lice. Ancient Greece and Romans used animal or plant origin materials and methods to control pests. Pyrethrum, derived from the flower of *Chrysanthemum cinerariaefolium, was* also used for pest control for more than 2000 years [3].

Use of chemical agents for pest control had been in practice for centuries, but not probably as widely as today [4]. Until the mid 18th century, derivatives of plant, animal, or mineral origin have also been used for pest control. The period between mid 18th century and mid 19th century marked the beginning of the use of industrial by-products and inorganic chemicals as pest control solutions such as sodium chlorate and sulfuric acid. The second half of the 19th century marked intensified research, production and application of chemical products as pesticides. Since the 1940s, use of synthetic chemical compounds for pest control emerged. Industrial pesticide production intensified after the Second World War, following huge production and application of DDT, BHC, aldrin, dieldrin, endrin, etc. These organochlorine compounds were discovered nearly a century before, but their pesticidal properties had been recognized later. These 'economic poisons' were used mostly for military and public health purposes [5]. Resistance to DDT, its adverse effects on non-target organisms and the problem of its residues were recorded in 1946, well before Dr. Paul Muller received the Nobel prize for the discovery of its insecticidal property in 1949 [6]. Called economic poisons, these compounds were extensively promoted in the agriculture sector as 'medicine for pest and disease', as demand for military use diminished after the World War. Usage in agriculture was thus encouraged to sustain the industry, and not per se about their functionality. From thereon, the spectrum of agrochemicals continuously expanded to maintain the momentum of business and industry profits. Over a period, genetic engineering and recently nanotechnology have been employed in production of pesticides even while biosafety concerns of these compounds remained unresolved. Currently thousands of different chemicals as pesticides are under use around the globe.

3. Intrinsic toxicity of pesticides

Pesticides are inherently toxic substances designed to kill or inhibit the growth of organisms. Toxicity points to the ability of a substance to produce or trigger an adverse effect in biology of living organisms. According to ICCPM, toxicity refers to a "*physiological or biological property that determines the capacity of a chemical to do harm or produce injury to a living organism by other than mechanical means*" [1]. Pesticides are characterized by their toxicity properties, which vary between different chemical groups, impacting on target and non-target organisms [7]. Pesticides impair vital life sustaining biological functions in organisms. It is worth noting that these biological functions are very similar among different life forms right from unicellular to multicellular. A pesticide designed to impact a particular biological function in a pest can have similar impacts on other life forms, when it gets in contact with them inadvertently. Since most life forms are interdependent, spraying of these toxic substances can impact more than one species, and not necessarily the target pest. The adverse effects of pesticides, as documented on human beings can manifest as mild symptoms such as headache or irritation, or severe symptoms such as convulsions and even death. The adverse outcomes depend upon how toxic the substance is, the amount of the substance and duration of exposure to the substance. Some of the effects can be reversible, while the others are irreversible leading to chronic or fatal outcomes. The toxicity of pesticides is further enhanced as formulated products in the markets. The inert ingredients or co-formulants in pesticide formulations may have higher toxicity than the active ingredients [8].

Pesticides are poisonous substances as they are designed to kill or harm organisms known as pests. Apparently, they are not only harmful to the so-called target pests, but are capable of harming or killing all types of life forms that come in contact with them. All pesticides are biocides and, hence, are hazardous substances.

4. Mechanisms of toxicity

The way by which pesticide exerts its harmful effect to kill an organism is known as its mode of action or mechanism of toxicity. Pesticides kill organisms by interfering with certain essential vital functions in their body. A variety of toxicity mechanisms have been noted, which includes inhibition of the activities of biomolecules such as enzymes, neurotransmitters, hormones, secondary messengers in the various organ systems, killing adults or preventing growth and development in young ones. For example, neurotoxic pesticides (eg. organophosphates, carbamates, pyrethroids and neonicotinoids) prevent the normal functioning of nerve cells by inhibiting enzymes involved in neurotransmitter metabolism or block nerve impulse transmission. As pesticides interfere with basic vital functions, in effect, they have the potential to kill not only the pests, but other animals, including mammals as well. However, depending upon the higher cellular organization and the amount of pesticide exposed to, the outcome of adverse effects in larger animals and human beings may vary.

Different chemical groups of pesticides are designed for different modes of action on target organisms. Most insecticides in general exert their effect on the nervous system of insects, inhibiting functions of neurotransmitters or communication channels. Insecticides that have different chemical nature exert their activity at different sites in the nervous system and result in a variety of adverse reactions.

For example, the most commonly used organophosphate and carbamate insecticides are toxic to the nervous system. They are extremely toxic to the nervous system as they irreversibly bind with the enzyme acetylcholinesterase, which is essential to hydrolyse the neurotransmitter called acetylcholine at neuronal junctions [9]. Organochlorine and phenylpyrazoles groups of insecticides block the neurotransmitter named gamma amino butyric acid (GABA)-activated chloride channel in the nervous system, resulting in hyper-excitation and convulsions [10]. GABA is the major inhibitory neurotransmitter in insects. Synthetic pyrethroid insecticides affect the sodium channels in the nervous system, continuously causing over-excitation [11]. Sodium channels are involved in the propagation of nerve impulses along neurons. The neonicotinoids affect the nicotinic acetylcholine receptor (nAChRs) and produce symptoms from hyper-excitation to lethargy, and even cause paralysis [12].

5. Toxicity classification

The toxicity of pesticides is divided into two types, acute toxicity and chronic toxicity. Acute toxicity refers to the short-term effects caused by pesticides following a single exposure happening in a relatively short period of time. The World Health Organization (WHO) recommended classification of pesticides based on acute toxicity, providing insights into the grouping of widely used pesticides for regulatory purposes. These classifications apply to technical grade active ingredients in pesticides and not for pesticide formulation. Toxicity of formulations vary depending on the type of formulation and the ingredients in it. This classification of pesticides has five categories namely, extremely toxic, highly toxic, moderately toxic, slightly toxic and technical grade active ingredient unlikely to present acute hazard in normal use [13]. Chronic toxicity refers to the adverse effects following repeated or long-term exposure. Such effects may be delayed or manifested after a long time, which includes neurobehavioral changes, cancers, reproductive and developmental effects, etc.

6. Highly hazardous pesticides (HHPs)

A small number of synthetic pesticides cause harm to human health and the environment disproportionately. They are termed as highly hazardous pesticides (HHPs). The Fourth International Conference of Chemicals Management (ICCM4) recognized HHPs 'as an issue of international concern' requiring concerted efforts to address them [14]. The term highly hazardous pesticides indicate without ambiguity that they are the most dangerous among pesticides. According to the ICCPM [15], HHPs are defined as "*pesticides that are acknowledged to present particularly high levels of acute or chronic hazards to health or environment according to internationally accepted classification systems such as WHO or Global Harmonized System (GHS) or their listing in relevant binding international agreements or conventions. In addition, pesticides that appear to cause severe or irreversible harm to health or the environment under conditions of use in a country may be considered to be and treated as highly hazardous*". HHPs have been recognized as a major public health concern by the World Health Organization [16], because of either severe acute effects or chronic effects or both, and pose the highest risk to farmers, workers, women and children. The United Nations considers HHPs as a global human rights concern [17] and calls for a global treaty to regulate them. The 2017 report of the UN's Special Rapporteur on the Right to Food noted "*reliance on hazardous pesticides is a short-term solution that undermines the rights to adequate food and health for present and future generations.*" The sixth UN Environmental Assembly, concluded on 1st March 2024, had called for concerted action towards eliminating HHPs by 2035 [18], and noted that there are unmanaged risks arising from the use of HHPs, and the progress to reduce the associated risk has been slow.

7. Adverse health effects of pesticides

Adverse effects caused by pesticides have been in discussion at global level, since the mid 20th century. The work of WHO on the application of pesticides for public health purposes, in the initial periods of their usage, gives ample insights to the danger posed by pesticides. The WHO had recognized and noted the hazards of pesticides to human beings in its very first World Health Assembly in 1948. A monograph published by the WHO in 1953 had compiled nearly 700 bibliographic references on the matter [19]. It had, since then, identified the potential adverse effects of pesticides to workers in the manufacturing unit, occupational dangers caused by pesticides during field application both for public health and agricultural purposes as well as hazards posed by pesticide contamination in food resources. Later the great work of American marine biologist and conservationist, Rachel Carson's *Silent Spring*, published in 1962 had highlighted the detrimental effects of pesticides on the environment. It identified the adverse effects caused by pesticides on birds and other organisms and the profound ecosystem damage [20]. Since then till date, thousands of peer reviewed articles and published reports are available across the globe that pointed to the adverse effects. In 2004, a joint report by the Food and Agricultural Organizations of the United Nations (UN FAO), the United Nations Environmental Programme (UNEP) and the WHO noted that all pesticides should be considered as potentially hazardous to human health [21].

Both the acute and chronic toxicity of pesticides are the outcomes of intrinsically toxic active ingredients. While the acute toxicity may manifest within a short span of time, mostly within 24 hours, chronic toxicity effects may take longer periods to appear. Toxicological studies on test animals conducted for pesticide active ingredients had indicated both acute effects and possible chronic effects of pesticides. Possible effects on human beings have also been identified [22]. The acute toxicity tests evaluate the adverse effects that occur to an organism following single or multiple exposures through oral, dermal or inhalation routes of exposure [23]. It determines the lethal dose at which death of the 50 percent of the test population happens, which is known as median lethal dose (LD50), and this figure is used towards estimating potential hazards to humans.

Even though, the acute toxicity test primarily considers death as a toxicological endpoint, it also can cause possible non-lethal adverse effects as well, which includes symptoms such as irritation, dizziness, sweating, restlessness, vomiting, weakness, rapid heart rate, diarrhoea, excessive salivation, muscular incoordination, thirst, loss of appetite, etc. Acute toxicity of pesticides happens when the quantity of pesticide in the body exceeds the tolerance level or the no observed adverse effects level (NOAEL), a measure developed to understand the toxicity. Exposure to pesticides, even at NOAEL or levels lower than NOAEL, can lead to slow poisoning effects or trigger chronic illness such as neurobehavioral problems and reproductive or hormonal effects, which may become evident quite some time after the exposure. Even at very low levels, exposure to pesticides may produce trans-generational impacts, as well [24].

Sub chronic toxicity study is conducted for a 30–90 days period and chronic toxicity tests are conducted for the lifetime (until the test organism survives) of the test animal. The latter generally involves tests for carcinogenicity and non-carcinogenic effects, reproductive and developmental effects and mutagenicity [25]. However, many of the chronic toxicity effects and symptoms may not appear for considerably longer periods - which may range from a few weeks through months or even years after the exposure, and this temporal fact makes it difficult to attribute health effects to pesticide exposure that appear much later in an individual's life. Studies show that chronic health effects include cancer and other tumours; brain and nervous system damage; birth defects; infertility and other reproductive problems; developmental toxicity, damage to the liver, kidneys, lungs and disruption of the hormone system functioning. Studies on effects of pesticides on humans, as well as epidemiological assessments, have attributed pesticides for a number of diseases and disorders including various cancers, immunotoxicity, hormone related ill health, reproductive health problems, miscarriage, developmental disorders including lowering of IQ among agricultural population [26].

The pesticide product available in the market is a mixture of active ingredients that give pesticidal properties as well as other chemicals that are added to increase its effectiveness [27]. Therefore, the toxicity of pesticide products to human beings, whether it is acute or chronic, available in the market often can have an entirely different toxicity outcome than the active ingredient, due to the presence of co-formulant chemicals or additives and their synergistic effects. The toxicity of some of the co-formulants can be higher or lower than that of the active ingredient as well [28]. Further, the prevailing agro climatic conditions, especially temperature and humidity, play a crucial role in determining the ultimate toxicity outcome and effects in the exposed individual. The outcome of adverse health effects caused by pesticides is diverse, categorized into neurotoxicity effect, neurobehavioral problems, carcinogenicity, endocrine disruption, reproductive toxicity, developmental toxicity, respiratory toxicity, immunotoxicity, ocular and dermal problems [29, 30].

7.1 Adverse effects of pesticides on children

Pesticide poisoning among children is an important concern. Long back in 2004, the UN organizations noted that infants and children are disproportionately affected by serious health problems caused by pesticides [21]. Exposure of children to pesticides can begin even before they are born, during conception along the pregnancy period. This exposure happens mainly through the parents, as a consequence of parental exposure to pesticides. Such exposure has the potential to harm the developing foetus and adversely affect the normal course of development and can result in neurobehavioral problems, birth defects and even stillbirth [26, 31].

Exposure of children to pesticides possibly happens after the birth as well. Infants are exposed to pesticides through breast milk, inhalation and intake of contaminated food and water. Many studies in India and the rest of the world have identified the presence of pesticides in breast milk [32–34]. Children living or playing in and near agricultural fields where pesticides are applied, children working in farms as labourers or as helping family labour are also getting exposed to pesticides [35]. Children's bodies are most vulnerable to chemicals and pesticides as their natural defence mechanism is poorly developed. The timing of exposure and toxicity of pesticides is critical. Rapidly developing biological systems of children can be dangerously derailed by even very low doses of pesticides that are 'generally considered safe' [21]. Continuous exposure to low doses of single and/or multiple pesticides over a long period of time can result in an explosion of adverse health effects in adulthood.

Exposure to pesticides in the adolescent period could worsen the adverse effects. Among adolescent children, both in boys and girls, pesticide exposure could lead to serious ill health such as cancers, behavioural problems, learning disabilities, poor memory and lowered IQ, as well as leading to explosion of adulthood conditions such as hormone imbalances, poor reproductive health, subfertility and infertility, miscarriages, cervical cancers, PCOD, endometriosis, neurodegenerative disorders, metabolic disorders, various cancers, etc. [21, 26]. Children in farming communities are more frequently directly exposed to pesticides, but non-farming communities are also exposed indirectly through air, water and food.

7.2 Adverse effects of pesticides on environment

Pesticides move from its site of application because of environmental factors resulting in spreading pollution and contamination [36]. They enter various environmental systems such as air, water and soil, when pesticides are sprayed in the open field or during the disposal of obsolete and banned products. During production and formulation processes of pesticides huge amounts of chemical pollutants get released into the environment. Once in the environment, pesticides may undergo processes such as transfer (or movement), deposition and degradation. Some pesticides do retain the structure under all circumstances, and are categorized as Persistent Organic Pollutants (POPs) or forever chemicals. Pesticide degradation produces by-products that sometimes can be more toxic than parent compounds [37]. Depending on the environmental conditions, soil parameters and chemical characteristics, degradation may take from hours to days or even years, and over these periods the parent compound or metabolite continue to pollute the environmental compartments. Therefore, residues of pesticides and their metabolites are often reported in air, water, soil and biota.

Generally, pesticides reach the environment while spraying, washing of spray tanks and other equipment used for mixing and loading, and releases of effluents from production facilities. Spray drift, runoff, adsorption, leaching, volatilization, are the processes by which contamination happens polluting both terrestrial and aquatic ecosystems and biotic components [38]. The environmental dangers posed by a pesticide depend on the quantity released into the environment, extent of time it takes the pesticide to break down, the substances it breaks down into, exposure routes, its ability to get accumulated in the tissues of organisms due to lipid affinity, and its toxicity to different organisms. Thus, pesticides can spread through the environment from production, use and disposal. Pesticide containers and leftover pesticides cause release of toxic chemicals into the air, soil, surface water or groundwater and contaminate them [39].

Residues of pesticides are omnipresent in agriculture systems. Residues of organochlorines have been noted across the globe with steady concentrations over longer periods indicating poor degradation. Residues of insecticides, fungicides and herbicides have been reported. Pesticide residues above the permissible limits have been reported in both drinking and ground water indicating widespread contamination at high concentrations [40].

Studies estimated that more than 99% of pesticides applied are wasted and only one percentage is contributing to pest control [41]. The wasted 99% is entering into various environmental compartments such as soil, air and water, and thereby contaminating the important life sustaining systems. Pimentel and Levitan reported that only a minor fraction (0.1%) of the applied pesticide actually reaches the target pest, while the lion's share moves to soil, air and water [42] causing a deadening effect on soil life and environment.

It is obvious that pesticide use in past decades has ultimately contaminated natural resources. Continued use of pesticides without a proper understanding of their intricate toxic mechanism has aggravated the problem of pest menace and has resulted in the resurgence of many pests. Large quantities of pesticide in the environment is causing severe pollution and contamination of the entire biota, impacting the delicate systems of the food web over the trophic levels leading to unintended, least understood consequences. Managing or reversing such contamination becomes tedious and impossible with little or inadequate solutions available.

The UN Special Rapporteur on the Right to Food, acknowledged that pesticides persist for longer periods - even decades - in the environment, threatening ecological systems around the globe including food production systems, and consequently results in pollution of soil and water sources, causing biodiversity loss, destroy growth of beneficial insects that provide natural pests control and reduce nutritional quality of food [43].

7.3 Factors contributing to adverse effects

The intrinsic poisonous nature of pesticide (hazard) is the cause of adverse effects. In addition, there are various scenarios and practices that contribute to exposure. Right from making a decision to purchase pesticides, use and disposal of pesticide and their containers, there exist multiple scenarios that could contribute to exposure to pesticides. It has to be noted that pesticide exposure and poisoning happens not only to farmers and farm workers, but also to non-farming communities as well. Studies have been pointing to the fact that 'safe use of pesticide' is a myth because pesticides are intrinsically poisoning substances, and they can cause health problems and environmental impacts both in short term and long term. Storage of pesticides in the vicinity of homes, preparation of spray solution, use of cocktails, spillage during mixing, loading and spraying, use of faulty spray equipments, entry of people into a field soon after pesticide application, increasing dosage of pesticides in successive applications or year after year, re-use of pesticide containers for domestic purposes, unsafe disposal methods, washing of equipment used for mixing, spraying, etc., and use of make shift measures assuming safety, and failure to maintain proper hygiene by farmers/workers while working with pesticides are contributing to higher risk of pesticide exposure in one way or the other [44]. Further, the environmental parameters such as wind direction, ambient temperature, and humidity affects rate of exposure and often worsen the intensity of poisoning and hence can lead to fatal outcomes. The proponents and regulators are failing to address these adverse effects. Lack of effective response is leading to ever increasing exposure and poisoning, complicating the impacts.

7.4 Undermined role of co-formulants in causing adverse effects

Co-formulants are a heterogeneous group of molecules encompassing solvents, surfactants, adjuvants, wetting or anti-foaming agents, stabilizers, emulsifiers, etc. These co-formulants are added to pesticides, sometimes at high concentrations, to enhance solubility of the active ingredient in water to increase effectiveness of pesticide formulation to enable it to spread, cover more surface area, increase absorption and penetration, and enhance the toxic activity of the product [45]. Co-formulants will ultimately shape the properties and applicability of the pesticides. Co-formulants used in pesticide products were confirmed as being harmful to health and the environment, in the recent past. Sometimes referred to as co-formulant, an adjuvant is a supplemental substance added to a formulation to enhance the performance and/or physical properties of the desired chemical. However, it is important to remember that adjuvants have labels indicating their hazardous nature.

Globally, regulation of co-formulants is still in its formative stage. They are not required to be registered by the U.S. Environmental Protection Agency (EPA), under the purported reason that adjuvants do not have pesticidal properties on their own. Distribution of adjuvants is rarely regulated by the states in the US. In 2021, the EU Commission adopted a first list of "unacceptable co-formulants" which includes 144 substances to be banned due to their inherent hazardous properties [46]. In the absence of a reporting system on adjuvant or co-formulants at the registration stage, determination of co-formulants in different pesticide formulations is becoming a challenge. To ensure human, animal and environmental safety, methods to assess and understand the toxicity of adjuvants are scarcely available. The health risk assessment of pesticides sans adjuvants in most countries, including in the European Union and in the United States. All over the world, pesticides regulation focuses almost exclusively on the stated active ingredient, and not on co-formulants.

Numerous negative health effects have been reported in humans and on the environment due to adjuvants, or co-formulants, indicating their toxicity. Despite the known toxicity of adjuvants, they are not regulated or under regulated. Their toxic effects are being generally ignored. Adjuvants are not subject to regulatory actions such as acceptable daily intake. They are not included in the health risk assessment of dietary exposures to pesticide residues.

Regulatory acknowledgement of the challenges and concerns associated with their use is not full and adequate. A regulatory framework of adjuvants is absent, especially in Low and Middle Income Countries. To ensure thorough understanding of pesticide toxicity, adjuvants or co-formulants have to be included in the regulatory scrutiny and their toxicological dynamics and synergic effects in the formulation has to be assessed.

7.5 Pesticide poisonings

Globally, unintentional acute poisoning (UAP) caused by pesticides is estimated to be about 385 million cases annually with 11,000 fatalities, which means 44% of the farming community is exposed to and poisoned every year [47]. This is a huge number of pesticide poisoning cases, largely ignored, by the knowledgeable medical community and lack of awareness among others about the correlation between growing chronic illness and pesticide usage.

Pesticide poisoning in India and other developing countries has been an unresolved concern for decades. The ICCPM has been one of the most important international initiatives on this concern. Acute poisoning globally has been due to organochlorines (OC), organophosphates (OP), carbamates and herbicides such as paraquat. Intentional, occupational and accidental poisonings among adults and children were documented in the past primarily due to insecticides that inhibit functioning of the nervous system [48–50]. For these reasons, some pesticides have been banned nationally and globally and certain pesticide formulations are enlisted with the Prior Informed Consent Procedure of the Rotterdam Convention. Some of these pesticides continue to be a frequent cause of pesticide poisoning in India. Monocrotophos, a highly hazardous pesticide recognised for its high acute toxicity since decades, has been implicated in unintentional poisonings of farmers, workers, children, and pilgrims in India. Its usage was prohibited for vegetables in 2005, and recently the government decided to ban this insecticide.

For example, the herbicide, paraquat dichloride is responsible for thousands of deaths worldwide, mostly intentional but also occupational and accidental among adults and children [51]. Paraquat dichloride, a severe poison that does not have an antidote, has been banned in more than 50 countries around the world. In India, it is being used extensively causing deaths and debilitation. Paraquat poisoning has become a clinical problem in the last 10 years. Various studies chronicled 16,659 poisoning deaths between 1999 and 2018 [52]. Most pesticides do not have antidotes to manage pesticide poisonings and hence, medical management of poisonings leaves behind little chances for survival, with paraquat being a leading example. There is enough evidence calling for well-improved, stringent regulation of pesticides in India, primarily to protect farming communities and consumers from the adverse effects caused by pesticides, as well as preventing environmental contamination and biodiversity loss [53]. However, regulatory action is too low, even while the concerns are growing, especially when more than three hundred pesticides are registered in India, and among them more than 100 qualify to be highly hazardous.

In India, between 1997 and 2002, 8040 patients were admitted to hospital in Warangal with pesticide poisoning [54]. A fact finding mission estimated more than 1000 poisonings with 50% fatality in the 2001–2002 cotton season in central India [55]. Hundreds of occupational pesticide poisoning incidents from farming communities was reported in the cotton belt of Maharashtra in 2017 primarily due to inhalation and contact exposure [56]. The State of Maharashtra accounts for the highest pesticide use in India, with particularly high use on cotton. Meagre literature is available on occupational pesticide exposure and poisoning of farmers and farm workers in India. While the literature on self-poisoning (or intentional poisoning) is growing, less information is available on unintentional occupational poisoning. A global assessment estimates about 44 percent of farming communities are annually poisoned around the globe. It further reported that the highest number of non-fatal poisoning occurs in southern Asia, followed by south-east Asia and east Africa [47]. The figure of unreported incidents of poisoning would have been manifold higher to this estimate. Symptoms of mild pesticide poisoning often resemble that of common illnesses such as cold or common flu, and hence such poisoning incidents are misdiagnosed or under-reported. If immediate symptoms in mild poisoning are not severe, medical intervention is not sought usually. Mild poisoning includes irritation of skin, eyes, and throat, headache, dizziness, thirst, nausea, fatigue and nervousness; moderate poisonings additionally will have vomiting, coughing, excessive salivation, blurred vision, abdominal cramps, etc.; and severe poisoning will manifest difficulty in breathing, pinpoint pupils, unconsciousness and even death depending upon different pesticide active ingredient, age of exposed persons and exposure levels [57]. Despite ill-health, most patients in India tend to avoid accessing formal health systems, due to inaccessibility, beyond economic means and inadequate correlation with the cause. Pesticide poisoned people, most of them being small scale farmers and workers, in poorly connected areas with inadequate health facilities, do not take precautionary medical advice either, in both acute and chronic impacts, except when it is very severe and life-threatening.

A study conducted in Austria reports higher levels of pesticide residue detection based on passive air sampling, indicating pollution [58]. It detected 63 different pesticide active ingredients belonging to insecticides, fungicides and herbicides. It further reported pesticides are having potential ecotoxicological exposures to mammals, birds, earthworms, fish and honey bees. Impacts of pesticides in the form of high acute toxicity, reproductive toxicity, carcinogenicity and endocrine disruption were also reported. Documented impacts indicate widespread pollution caused by pesticides and inadequate protection to environment and organisms from their toxicity, which cannot be managed unless the hazard is eliminated by banning these poisons.

7.6 Pesticide residues in food

Frequent detection of pesticide residues in food has been raising concerns and is a global food safety risk as well. In India, national authorities are mandated to apply Maximum Residue Level (MRLs) standards to ensure food safety. It is a reality that conventional production systems yield food products contaminated with residues of pesticides. Official information from the All India Network Project on Pesticide Residues for the period between 2018 and 2019 to 2022–2023 shows an increase in the percentage of samples detected with pesticide residues. Samples having pesticide residues increased from 22.6% in 2018–2019 to 35.9% in 2022–2023. Additionally, there has been a gradual increase in the percentage of samples detected with residues exceeding MRL, pointing to the fact that the food resources are increasingly becoming poisonous.

8. Use conditions and suitability of risk mitigation measures including personal protective equipments (PPE)

Pesticide use conditions include literacy levels, knowledge levels, perception of adverse effects, behaviour of farming community, environmental parameters such as ambient temperature, humidity and wind, as well the governance factors such as policy framework, rules and regulations. All these affect exposure and poisoning. Imparting education and training to the farming community, use of PPE is the most common and prevalent strategies recommended by industry and regulatory authorities for pesticide risk reduction. However, pesticide regulation does not include an assessment of the suitability and practicality of suggested measures to prevent the risk of exposures, given the pesticide use conditions and environmental parameters in India. Often, use of personal protective equipment (PPE) is one and only commonly suggested method to prevent exposure.

The suitability of PPE use has to be reviewed and feasibility of such use has to be assessed carefully. It is a reality that the majority of farmers and farm workers do not use and follow recommended safety precautions and practices. Use of recommended PPE is low, if not zero. A study reports that a high proportion of farmers in Bangladesh, India, Philippines and Sri Lanka did not use even minimal PPE while working with pesticides [59]. Rudimentary measures, using locally available material, have been reported [44].

Studies further showed that most farmers are unaware of the types, poisoning, safety precautions and potential adverse effects of pesticides [60]. Studies reporting poor PPE use underscore lower awareness on pesticides and putting the responsibility of dealing with potential adverse effects and precautionary measures on users. Recommended PPE are not available, unaffordable, unsuitable to the work conditions, especially in a tropical climate. In this scenario, it is relevant to note the Article 3.6 of the International Code of Conduct on Pesticides Management, that recommends avoiding use of pesticides that require PPE, specifically in hot weather. Humid conditions prevent comfortable use of the recommended protective gear. Recommended protective gear is unsuitable for usage in tropical, hot and humid conditions.

Generally, the PPE recommended by regulators and manufacturers are neither available nor affordable, and inappropriate for the farming community. The Indian Insecticide Rules, 1971, without ambiguity, specified PPE usage when working with pesticides. Rule 39 specifies that the PPE shall be made of washable materials that resist the penetration of insecticides formulations [61]. In practice, farmers and workers use make-shift measures and or pick components of PPE that are available in the market at cheap rates, such as gloves. These cheap plastic materials usually do not comply with the specifications.

A comprehensive PPE suit, such as protective outer garment or overalls, hood or hat, protective gloves covering the fore-arm, dust-proof goggles, face shield and boots, should be made of materials like polycarbonate that are impermeable to chemicals and liquids. Along with this suit, recommended respiratory devices or masks also need to be used. Specifically designed protective equipment and clothing reserved for pesticide application and handling is often made of water resistant, chemical resistant materials such as polycarbonate [62]. If the authorities or industry are unable to provide the recommended adequate PPE, and/or the farming community cannot wear it due to inappropriateness, un-comfortableness and suffocation, then the regulatory authority has to ban pesticides as recommended by Article 3.6 of the ICCPM.

Pesticide manufacturers and regulatory authorities recommend wearing PPE. Further, the PPE information given on the label of pesticides is insufficient, often stating "wear protective clothing like apron, gloves, face shield and boots" or "wear full protective clothing while spraying or broadcasting" without referring to the recommended full PPE. Additionally, the language and text in the label information is incomprehensible to the general users and in case of poor literacy conditions the label becomes useless [44].

The hierarchy of exposure control [63] emphasizes that wearing PPE is one of the least effective strategies to reduce exposures [62]. The hierarchy of control focuses on elimination of hazard, followed by hazard replacement, engineering controls (basically isolate people from the hazard), administrative controls (change the way and behaviour of people) and, lastly, use of PPE to protect users from adverse effects. Elimination is the most effective precaution to prevent pesticide exposures and use of PPE is the least effective precaution.

When mishaps such as poisoning happen, the blame is put on poor practices of the farming community. Poor awareness, lack of training, improper use of pesticides and not using PPE are the reasons cited for exposure and poisoning, often pushing the responsibility on the pesticide user. This is unacceptable, given that manufacturers and regulators are fully aware that pesticides are inherently toxic substances. By resorting to such 'blame game', promoters of pesticides intend to shift the focus from the fact that pesticides are toxic substances and have the potential to cause poisoning and adverse effects including death, and undermine the fact that using PPE as a barrier to prevent exposure and attain safety is practically impossible. An Iranian study has noted that farmers tried but were not able to use PPE due to cost, suffocation and discomfort [64]. Studies conducted in Nepal and parts of South India reported that inappropriateness, discomfort were the major barriers for not using PPE [65, 66]. A farmers' knowledge perception study indicated that awareness on pesticide adverse effects by itself does not make wearing PPE suitable to prevailing work conditions [67].

An assessment about the role of PPE in reducing the risk of pesticide exposure noted that the discomfort, cost and availability render use of recommended PPE impractical [68]. On the other hand, hazardous pesticides receive regulatory approval only because of the assumed protection from use of PPE. A comfortable PPE has not been developed. Not much work or research has been done to develop PPEs that are suitable to Indian humid and field conditions. In India, affordability is the key parameter for all applications. Small and marginal farmers need PPE that is affordable. PPEs for pesticide use should be able to protect the farmers and farm workers from contamination. There is no PPE that has demonstrated that it is protective. Above cited paper says, numerous studies have found significant gaps between the ideal theoretical model of PPE and reality. It further noted significant data gaps between the assumed theoretical safety of PPE and practical complexities of wearing it, and these gaps magnify the risk of exposure. Wearing the recommended PPE would be difficult due to a high level of discomfort, suffocation, and overheating to the user in tropical agro climatic conditions. Considering these facts, and given the climate change and increased temperatures extremes, the false or impractical promise of assumed protection by PPE should not become the ruse for pesticide approval [69].

9. Pesticides and climate change

A recent report by PAN UK [70] highlights that about 99% of synthetic pesticides are derived from fossil fuels. In the entire life cycle right from manufacturing, packaging, transportation, application, and disposal, pesticides contribute considerably to the release of greenhouse gas emissions, exacerbating the climate emergency, in addition to the detrimental effects on health and environment.

Synthetic pesticides disrupt the delicate and balanced ecosystem function of soil, the complex web of life in the soil and soil fertility, by killing soil organisms which are critical for maintaining soil health. Use of pesticides turns soil devoid of life and makes it into a reservoir of toxic pollutants preventing regeneration of life sustaining activities in soil ecosystems, and thus reducing its ability to sequester carbon, coupled with increased carbon emission. In terms of productivity, life-less dead soil requires application of more synthetic fertilizers to support agriculture production, while the highly fertilized crops will become more susceptible to pest and disease infestation demanding application of more pesticides, further accelerating adverse effects [71].

10. Regulation and regulatory requirements

Stringent regulation of pesticides is essential. The process of pesticide registration provides an opportunity for selection of pesticides with lowest risk and to exclude products that pose unacceptable risk under the conditions of use in India. The ICCPM [15] has defined Registration. It is the process followed by the national government based on which the sale and use of a pesticide is approved. This process includes the evaluation of scientific data that demonstrates that the product is effective in its use. Further, it has to be ascertained scientifically that this pesticide does not pose risk to humans, animals and the environment. Yet, given the inherent risk of pesticides, the Food and Agriculture Organization of the United Nations [72], suggests three important steps for pesticide reduction: reduce dependence on pesticides, identify pesticides with lowest risk and ensure proper use. Hence it is important to follow a standard process employing relevant criteria for scrutinizing and decision-making based on internationally accepted criteria before registering pesticides. Not to register products belonging to Highly Hazardous Pesticides as specified by the criteria developed by the FAO-WHO Joint Meeting on Pesticides Management (JMPM) [73] is one of the options to keep away the most dangerous products. Further, those pesticides belonging to WHO Class II and or Class III [74], which so far resulted in pesticide poisoning in India, are also needs to be kept away.

It is of great importance to follow the ICCPM provisions especially article 3.6 and 6.1.1 while deciding on pesticide registration. In the case of small scale users and farm workers in hot climates, Article 3.6 of ICCPM mandates avoidance of pesticides that require PPEs, while handling and application, in three scenarios: PPE kits are uncomfortable, are expensive or are not readily available. Further, article 6.1.1 mandates the governments at the time of approval of pesticides to assess local needs, socio-economic conditions, literacy levels, climatic conditions, availability and affordability of appropriate pesticide application equipment, as well as personal protective equipment.

The following session presents examples of regulatory flaws on pesticides from an Indian context. In India, the Insecticide Act of 1968 and the Insecticide Rules 1971 governs pesticide regulation. It is almost 56 years old legislation. Some of the provisions have continued relevance in present conditions. However, it needs to be reformed to reflect the current conditions of pesticide regulation, especially when a lot of knowledge has been developed in the assessments of pesticides [75]. One of them is a mandatory review - mandates a review of pesticides in general to guarantee their safety and effectiveness. This is not done currently in India, as legislation backed process. An important drawback of the current legislation is that few pesticides, being used prior to 1968, before the official process of registration started in the country, were allowed to be used without evaluation, in India, in the category of *deemed to be registered pesticides* (DRPs) [76]. Later, after 1968, using this provision, several pesticides were registered under the DRP category, without mandatory evaluation. These pesticides have not been assessed properly, owing to data gaps pertaining to bio efficacy, effectiveness and safety concerns, but approved for use. By 2018, 62 pesticides are regarded as DRPs in India. Alarmingly, highly hazardous pesticides like Monocrotophos and endosulfan were among the DRPs, and were implicated in public health disasters as well as several poisoning incidents in India. Moreover, many pesticides, most of them are HHPs, that are banned in other countries have been allowed for use in India. After pressures from civil society, a committee was formed to review those 66 pesticides in 2013, and it submitted a report. However, even after 10 years many of the critical recommendations of the committee have not been fulfilled [77].

Pesticide poisoning remains unacknowledged at the regulatory level, even as fatal pesticide poisoning is increasingly becoming common in India. Foraging animal poisoning due to herbicides is also on the rise. Nearly seven goats in Kerala's Wayanad district died after getting poisoned by herbicide-treated grass. Rising cases of pesticide poisoning establish the need for responsive regulatory interventions. This lack of responsiveness of the existing regulation to the emerging challenges is leading to some State-level interventions. A few States have been repeatedly restricting selected pesticides, within the purview of their powers obtained under the Insecticides Act, 1968. Misbranded, under-quality or fake pesticides menace is a challenge before the State governments. Despite bans and restrictions, this problem is not going away. Stepping up its efforts, recently, a bill to amend the Insecticide Act, 1968 in its application in the State was introduced by the Maharashtra government in the legislative assembly on 8th August 2023. It sought to punish manufacture and sale of adulterated, non-standard or misbranded insecticides. This simple two-line amendment wants punishment to be cognizable and non-bailable. Through this amendment, the Government admits that misbranded or substandard pesticides are being manufactured and sold rampantly in the state. Significantly, Maharashtra government did not expand this amendment, to include other challenges and most importantly pesticide poisonings. In 2018, Maharashtra recommended ban on five pesticides named in the 2017 pesticide-poisoning incident in Yavatmal. In Odisha, around 200 people, largely farmers, died as a result of herbicide-related tragedies according to the registered death reports since September 2017. Recently, the Odisha government banned paraquat for a limited period, and has recommended its permanent ban. There is no response yet from the central pesticide regulatory mechanism.

The insecticide monocrotophos, an HHP, is extensively used in India. In several pesticide poisoning cases across India, monocrotophos was mentioned, including in the 2017 Yavatmal pesticide poisonings. Maharashtra Association of Pesticide Poisoned Persons (MAPPP) advocated a ban on this, along with other pesticides linked in the deaths and chronic injuries to farmers and farm workers. In the aftermath of pesticide poisoning in Yavatmal, the Maharashtra government wrote a letter to the national government to ban monocrotophos and four other pesticides. The Government of India ignored that letter. Later in 2023, a government of India Notification (S.O. 4294(E) dated 29th September 2023) banned four pesticides (Dicofol, Dinocap, Methomyl and Monocrotophos). As per this notification, India has not banned monocrotophos outright. There are some conditionalities built into it. A window period of 1 year was given for users to move towards alternatives. However, this delayed ban benefits the industry. The notification says, "sale, distribution or use of Monocrotophos 36% SL shall be allowed only for clearance of existing stock till its expiry period." The language used here is also ambiguous enough for the industry to build stocks in this window period of 1 year. As a result, Monocrotophos can continue to be used beyond the 1 year period until the stocks are cleared. Specific ban on the manufacture of Monocrotophos (all its formulations) is missing in this notification [78].

The Ministry of Agriculture and Farmers Welfare in India issued a draft notification dated 18th May 2020, proposing ban on 27 pesticides. A final notification, after considerable period, was issued, which dilutes the content in the draft notification. Only 11 pesticides were listed. Among the 11 pesticides, only three are banned. A qualified ban on one pesticide was notified, while allowing restricted usage of seven other pesticides. This comes after the Supreme Court of India observed, in July, 2023, about the penchant of the government to constitute a series of committees until it gets a favourable response in the review of the proposal of banning 27 pesticides in 2020. The remaining 16 pesticides, out of the 27 pesticides in the previous draft notification, do not find any mention in the final notification. The Agriculture Ministry did not deem it necessary to explain why 16 pesticides named in the 2020 notification have been left out.

In India, information on pesticides is very low, which is not helping in increasing awareness and knowledge of various stakeholder groups. Studies on pesticides, and their impacts, are far and rare. Basic information on pesticide poisoning in most studies is not found. Information and statistics on pesticide usage and residues continues to be the prerogative of the government. Mandatory provisions in the Insecticide Act, 1968, to collect data from the pesticide retail network are frazzled, inefficient and discontinuous. Discussion on Highly Hazardous Pesticides is at best rudimentary.

The Government of India does not acknowledge and recognize the hazardous potential of pesticides. Legislation needs to be attuned as per the harm caused by these pesticides to ecology and environment. HHPs have to be regulated. The pesticide management bill 2020 (draft) does not refer to HHPs and their usage. Regulation is lax, while monitoring the usage of HHPs is not even in place. Agricultural extension systems continue to recommend HHPs, on crops, without reference to the registered use. The gap between extension and registration is huge, with either of them not making efforts to bridge this gap. This is probably because the process of regulation is heavily influenced by the pesticide industry, which is focussed on preventing review, evaluation and scientific studies of pesticides. Most of the HHPs have been registered without meeting complete data requirements, or rudimentary information generated elsewhere. Effectiveness of HHPs is not demonstrated fully. Field data on the impact of HHPs on the targeted pests is completely absent. There is no independent verification process either. Meanwhile, residues of HHPs were detected in food and vegetables. The regulators initiated no further steps whenever such residues were detected, indicating flaws in regulation. Evaluation of impact of HHPs on human health, especially on farmers and their families, has to be done periodically.

There is inadequate dynamism in pesticide regulation in India. Despite increasing reports of pesticide residues in food, pesticide poisoning of farmers and farm workers in rural areas, and loss of biodiversity, pesticide usage is continuing unregulated.

10.1 Regulation of pesticides - State efforts in India

Earlier in 2011, following the global endosulfan ban and its aftermath, the Government of Kerala had banned 14 pesticides due to their potential adverse health effects. In 2017, Maharashtra government prohibited five pesticide formulations after they were identified in the wave of cotton farmer and farm worker occupational poisonings.

While there has been significant effort from various state governments in India to prohibit pesticides due to their potential to have harmful effects, a similar level of effort and response has not been seen from the federal government. As of March 2024, 49 pesticide active ingredients and five pesticide formulations are banned in India, while the number of registered pesticides rose to 339.

10.2 Pesticide management bill - A proposed new legislation in India

The Pesticide Management Bill 2020 [79] was introduced in the Rajyasabha on 23rd March 2020. Previously, the Union Cabinet approved this draft. This bill was slated to replace the Insecticide Act, 1968. The objective of the Bill is to regulate pesticides including their manufacture, import, packaging, labelling, pricing, storage, advertisement, sales, transport, distribution, use and disposal. Basically, this regulation is being brought to ensure availability of safe and effective pesticides, to minimize risk to human beings, animals, living organisms other than pests, and the environment. It wants to promote pesticides that are biological too, based on traditional knowledge. Originally a bill to replace the Insecticide Act, 1968, was introduced in 2008. Due to opposition, the content of this replacement has undergone several changes. Even the contents of the 2020 bill do not reflect a number of prevailing important issues. It does not propose decentralization of pesticide regulation. Higher role for State governments has been felt intensely in recent years. Advocacy groups have flagged lacunae in price and quality control mechanisms.

The 2020 bill does not address the pesticide use scenario in India in its true sense. In India pesticides are registered for specific uses, according to crop-pest combinations. Based on the scrutiny of the industry claims on efficacy and safety, the Central Insecticides Board and Registration Committee approves uses of registered pesticides (Insecticides, fungicides, herbicides/weedicides, plant growth regulators) associating particular crops, pests, dosage to be applied per unit area (ha), dilution concentration/water to be mixed with, and waiting period for each and every formulation of all the registered pesticides. The role of CIBRC ends at registering pesticides and issuance of approved uses, among other matters. On the other hand, the State agriculture University/Department issues crop production guides or package of practices for crop production. These guides recommend specific pesticides to be used for controlling specific pests for specific crops. Further Commodity boards such as Coffee Board, Tea Board, Spices Board and other public institutions also give recommendations of use. However, recommended uses are often different from the approved uses of pesticides in India.

Field level application of pesticides is not in harmony with the use approved by the CIBRC. A 2015 report titled 'Conditions of Paraquat use in India' showed that the weedicide paraquat dichloride was approved for nine crops, while State Agriculture Universities/Department or Commodity Boards had recommended it for using in 17 crops whereas, manufacturers had recommended it for 13 crops/uses; and field data showed that it was used for 25 crops. The Bill does not have provisions to harmonize approved use and recommended use. This is really important considering the matter of pesticide residues and food safety, as residue standards and related regulations are developed based on approved uses. Regulatory authorities have note taken seriously the usage happening beyond the approved and recommended use. Residue monitoring does not link up unauthorized usage, ignores related food safety concerns, with persistent impacts on public health and agriculture exports.

Analysts felt that the provisions of the draft bill are oriented for promoting and protecting business and the pesticide industry. There is no provision that is mentioned for protecting human beings, animals and environment from harmful effects of pesticides. The term 'to strive to minimize risk' is a weakly couched objective. There is no affirmative objective in the Bill. The objective could be reworded as 'elimination of risk to human beings.....and environment'.

10.3 Double standards in pesticide regulation

Acknowledging adverse effects of certain pesticides, many developed countries have come up with stronger regulations including banning or denying approval of hazardous pesticides. However, this response is limited to protection of the intra-country environment. Pesticides that are banned in European Union (EU) countries are being manufactured and exported to other countries, such as India. Due to a strong multinational pesticide lobby, regulation of such imports remains inadequate. On the other hand, countries that banned certain pesticides do not specify prohibition of export-linked production. This allows companies to manufacture and export banned or unapproved pesticides for use in another part of the world, threatening human health and environmental well-being. CIEL reported in 2023 [80] that high-income countries export pesticides, which they banned, to get profits. Unwanted hazardous substances are exported by manufacturers to low and middle income countries, where protective regulations are absent and with low technical capacity to regulate usage of hazardous substances. As a result, communities, farmers, and workers in the importing countries are continuously exposed to toxic agrochemicals. The 'toxic export' of banned pesticides in the exporting country impacts people and the environment in importing countries that have poor resources to assess adverse effects and regulatory systems. This kind of 'double standards' of the exporting countries is widely recognized as a human rights violation. For example, in the case of the EU, many hazardous agrochemicals are not approved or banned for use because of their adverse effects on human health and/or on the environment, yet manufacturers in certain EU countries are still producing them to export. India also is guilty of such double standards, allowing export of certain pesticides banned domestically. Pesticide regulation is harmoniously lax across the countries. In India, there is no process or procedure that calls out this kind of double standards.

11. Reducing pesticide risk

Scientifically proven and widely accepted fact that pesticides are toxic and harmful substances that had potential to cause irreversible damages to human health and environment, should have been a strong enough reason for international organizations, regulatory frameworks and treaties to reduce the use of toxic pesticides. International agencies such as World Health Organization, Food and Agriculture Organization, as well as National Governments have recommended several safety precautions to be taken while working with pesticides, in order to prevent exposure to these poisons. However, these precautions are not sufficient and are inadequate in assuring safety and prevention of poisonings and contamination. The risk arising from pesticide use is directly proportional to the intrinsic toxicological property of pesticide (hazard) and exposure. A reduction in usage or exposure can in turn reduce the risk from pesticides. The Food and Agriculture Organization of the United Nations released a guide wherein it calls for reduced reliance on pesticides, usage of pesticides with lowest risk and proper use [72]. Avoiding pesticides having high toxicity, adopting non-chemical pest management practices, selecting pesticides that have lowest toxicity, strictly complying with recommended safety precautions and PPE can help reduce risks of pesticides. These steps have to be followed step-wise over a period of time, towards phasing out pesticides. Agricultural system has to have a transition and paradigm shift into sustainable non-chemical methods. It has to be noted that, though safety measures are followed, the safe use of synthetic pesticides can never happen, as it involves risks at varying levels to health and environment, which can not be eliminated or mitigated.

Addressing the concerns over non-chemical method of food production, the UN special rapporteur, in the report "Agro-ecology and the right to food" presented before the UN Human Rights Council, did an extensive review of scientific literature. It concluded that with adequate policy and resources support, methods of agroecology can double food production in the world. It can also mitigate climate change and alleviate rural poverty [81].

Recognizing the multifunctionality of farming systems [82] and realizing the fact that pesticides destroy the multifunctional interactions, stringent efforts are needed to eliminate pesticides. Chemical intensive agriculture has been recognized as an accelerating factor to the triple planetary crisis of biodiversity loss, pollution and climate change. The identified solution to this enveloping crisis is pursuing agroecology. Integrating the commitments of Global Biodiversity Framework (GBF) under the Convention on Biodiversity as well as Global Framework on Chemicals (GFC) and other UN policy frameworks provides significant opportunities towards phasing out of HHPs and 50% lesser risk from pesticides by adopting agroecological options [83]. This necessitates policy reforms as well as national commitments to spur increased adoption of agroecology and related reduction of HHP usage.

12. Conclusion

Adverse effects of pesticides on children, farmers, farm workers, biodiversity and wildlife is becoming a serious public health concern in India and globally. The acute, chronic and trans-generational effects of hazardous pesticides on life are understudied and as such remains unrecognized. Pesticide regulation in India is basic, and is not geared to address various challenges arising out of intensive usage of highly hazardous pesticides in rural environments. Review capacity of the regulatory institutions, especially given the new active principles and co-adjuvants, remains undemonstrated. Text in the proposed new Bill does not have provisions to build a robust structure of regulation that addresses the challenges emerging from widespread production, usage and trade of agrochemicals. Current Indian pesticide regulation is inadequate and way behind in dealing with emerging challenges.

Achieving food security is the usual excuse for use of hazardous pesticides. This lie has been taken to the level of propaganda by the profiteering agrochemical industry simply to continue its business as usual, despite mounting evidence of chemicalisation of farmers, farm workers, soil, water and other environmental resources resulting in life threatening outcomes. Regulatory efforts towards pesticides management merely legitimize the use of poisonous substances whose potential hazards and risk cannot be minimized or managed effectively.

Year after year, the scenario of agriculture workers getting poisoned due to pesticides is becoming worse. Lack of regulatory efficiency and monitoring are contributing to illegal and dangerous practices of pesticide trade and use, which finally ends up in agriculture workers suffering from poisoning and loss of lives. Agriculture has become one of the most dangerous occupations owing to the intensive use of hazardous pesticides and agrochemicals. Marginal and small-scale farmers as well as farm workers mostly bear the harmful effects of such poisons. While National Crime Records Bureau (NCRB) in India chronicles poisoning of thousands of agriculture workers by pesticides, an equal or more in numbers of pesticide poisoned persons go unnoticed. Several cases of pesticide poisonings and deaths as well as pesticide induced chronic health effects have been reported across Indian states such as Kerala, Tamil Nadu, Andhra Pradesh, Telangana, Maharashtra, Punjab, and West Bengal.

Following exposure, thousands of agriculture workers lead incomparable sufferings and miserable lives. Poverty, malnutrition, unsustainable livelihood, lack of recognizable social status and decent living, poor health care, etc. are the major challenges faced by agriculture workers in India. Further, farm workers being the poorest of the poor, are unable to find resources to educate their children, for better employment prospects. In this situation, acute and chronic illness caused by pesticide poisoning has a devastating effect. Most agriculture workers live close to agriculture fields or spend considerable time in agricultural fields; they are exposed to an environment - air, water and soil as well as food resources - polluted by agrochemicals, which further deteriorate their immunity and health. All these facts necessitate the urgent need to dechemicalise agriculture and moving to agroecological practices. Concerted efforts are required from the International regulatory bodies, National and State governments to move into agroecological farming methods considering safety of agriculture workers and in view of agriculture sustainability, food and nutrition security, enhanced livelihood options, climate change impacts, environmental safety and better public health.

Acknowledgements

Authors acknowledge the encouragement and support provided by the Board of Directors of Pesticide Action Network (PAN) India to work out the chapter.

Conflict of interest

The authors declare no conflict of interest.

Author details

A.D. Dileep Kumar[1*] and Narasimha Reddy Donthi[2]

1 Pesticide Action Network (PAN) India, Thrissur, Kerala, India

2 Public Policy Expert, Hyderabad, India

*Address all correspondence to: dileep@pan-india.org; dileepnrm@gmail.com

References

[1] FAO and WHO. The international code of conduct on pesticide management. Published online 2014. Available from: www.fao.org/fileadmin/templates/agphome/documents/Pests_Pesticides/Code/CODE_2014Sep_ENG.pdf

[2] Vettorazzi G. The role of the World Health Organization in pesticide research. Toxicology. 1975;**4**(1):31-40. DOI: 10.1016/0300-483x(75)90019-0

[3] Unsworth J. History of pesticide use. 2010. Available from https://agrochemicals.iupac.org/index.php?option=com_sobi2&sobi2Task=sobi2Details&catid=3&sobi2Id=31 [Accessed: May 29, 2024]

[4] Bertomeu-Sánche JR. Before the silent Sprint: Pesticides in twentieth century Europe. Pesticides: Past and present. HoST - Journal of History of Science and Technology. 2019;**13**(1):1-27. DOI: 10.2478/host-2019-0001

[5] Unsworth J, Nakagawa Y, Harris C, Kleter G. The Battelle for a sustainable food supply Chapter 2. In: Vaz S Jr, editor. Sustainable Agrochemistry, a Compendium of Technologies. Switzerland: Springer Nature; 2019. pp. 13-84. ISBN 978-3-030-1780-1. ISBN 978-3-030-17891-8 (e-book)

[6] Delaplane KS. Pesticide Usage in the United States: History, Benefits, Risks, and Trends; Bulletin 1121. Athens, Georgia: Cooperative Extension Service, The University of Georgia College of Agricultural and Environmental Sciences; 2000. Available from: http://pubs.caes.uga.edu/caespubs/pubs/PDF/B1121.pdf

[7] Bolognesi C, Merlo FD. Pesticides: Human health effects. In: Nriagu JO, editor. Encyclopedia of Environmental Health. Burlington: Elsevier; 2011. pp. 438-453

[8] WHO & UNEP. Public health impact of pesticides used in agriculture. Published online 1990. Available from: https://apps.who.int/iris/handle/10665/39772

[9] Fukuto TR. Mechanism of action of organophosphorus and carbamate insecticides. Environmental Health Perspectives. 1990;**87**:245-254. DOI: 10.1289/ehp.9087245

[10] Jayaraj R, Megha P, Sreedev P. Organochlorine pesticides, their toxic effects on living organisms and their fate in the environment. Interdisciplinary Toxicology. 2016;**9**(3-4):90-100. DOI: 10.1515/intox-2016-0012

[11] Soderlund DM. Molecular mechanisms of pyrethroid insecticide neurotoxicity: Recent advances. Archives of Toxicology. 2012;**86**(2):165-181. DOI: 10.1007/s00204-011-0726-x

[12] Buszewski B, Bukowska M, Ligor M, Staneczko-Baranowska I. A holistic study of neonicotinoids neuroactive insecticides-properties, applications, occurrence, and analysis. Environmental Science and Pollution Research International. 2019;**26**(34):34723-34740. DOI: 10.1007/s11356-019-06114-w

[13] World Health Organization Recommended Classification of Pesticides by Hazard and Guidelines to Classification. 2019th ed. Geneva: World Health Organization; 2020. Licence: CC BY-NC-SA 3.0 IGO. Available from: https://iris.who.int/bitstream/handle/10665/332193/9789240005662-eng.pdf?sequence=1

[14] UNEP. Highly hazardous Pesticides (HHPs). United Nations Environmental Programme. Available from: https://www.unep.org/topics/chemicals-and-pollution-action/pollution-and-health/highly-hazardous-pesticides-hhps [Accessed: June 16, 2024]

[15] FAO and WHO. The international code of conduct on pesticide management. Rome: World Health Organization Food and Agriculture Organization of the United Nations; 2014. Available from: www.fao.org/fileadmin/templates/agphome/documents/Pests_Pesticides/Code/CODE_2014Sep_ENG.pdf

[16] The World Health Organization. 10 chemicals of public health concern. 2020. Available from: https://www.who.int/news-room/photo-story/photo-story-detail/10-chemicals-of-public-health-concern [Accessed: May 24, 2024]

[17] United Nations. Pesticides are "global human rights concern", say UN experts urging new treaty. 2017. Available from: https://www.ohchr.org/en/press-releases/2017/03/pesticides-are-global-human-rights-concern-say-un-experts-urging-new-treaty [Accessed: May 17, 2024]

[18] UNEP. Resolution adopted by the United Nations Environment Assembly on 1 March 2024. 6/11 highly hazardous pesticides. Available from: https://documents.un.org/doc/undoc/gen/k24/008/43/pdf/k2400843.pdf?token=jgHV0qpVHLE7NMwVmj&fe=true [Accessed: May 06, 2024]

[19] Barnes JM, World Health Organization. Toxic Hazards of Certain Pesticides to Man : Together with a Select Bibliography on the Toxicology of Pesticides in Man and Mammals. Geneva: World Health Organization; 1953. Available from: https://iris.who.int/handle/10665/41134 [Accessed: May 07, 2024]

[20] Carson R. Silent Spring. Boston: Houghton Mifflin Company; 1962

[21] Goldmann L, Murray B, Wahlström B, Pronczuk J. Childhood Pesticide Poisoning Information for Advocacy and Action. WHO, FAO: UNEP; 2004

[22] Lewis KA, Tzilivakis J, Warner D, Green A. An international database for pesticide risk assessments and management. Human and Ecological Risk Assessment: An International Journal. 2016;**22**(4):1050-1064. DOI: 10.1080/10807039.2015.1133242

[23] Saganuwan SA. Toxicity study of drugs and chemicals in animals: An overview. Bulgarian Journal of Veterinary Medicine. 2017;**20**(4):291-318. ISSN 1311-1477. DOI: 10.15547/bjvm.98

[24] Shoda A, Murata M, Kimura M, Hara Y, Yonoichi S, Ishida Y, et al. Transgenerational effects of developmental neurotoxicity induced by exposure to a no-observed-adverse-effect level (NOAEL) of neonicotinoid pesticide clothianidin. The Journal of Veterinary Medical Science. 2023;**85**(9):1023-1029. DOI: 10.1292/jvms.23-0101. Epub 2023 Aug 4

[25] Toxicity tests for human health assessments for pesticides. Available from: https://www.epa.gov/pesticide-science-and-assessing-pesticide-risks/toxicity-tests-human-health-assessments-pesticides [Accessed: May 22, 2024]

[26] Watts M. Poisoning our future: Children and pesticides. Penang: Pesticide Action Network Asia and the Pacific (PAN AP); 2013

[27] USEPA. Basic information about pesticide ingredients. Available from: https://www.epa.gov/ingredients-used-pesticide-products/basic-information-about-pesticide-ingredients [Accessed: May 02, 2024]

[28] Watts M. Glyphosate Monograph. PAN Asia Pacific, Penang: PAN International; 2016. Available from: http://pan-international.org/wp-content/uploads/Glyphosate-monograph.pdf [Accessed: May 07, 2024]

[29] Kesavachandran CN, Fareed M, Pathak MK, Bihari V, Mathur N, Srivastava AK. Adverse health effects of pesticides in agrarian populations of developing countries. Reviews of Environmental Contamination and Toxicology. 2009;**200**:33-52. DOI: 10.1007/978-1-4419-0028-9_2

[30] Keifer MC, Firestone J. Neurotoxicity of pesticides. Journal of Agromedicine. 2007;**12**(1):17-25. DOI: 10.1300/J096v12n01_03

[31] Roberts JR, Karr CJ. Council on environmental health. Pesticide exposure in children. Pediatrics. 2012;**130**(6):e1765-e1788. DOI: 10.1542/peds.2012-2758. Epub 2012 Nov 26. Erratum in: Pediatrics. 2013 May;131(5):1013-4

[32] Sharma A, Gill JPS, Bedi JS, et al. Monitoring of pesticide residues in human breast milk from Punjab, India and its correlation with health associated parameters. Bulletin of Environmental Contamination and Toxicology. 2014;**93**:465-471. DOI: 10.1007/s00128-014-1326-2

[33] Witczak A, Pohoryło A, Abdel-Gawad H. Endocrine-disrupting organochlorine pesticides in human breast milk: Changes during lactation. Nutrients. 2021;**13**(1):229. DOI: 10.3390/nu13010229

[34] Mehta R, Sreenivasa MA, Mathew M, et al. A mixed-methods study of pesticide exposures in breastmilk and community & lactating women's perspectives from Haryana, India. BMC Public Health. 2020;**20**:1877. DOI: 10.1186/s12889-020-09966-x

[35] Varghese P, Erickson TB. Pesticide poisoning among children in India: The need for an urgent solution. Global Pediatric Health. 2022;**9**:2333794X221086577. DOI: 10.1177/2333794X221086577

[36] Tudi M, Daniel Ruan H, Wang L, Lyu J, Sadler R, Connell D, et al. Agriculture development, pesticide application and its impact on the environment. International Journal of Environmental Research and Public Health. 2021;**18**(3):1112. DOI: 10.3390/ijerph18031112

[37] Levengood JM, Beasley VR. Principles of ecotoxicology. In: Gupta RC, editor. Veterinary Toxicology: Basic and Clinical Principles. Amsterdam: Academic Press; 2007. pp. 689-708

[38] Bedos C, Cellier P, Calvet R, Barriuso E, Gabrielle B. Mass transfer of pesticides into the atmosphere by volatilization from soils and plants: Overview. Agronomie. 2002;**22**:21-33

[39] Agriculture and Natural Resources. Water quality: Controlling nonpoint source (NPS) pollution. Pesticide management to protect water quality. Understanding pesticides and how they affect water quality. 1999. Available from: http://www.aces.edu/pubs/docs/A/ANR-0790/WQ4.5.1.pdf [Accessed: May 06, 2024]

[40] United Nations Environment Programme, World Health Organization, Food and Agriculture Organization of the United Nations. Chapter 4/12:

Environmental and Health Effects of Pesticide Use. Nairobi: UNEP, WHO & FAO; 2022. Available from: https://wedocs.unep.org/20.500.11822/40353 [Accessed: May 07, 2024]

[41] Katyal JC, Chaudhari SK, Dwivedi BS, Biswas DR, Rattan RK, Majumdar K, editors. Soil health: Concept, status and monitoring. In: Bulletin of the Indian Society of Soil Science. Vol. 30. New Delhi: Indian Society of Soil Science; ISSS Bulletin 30; 2016. pp. 1-98

[42] Pimentel D, Levitan L. Pesticides: Amounts applied and amounts reaching pests. Bioscience. 1986;**36**(1986):86-91

[43] UN Special Rapporteur on the right to food. (A/HRC/34/48). 2017. Available from: https://documents.un.org/doc/undoc/gen/g17/017/85/pdf/g1701785.pdf?token=ZGZpmjsFXpvqqPNc1L&fe=true [Accessed: May 07, 2024]

[44] Dileep Kumar. Highly hazardous pesticide series: State of chlorpyrifos, fipronil, atrazine and paraquat dichloride in India. 2022. Available from: https://pan-india.org/wp-content/uploads/2022/08/HHP_Chl-Fip-Atr-Pqt_Report-Final-web_PAN-India.pdf [Accessed: May 07, 2024]

[45] De Ruiter H, Holterman HJ, Kempenaar C, Mol HGJ, De Vlieger JJ, van de Zande JC. Influence of Adjuvants and Formulations on the Emission of Pesticides to the Atmosphere. Wageningen: Plant Research International B.V.; 2003

[46] EU. Regulation (EC) 2021/383 amending Annex III to Regulation (EC) No 1107/2009

[47] Boedeker W, Watts M, Clausing P, et al. The global distribution of acute unintentional pesticide poisoning: Estimations based on a systematic review. BMC Public Health. 2020;**20**:1875. DOI: 10.1186/s12889-020-09939-0

[48] Baker EL, Zack M, Miles JW, Lee A, Warren M, Dobbin RD, et al. Epidemic malathion poisoning in Pakistan malaria workers. The Lancet. 1978;**311**(8054):31-34. ISSN 0140-6736. DOI: 10.1016/S0140-6736(78)90375-6

[49] Eddleston M. Patterns and problems of deliberate self-poisoning in the developing world. QJM: An International Journal of Medicine. 2000;**93**(11):715-731. DOI: 10.1093/qjmed/93.11.715

[50] Jeyaratnam J, de Alwis Seneviratne RS, Copplestone JF. Survey of pesticide poisoning in Sri Lanka. Bulletin of the World Health Organization. 1982;**60**(4):615-619

[51] Wesseling C, Aragon A, Castillo L, Corriols M, Chaverri F, de la Cruz E, et al. Hazardous pesticides in Central America. International Journal of Occupational and Environmental Health. 2001;7:287-294

[52] Karunarathne A, Bhalla A, Sethi A, et al. Importance of pesticides for lethal poisoning in India during 1999 to 2018: A systematic review. BMC Public Health. 2021;**21**:1441. DOI: 10.1186/s12889-021-11156-2

[53] Kumar D. Conditions of Paraquat Use in India. Zurich: IUF, Berne Declaration, PAN AP and PAN India; 2015. Available from: http://www.pan-india.org/wp-content/uploads/2017/03/BD_paraquat_4-15_def-WEB.pdf

[54] Srinivas Rao C, Venkateswarlu V, Surender T, Eddleston M, Buckley NA. Pesticide poisoning in South India: Opportunities for prevention and improved medical management. Tropical Medicine & International

Health. 2005;**10**(6):581-588. DOI: 10.1111/j.1365-3156.2005.01412.x

[55] Damoder P, Rangarajan R, Narasimhareddy D, Dutta M , Patil R, Nagasena R, Abitha S. The Killing Fields Farmer deaths due to exposure to pesticides in Warangal District. New Delhi: ToxicLink; 2002. Available from: https://toxicslink.org/wp-content/uploads/2022/08/The_Killing%20Fiealds_Warangal%20report_2002.pdf

[56] Narasimha R, Dileep Kumar AD. Pesticide poisonings in Yavatmal District in Maharashtra: Untold realities. Published online 2017. Available from: http://www.pan-india.org/wp-content/uploads/2017/10/Yavatmal-Report_PAN-India_Oct-2017_web.pdf

[57] CCOHS. Pesticides - Health Effects. Canadian centre for occupational health and safety. Available from: https://www.ccohs.ca/oshanswers/chemicals/pesticides/health_effects.htm [Accessed: May 07, 2024]

[58] Zaller JG, Kruse-Plaß M, Schlechtriemen U, Gruber E, Peer M, Nadeem I, et al. Pesticides in ambient air, influenced by surrounding land use and weather, pose a potential threat to biodiversity and humans. The Science of the Total Environment. 2022;**838**(Pt. 2):156012. DOI: 10.1016/j.scitotenv.2022.156012. Epub 2022 May 18

[59] Matthews GA. Attitudes and behaviours regarding use of crop protection products—A survey of more than 8500 smallholders in 26 countries. Crop Protection. 2008;**27**(3-5):834-846. DOI: 10.1016/j.cropro.2007.10.013

[60] Sharma D, Thapa R, Manandhar H, Shrestha S, Pradhan S. Use of pesticides in Nepal and impacts on human health and environment. Journal of Agriculture and Environment. 2013;**13**:67-74. DOI: 10.3126/aej.v13i0.7590

[61] GoI. Insecticides Rules 1971. (GSR 1650, DT. 9-10-1971). New Delhi: Government of India; 1971. Available from: https://ppqs.gov.in/sites/default/files/insecticides_rules_1971.pdf [Accessed: May 14, 2024]

[62] FAO and WHO. Guidelines for Personal Protection When Handling and Applying Pesticide. Rome: International Code of Conduct on Pesticide Management; 2020. Available from: https://www.fao.org/3/ca7430en/ca7430en.pdf

[63] NIOHS. Hierarchy of controls. Published online 2023. Available from: https://www.cdc.gov/niosh/topics/hierarchy/default.html [Accessed: May 22, 2024]

[64] Afshari M, Poorolajal J, Rezapur-Shahkolai F, Assari MJ, Karimi-Shahanjarini A. Which factors influence farmers' use of protective measures during pesticides exposure? Workplace Health & Safety. 2019;**67**(7):338-349. DOI: 10.1177/2165079919827042

[65] Bhandari G, Atreya K, Yang X, Fan L, Geissen V. Factors affecting pesticide safety behaviour: The perceptions of Nepalese farmers and retailers. The Science of the Total Environment. 2018;**631-632**:1560-1571. DOI: 10.1016/j.scitotenv.2018.03.144

[66] Indira Devi P. In: Bellamy R, Shyamsundar P, editors. Facing Hazards at Work: Agricultural Workers and Pesticide Exposure in Kuttanad, Kerala. Nepal: SANDEE; March 2007

[67] Satya Sai M, Revati G, Ramya R, Swaroop A, Maheswari E, Kumar M. Knowledge and perception of farmers regarding pesticide usage in a rural farming village, Southern India. Indian Journal of Occupational and Environmental Medicine. 2019;**23**(1):32. DOI: 10.4103/ijoem.IJOEM_121_18

[68] Garrigou A, Laurent C, Berthet A, et al. Critical review of the role of PPE in the prevention of risks related to agricultural pesticide use. Safety Science. 2020;**123**:104527. DOI: 10.1016/j.ssci.2019.104527

[69] Prakash M, Donthi NR. Pesticides regulations in India and the need for protection against health risks. Vol. 57, Issue No 25. Mumbai: Economic and Political Weekly; 2022. Available from: https://www.epw.in/engage/article/pesticides-regulations-india-and-need-protection

[70] PAN UK. Pesticides and the climate crisis. 2023. Available from: https://www.pan-uk.org/pesticides-and-the-climate-crisis/ [Accessed: May 07, 2024]

[71] PAN UK. Pesticides and climate change. 2021. Available from: https://www.pan-uk.org/pesticides-and-climate-change/ [Accessed: May 07, 2024]

[72] FAO and WHO. The International Code of Conduct on Pesticide Management. Rome: World Health Organization and Food and Agriculture Organization of the United Nations; 2014. Available from: http://www.fao.org/fileadmin/templates/agphome/documents/Pests_Pesticides/Code/CODE_2014Sep_ENG.pdf

[73] FAO. Guideline on highly hazardous pesticides. In: International Code of Conduct on Pesticides Management. Rome: Food and Agriculture Organization of the United Nations and World Health Organization; 2016. Available from: https://www.fao.org/3/a-i5566e.pdf

[74] World Health Organization. WHO Recommended Classification of Pesticides by Hazard and Guidelines to Classification: 2009. Geneva: WHO; 2010. Available from: www.inchem.org/documents/pds/pdsother/class_2009.pdf

[75] PAN India. Pesticide management bill 2020 need critical amendments. 2020. Available from: https://pan-india.org/3596-2/

[76] Dileep, Kumar AD, Athira PS. Deemed to be registered pesticides in India. Kerala: Pesticide Action Network India; 2022. Available from: https://pan-india.org/wp-content/uploads/2022/09/Deemed-to-be-Registered-Pesticides_report_PAN-India-Web.pdf

[77] CIBRC. Minutes of 361st Special Meeting of Registration Committee Held on 22nd December, 2015. New Delhi: Central Insecticides Board and Registration Committee, Ministry of Agriculture and Farmers Welfare, Government of India; 2015. Available from: https://ppqs.gov.in/sites/default/files/361rc2015.pdf; [Accessed: May 07, 2024]

[78] PAN India. Government of India makes U turn, banning only three out of the proposed 27 hazardous pesticides in 2020. Press Release. 2023. Available from: https://pan-india.org/government-of-india-makes-u-turn-banning-only-three-out-of-the-proposed-27-hazardous-pesticides-in-2020/

[79] The Pesticide Management Bill, 2020. Bill No. XXII of 2020. Introduced in Rajya Sabha Parliament of India

[80] CIEL. A Toxic Double Standard: Analyzing the Legality of the Export of Banned Pesticides. 2023. Available from: https://www.ciel.org/a-toxic-double-standard-analyzing-the-legality-of-the-export-of-banned-pesticides/ [Accessed: May 07, 2024]

[81] United Nations. In: De Schutter O, editor. Report Submitted by the Special Rapporteur on the Right to Food. Geneva: UN; 2010

[82] The International Assessment of Agricultural Knowledge, Science and Technology for Development (IAASTD). UNDP, FAO, UNEP, UNESCO, World Bank, WHO & GEF; 2008. Available from: https://wedocs.unep.org/bitstream/handle/20.500.11822/7880/-Agriculture%20at%20a%20crossroads%20-%20Executive%20Summary%20of%20the%20Synthesis%20Report-2009Agriculture_at_Crossroads_Synthesis_Report_Executive_Summary.pdf?sequence=6&amp%3BisAllowed=y%2C%20Chinese%7C%7Chttps%3A//wedocs.un [Accessed: Jun 09, 2024]

[83] PAN UK. Leveraging NBSAPS beyond the CBD-how action on agroecology and pesticides delivers across multiple UN policy frameworks. 2024. Available from: https://www.pan-uk.org/site/wp-content/uploads/Leveraging-NBSAPS-beyond-the-CBD.pdf [Accessed: Jun 14, 2024]

Section 3

Biorational Strategies in Insect Pest Management

Chapter 8

Botanical Insecticides for Crop Protection: Major Classes and Possible Mechanisms of Action

Sorabh Chaudhary, Sunil Kumar Yadav, Pankaj Verma, Sushma Sagar and Mehi Lal

Abstract

Botanical insecticides, derived from plant-based bioactive compounds with insecticidal activities, have been used to protect crops for more than a century. Synthetic insecticides have become essential in modern agricultural practices because they act quickly, are cost-effective, easy to use, and effective against a wide range of harmful species of insects and pests. Synthetic insecticides, once beneficial, have now become a threat to human health and the environment due to their toxic nature and environmental impact, and hence, botanical pesticides can be utilized as a substitute to synthetic chemical pesticides. Botanicals are easily available, biodegradable, have a broad spectrum of activity, are inexpensive, and have low toxicity to humans and non-target organisms. Various plants species including neem, pyrethrum, garlic, citrus, etc., contains various types of plant secondary metabolites including alkaloids, flavonoids, terpenes, essential oils, glycosides, esters, etc., were popular botanicals. These metabolites exhibited different physical/chemical characteristics and mechanisms of action and affect the insects in various ways such as insecticides, repellents, antifeedants, toxicants, growth retardants, herbicides, and attractants. This chapter describes the in-depth knowledge of various classes of bioinsecticides and their possible mode of actions.

Keywords: botanical insecticides, essential oils, neem, repellent, phototoxin, integrated pest management

1. Introduction

The global human population is increasing exponentially (about 75 million annually), and if this pattern persists, it is projected that the global population will exceed 10 billion by the end of the century [1, 2]. The rapidly growing population further put greater pressure on land, water, and resources, consequently requiring farmers in developing nations to significantly increase their agricultural output to sustain the growing individuals [3]. Pests and diseases considered as the major obstacles to achieving successful crop production. It is estimated that around 35–40% of global crop production is lost every year due to pests before harvest for all potential cereal, vegetable, and fiber crops [4]. Hence, it is crucial to implement cutting-edge, affordable agricultural methods to enhance food production, control pests, and manage diseases both before and after harvest in order to boost crop yields and promote sustainable farming. Over the past, conventional synthetic chemical insecticides have been effectively used to control pests and increase agricultural productivity. The major categories of chemical insecticides are belonged to chlorinated hydrocarbons, organophosphates, carbamates, and pyrethroids. While agrochemicals have increased agricultural production, their widespread and uncontrolled use has led to significant issues including environmental damage from water, air, and soil contamination, harm to non-target organisms, health risks from agrochemical residues in food, and decreased effectiveness due to pest resistance [5, 6]. Agrochemicals can enter the soil through direct use like applying them to control weeds and treating seeds and indirectly through spraying plant parts, the dropping of treated leaves or fruits, and movement of contaminated water on the surface and within the soil [7, 8]. When these chemical compounds accumulated in the soil, they can be transported by leaching and surface runoff and can decompose through chemical processes such as hydrolysis, photolysis, and chemical degradation. Additionally, they can interact with the soil's microbial community and break down naturally [9, 10]. The excessive use of agrochemicals leads to increased persistence in the soil, which can have a detrimental impact on soil microbial consortium. Different types of pesticides elicit varied responses from the microorganisms within the microbial consortium, potentially leading to increased or inhibited growth and metabolism [7, 11]. Previous studies found that extensive use of pesticides cause qualitative and quantitative changes in the soil microbial consortia [12, 13], alteration of nitrogen cycling [14], changes in soil enzymatic activity, and disruption of the symbiosis between mycorrhizae and root nodules in legumes. These factors collectively altered the soil fertility and, consequently, plant growth [15]. Therefore, in order to meet the increasing need for food, sustainable agriculture must effectively improve soil quality through methods like employing biopesticides and biofertilizers, crop diversification, crop rotation, and utilizing microbiota [16, 17]. The quest for better and environmentally friendly ways to manage pests has grown, with a focus on using natural enemies found in soil and plant-based products for biological control. Biopesticides provide a promising substitute for conventional chemicals in controlling the crop diseases for sustainable agriculture [18–20]. The use of biopesticides offers several advantages, including minimal persistence and residuality, which help prevent environmental pollution and minimize adverse effects on living organisms [21]. Additionally, biopesticides exhibit high specificity towards hosts, resulting in a delayed knockdown, and are less prone to encountering pest resistance because of their diverse chemical structures (**Figure 1**), which are attributed to the remarkable biosynthetic capabilities of plants.

Since ancient times, besides to manage animal pests, natural products have been utilized for the management of various plant diseases and to control weeds. Plants have served as the primary sources of natural pesticides for centuries. With the advancement of modern analytical methods, preparations standardized for the active ingredient(s) have become available in recent decades. This development has enabled the production of reliable products. Botanical insecticides, derived from plants and

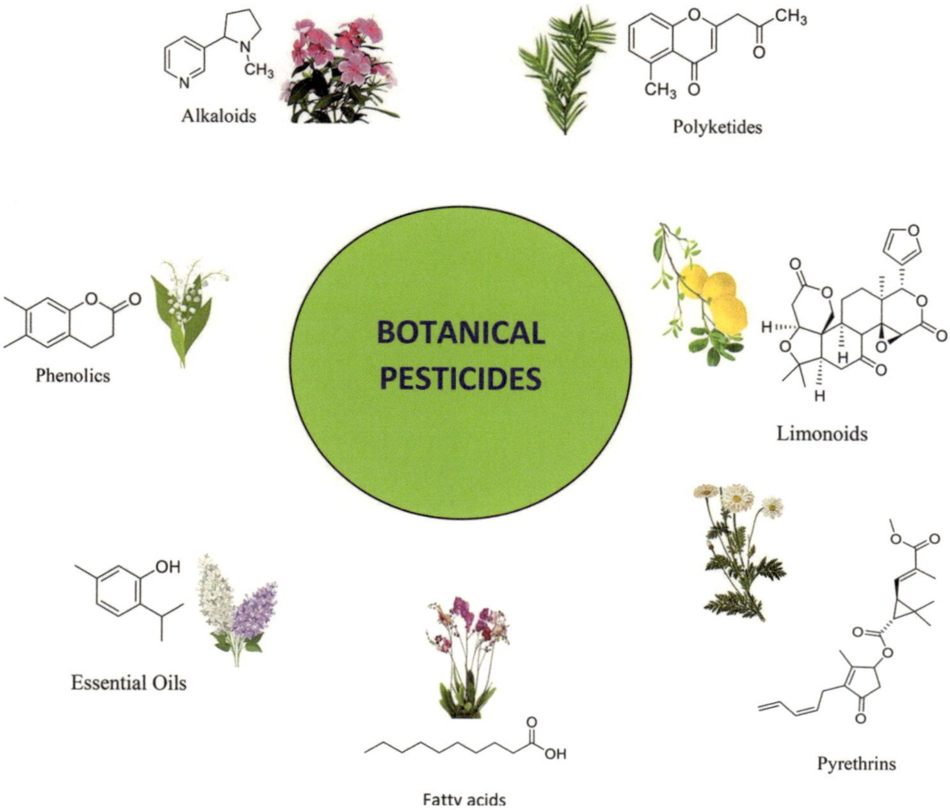

Figure 1.
Structural diversity of bioactive compounds in plant-based pesticides.

their extracts, are eco-friendly pest control solutions that are biodegradable and provide sustainable crop protection. These compounds have been used for centuries in various cultures for pest control, and their resurgence in modern agriculture reflects a growing demand for sustainable farming practices. These alternatives to synthetic insecticides have gained popularity due to advancements in pest control and the growing demand for Integrated Pest Management (IPM) strategies [22]. Among the botanical insecticides rotenone from *Derris elliptica*, nicotine from tobacco leaf, pyrethrins from pyrethrum flowers (*Chrysanthemum cinerariaefolium*), and azadirachtin from neem (*Azadirachta indica*) have attained commercial importance. Botanical insecticides are derived from plant sources and have insecticidal properties. These natural extracts are versatile, with low toxicity, high efficiency, and easy degradability. They leave behind no residual insecticide and do not contribute to insect resistance. Botanical insecticides are capable of effectively controlling various plant pests without causing harm to humans, animals, or the environment. Their composition, production methods, mechanism of action, and mode of operation differ from those of chemical insecticides. Some examples of botanical insecticides with their source plant, active ingredients, and manufacturing company are listed in **Table 1**. The present chapter describes an overview of different classes of botanical insecticides with their possible mode of actions for controlling the insect/pest for sustainable agricultural productivity.

Species (family)	Active compounds	Commercial product (company)
Allium sativum L. (Liliaceae)	Sulphur containing compounds e.g. diallyl trisulfide, diallyl disulfide, methyl allyl trisulfide	AjoNey (Invernaderos Hidroponicos Neisi, Mexico); EcoA-Z®, L'EcoMix® or CapsiAlil® (EcofloraAgro, Colombia)
Annona squamosa L. (Annonaceae)	Squamocin (annonin), debitterized annona oil	ANOSOM® (Agri Life, India)
Azadirachta indica Juss. (Meleaceae)	Azadirachtin, salanin, nimbin	MARGOSOM® (Agri Life, India); Molt-X® (BioWorks, Inc., USA); NeemAzal T/S® (Trifolio-M, Germany); AZERA™ (MGK®, USA); Neemix 4.5® (Certis, USA); Azatin XL® (OHP Inc., USA); Azamax® (UPL Ltd., Brazil); Fortune Aza 3%EC® (Fortune Biotech, USA)
Capsicum annum L. (Solanaceae)	Protoalkaloids e.g. capsaicin	Hot Pepper Wax (Rincon-Vitova Insectaries, USA); ChileNey (Invernaderos Hidroponicos Neisi, Mexico)
Celastrus angulatus Maxim. (Celastraceae)	Sesquiterpene pyridine alkaloids	CELAN-X SL (Marketing Arm International, Inc., USA)
Citrus sinensis (L.) Osbeck (Rutaceae)	Limonene and linalool	Demize EC (Paragon Professional Pest Control Products, U.SA); Prev-Am (Oro Agri SA (Pty) Ltd., South Africa)
Chrysanthemum cinerariaefolium (Trevir.) Vis. (Asteraceae)	Pyrethrins (cinerins; jamolins and pyerethrins)	Pruzit® (Neudorff, Germany); PyGanic® Crop Protection EC 5.0 or AZERA™ (MGK®, USA); 1.5% Aphkiller AS (Beijing Kingbo Biotech Co., Ltd, China)
Lonchocarpus spp., Derris spp. (Fabaceae)	Rotenone	5.% Rotenone ME (Beijing Kingbo Biotech Co., Ltd, China); Rotenone Dust (Bonide Products, Inc, USA)
Nictiana tabacum L. (Solanaceae)	Nicotine	Nico Dust or Nico Neem (Nico Orgo Manures, India); 10% Nicotine AS (Beijing Kingbo Biotech Co., Ltd., China)
Pongamia pinnata (L.) Pierre (Fabaceae)	Karanjin, debitterised karanjin oil	DERISOM® (Agri Life, India); RockEffect (Agro CS a.s., Czech Republic)
Schoenocaulon officinale A. Gray (Melanthiaceae)	Cevadine; veratridine	VERATRAN D® (MGK®, USA)
Thymus vulgaris L.	Phen(yl)ethyl propionate; Thyme oil	EcoVia WD® (Rockwell Labs Ltd., USA)
Origanum vulgare	Oregano oil	By-O-reg +® (By-O-reg +, USA)

Table 1.
Most common commercially used botanical insecticides [23].

2. Classes of botanical insecticides

Plant-based botanicals exhibited various biological activities including, repellents, insecticides, herbicides, fungicides, nematicides, and bactericides with different modes of action [24]. Various botanical insecticides with their plant source, active ingredients, chemical structure, plant source, mode of action, and targets are listed in **Table 2**.

DOI: http://dx.doi.org/10.5772/intechopen.1006743

Plant species	Compound	Chemical structure	Class	Mode of action	Target insect/pest	References
Tanacetum cinerarifolium	Pyrethrin I, R = CH_3 Pyrethrin II, R = CO_2CH_3	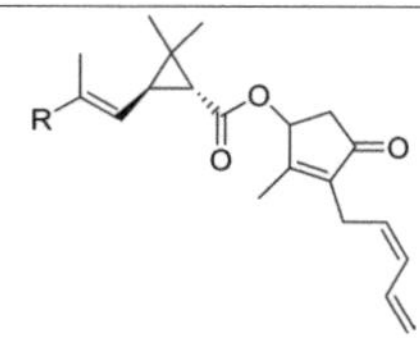	Cyclopropylmono-terpene esters	Agonists of the voltage-gated Na+ channels of insects, causing their sensory nerve fibers and motor nerve cells to fire repeatedly, causing the insect excited, paralyzed and eventually dead.	Houseflies, beetles, fleas, spider mites	[25, 26]
Cedrus deodara	Himachalol	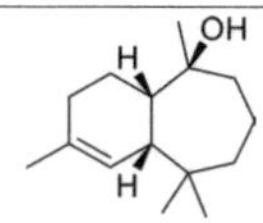	Sesquiterpene	Agonists of voltage-gated Na+ channels	Pulse beetle and Houseflies	[27]
Decalepis hamiltonii	Decaleside	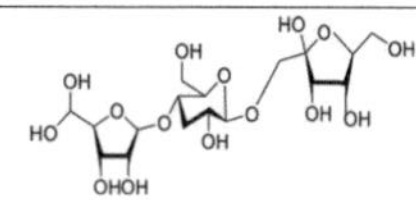	Trisaccharide	Inhibited the enzyme involving the ATP binding sites	Cockroaches, houseflies, stored grain pests	[28]
Schoenocaulon officinale	Cevadine R=(z)-$CH_3CH{=}C(CH_3)CO$ Veratridine R= 3,4-$(CH_3O)_2PhCO$	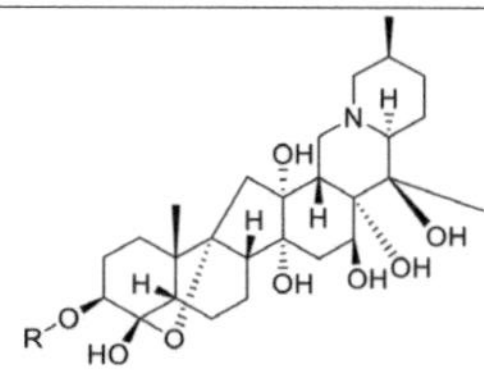	Sabadilla alkaloid or *Veratrum* alkaloid	Activate voltage-sensitive Na+ channels of nervous system	Leafhoppers, caterpillars, houseflies, stinks, thrips	[29–32]
Ryania speciosa	Ryanodine	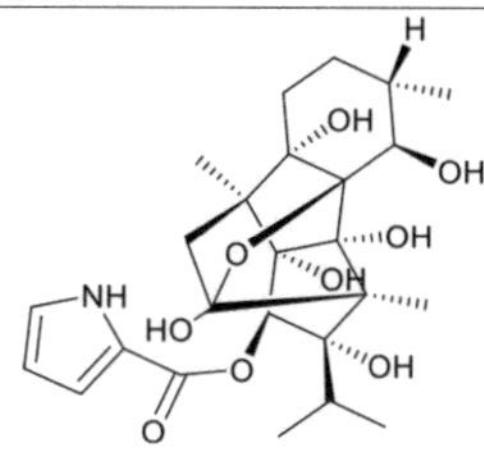	Diterpene	Effects nervous system by binding to the intracellular Ca^{2+} release channel proteins	Potato beetle, caterpillars, worms, lace bugs, aphids, squash bugs	[29, 33]
Lantana camara	(Coumaran)2,3-dihydrobenzofuran	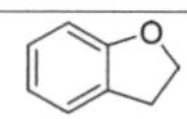	Benzofuran	Inhibit Acetylcholin-esterase (AChE) enzyme activity	Stored grain pests, houseflies	[34]

Plant species	Compound	Chemical structure	Class	Mode of action	Target insect/pest	References
Eucalyptus globulus Labill.	1,8-Cineole (Eucalyptol)	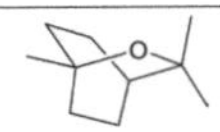	Monoterpene	Inhibit Acetylcholin-esterase (AChE) enzyme activity	Head louse (*Pediculus humanus capitis*)	[35]
Chenopodium ambrosioides L.	δ-3-carene	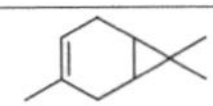	Monoterpene	Inhibit Acetylcholin-esterase (AChE) enzyme activity	*Tribolium confusum*, *Callosobruchus maculatus*	[36]
Nicotiana tabacum, *Stemona japonicum*, *Haloylon salicornicum*	Nicotine	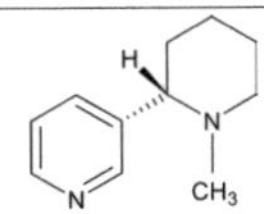	Pyridine alkaloid	Interact with the nicotinic acetylcholine (nACh) receptors at neuron synapses, resulting unregulated nerve firing causing malfunctions of the neurons.	Leaf hoppers, aphids, thrips, mites, spider mites	[26]
Artemisia absinthium, *Juniperus* sp., *Cedrus* sp.	Thujone	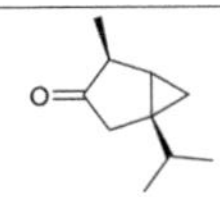	Monoterpene	Reversible modulator of $GABA_A$ receptor	Root worm larvae (*Diabrotica virgifera*), Fruit fly	[37]
Origanum vulgare	Carvacrol	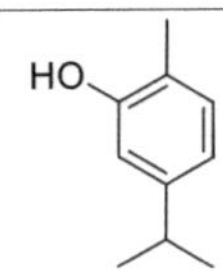	Monoterpene	Binds to nicotinic acetylcholine receptors	American cockroach (*Periplaneta Americana*)	[38]
Thymus vulgaris	Thymol	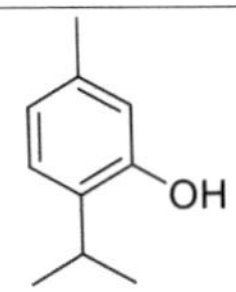	Monoterpene	Prevent octopamine receptors via tyramine receptor cascade.	American cockroach (*Periplaneta Americana*)	
Syzygium aromaticum	Eugenol	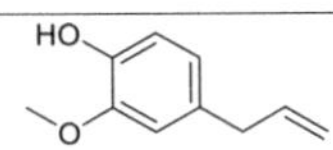	Monoterpene	Inhibits octopamine receptor (OAR) system	Bug (Triatoma infestans), fruit fly, American cockroach	[39–41]

Plant species	Compound	Chemical structure	Class	Mode of action	Target insect/pest	References
Aniba rosaeodora	Linalool	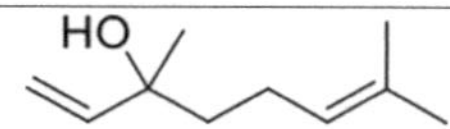	Monoterpene	Competitive and reversible inhibition of acetyl-cholinesterase activity	Fleas, houseflies, mosquitoes, spider mites	[42, 43]
Lonchocarpus Kunth, *Derris* Lour, *Rhododendron* L.	Rotenone	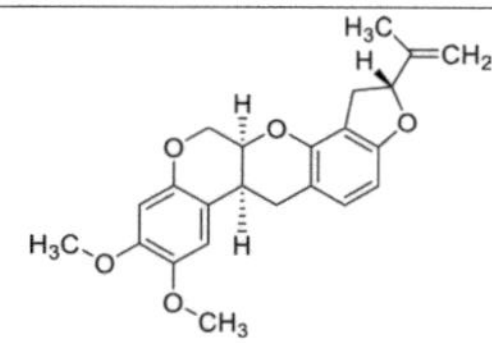	Isoflavonoid/ Rotenoid	Interact with NADH dehydrogenase resulting inhibit the oxidation of NADH to NAD	Caterpillars, lice, fleas, beetles, mosquitoes, fire ants	[26, 44]
Azadirachta indica	Azadirachtin	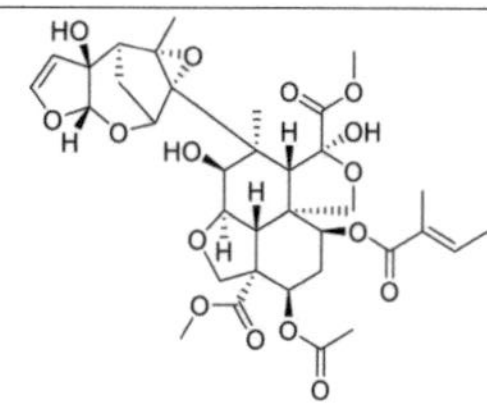	Tetranortriterpenoid limonoid	Acts as antifeedant, sterilant and repellent; disrupt insect growth by blocking the release of prothroracicotropic hormone (PTTH); distort phagostimulant disruptor by cholinergic transmission.	Stored grain pests, aphids, caterpillars, thrips, merely bugs	[30, 45–49]
Ageratum conyzoides	Precocene I (7-methoxy-2,2-dimethylchromene)	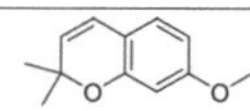	Chromene	Working as insect growth regulator	Grain beetle, Milkweed bug, Noctuid moth, Parasitic wasp	[50–53]
Catharanthus roseus	α-amyrin acetate	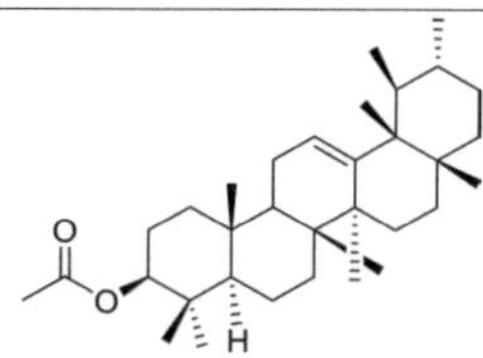	Steroid	Working as insect growth regulator	*Helicoverpa armigera*	[54]

Plant species	Compound	Chemical structure	Class	Mode of action	Target insect/pest	References
Xylopia aethiopica	(-) Kau-16-en-19-oic acid	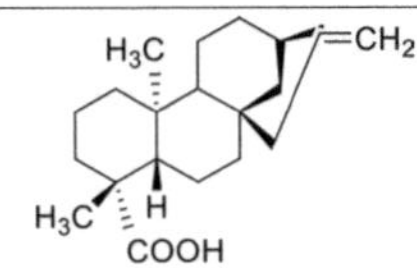	Kaurane diterpene	Antifeedant	Termites (*Reticulitermes speratus*)	[55]
Jatropha podagrica	15-epi-4E-jatrogrossi-dentadione	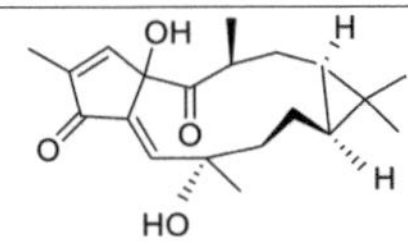	Diterpene	Antifeedant	Moth (*Chilo partellu*)	[45]
Senecio palmensis	11-Acetoy-5-isobutyryloysilphinen-3-one	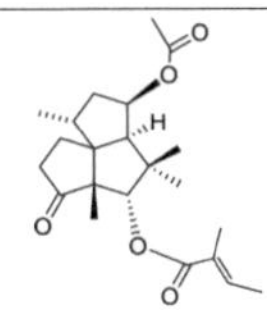	Silphinene sesquiterpene	Antifeedant	Colorado potato beetle, Aphids	[56]
Lotus corniculatus; *Trifolium repens*	Linamarin	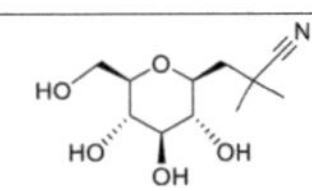	Cyanogenic glycoside	Antifeedant	Snails, slugs, lemmings, aphids	[57]
Pterocarpus macrocarpus	(-)-homoptreocarpin	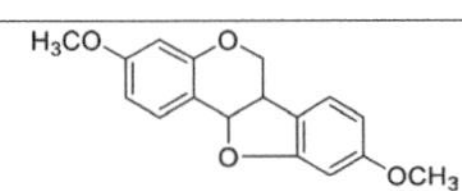	Isoflavonoid/ Pterocarpan	Antifeedant	Common cutworm, subterranearn termite	[58]
Nepeta cataria	Nepetalactone	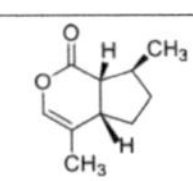	Monoterpene lactone	Repellent	Beetle, bees, flies, termites, lady beetle, mosquitoes	[59–61]
Tectona grandis	Tectoquinone	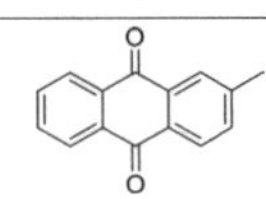	Anthraquinone	Repellent	Termites	[62–64]

DOI: http://dx.doi.org/10.5772/intechopen.1006743

Plant species	Compound	Chemical structure	Class	Mode of action	Target insect/pest	References
Alstonia boonei	Alstonine	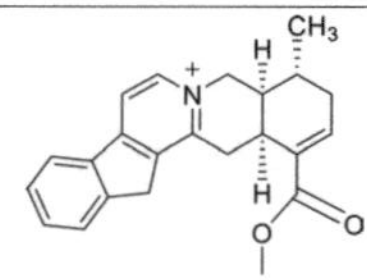	Indoloquinolizidi ne alakaloid	Repellent, larvicidal	Mosquitoes	[65]
Pinus strobes	Limonene	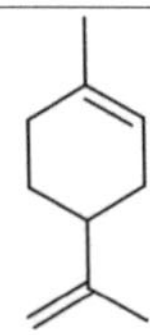	Monoterpene	Attractant	White pine cone beetle	[66]
Zea mays	(E)-β-caryophyllene	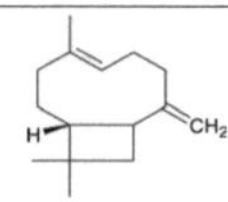	Sesquiterpene	Attractant	Nematodes (*Heterorhabditis megidis*), corn root worm (*Diabrotica virgifera*)	[67]
Juglans nigra	Juglone	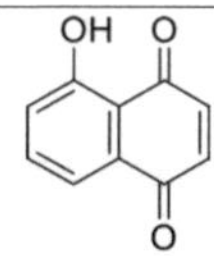	Naphtoquinone	Working as herbicide, inhibiting certain enzymes needed for the photosynthesis.	*Echinochloa crus-galli, Amaranthus retroflexus, Abutilon theopharasti*	[68, 69]
Eupatorium adenophorum	5,6-dihydroxycadinan-3-ene-2,7-dione	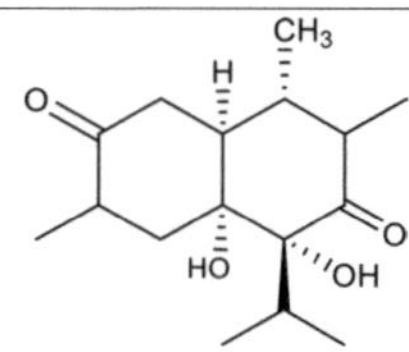	Cadinane sesquiterpene	Working as herbicide	*Arabidopsis thaliana*	[70]

Plant species	Compound	Chemical structure	Class	Mode of action	Target insect/pest	References
Poaceae spp.	m-Tyrosine	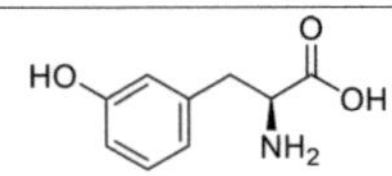	Amino acid	Working as herbicide; inhibit root growth by direct interference with amino acid metabolism, inhibit cell wall synthesis, and alter the plant hormone signalling.	Weeds	[71]
Prosopis juliflora	Tryptophan	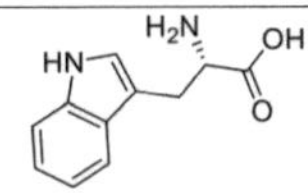	Amino acid	Working as herbicide	Barnyard grass (*Echinochloa crus-galli*)	[72]
Centaurea stoebe	(-)-Catechin	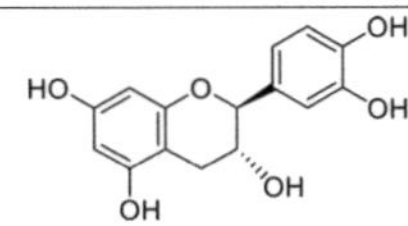	Flavanol	Working as herbicide; inhibits seed germination and root growth by rapid Ca^{2+} signalling cascade leading to cell death	*Koeleria macrantha*, and *Festuca idahoensis*	[73]
Cymbopogon citratus	Citronellol	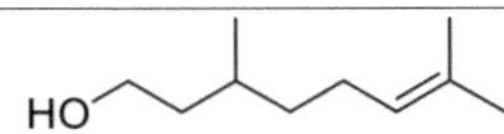	Monoterpene	Herbicide; showed toxic effects on plant growth by reducing chlorophyll and protein contents.	*Echinochloa crus-galli, Amaranthus tricolor*	[74]

Table 2.
Botanical insecticides with their origin and insecticidal activity.

2.1 Botanical insecticides affecting the nervous system

Researches suggests that plant-based insecticides can exhibit various modes of action in insects and mammals, particularly targeting the nervous system. These insecticides can affect γ-aminoutyric acid (GABA) gated chloride channels, octapamine receptors, nicotinic acetyl-choline receptors (nAChR) and acetylcholinesterase enzyme (AChE), voltage-gated sodium channels, and voltage-gated calcium channels.

2.1.1 Inhibition of GABA (γ-aminoutyric acid)-gated chloride channels

GABA is an inhibitory neurotransmitter present in the nervous systems of nearly all insect species, and it is the primary targets for pesticides, including bioinsecticides [75]. On binding of GABA with its receptors, the chloride (Cl^-) channels are activated, allowing chloride ions to flow into neurons, which leads to inhibition of the nervous system [76]. The active compounds present in various bio-insecticides can either inhibit or bind to the GABA receptors, resulting to the insect's death by hyper-excitation or by disturbance in neuronal activity [38].

The thujone found in *Artemesia absinthium* L. and the picrotoxin from *Anamirta cocculus* L. can block GABA-gated chloride channels, leading to reduced neuronal inhibition, hyper-excitation of the nervous system and potentially resulting in convulsions and death. Previous studies have indicated that thujone acts as a positive allosteric modulator of GABAA receptors and as a competitive inhibitor of [3H]Ethynylbicycloorthobenzoate ([3H]EBOB) binding [37]. Moreover, the GABA receptors may be inhibited by monoterpenoids carvacrol, pulegone, and thymol through [3H]EBOB binding [38]. Likewise, botanical compounds silphinene sesquiterpenes inhibit the aminobutyricacid (GABA) receptors by stabilizing non-conducting conformations of the chloride channels [26, 37]. GABA is considered as endogenous ligand that stimulates feeding and triggers taste cell responses in many plant-eating insects and suggests that compounds which inhibit GABA receptors could serve as antifeedants or deterrents, particularly affecting aphids, lepidopterans, and beetles [56, 77].

2.1.2 Inhibition of octopamine receptors

Octopamine play crucial role as endogenous amines [78] in the nervous system of arthropods, including insects [79]. They are derived from tyrosine [79] and function as neurotransmitters, neurohormones, and neuromodulators [39, 80]. The receptors are found throughout the nervous systems of insects, both in the central nervous system (CNS) and peripheral nervous system (PNS) areas. The octopaminergic system comprises different types of octopamine receptors that are associated with various second messenger systems, playing a pivotal role in controlling physiological functions and behaviours [81–83]. In honeybees (*Apis mellifera*), octopamine can influence memory and learning processes and modulate muscle performance, lipid metabolism, heart rate, and respiration in insects [79]. For instance, the octopamine1 receptor influences the rhythmic contraction of the locust extensor-tibiae muscle by altering the levels of calcium inside the cell. On the other hand, the effects of the octopamine2A and octopamine2B receptors are mediated by activating adenylate cyclase. Additionally, the octopamine3 receptors are responsible for regulating the levels of intracellular calcium and cAMP in the central nervous system of the locust [39]. The quick response of monoterpenes against certain insect/pest implies a neurotoxic mechanism of action. Reynoso and colleagues [40] demonstrated that how eugenol can repel and kill the blood-sucking bug *Triatoma infestans* by activating the octopamine receptor.

According to previous researches, octopamine receptors have been found in a wide range of insects, such as fireflies, flies, nymphs, cockroaches, and lepidopterans [39–41]. Enan [41] research highlighted that eugenol's insecticidal effects are linked to the octopamine receptor system (OAR), emphasizing that phenolic compounds exhibit greater toxicity than other monoterpenoids. Furthermore, Enan pointed out that the positioning and spacing of a hydroxyl group (OH) on the benzene ring dictate the level of toxicity of the active compound found in essential oils through the OAR system. The monoterpenoids α-terpineol and cinnamic alcohol inhibit octopamine receptors in the species *Camponotus pennsylvanicus*, *Periplaneta americana*, and *Blattella germanica* at an inhibitory concentration (IC_{50}) of 9 nmol/mL [41].

Monoterpenoids can play another role in their interaction with OAR, such as menthol enhancing the toxic effects of carbamate, a synthetic pesticide, by activating OAR, thereby increasing the efficacy at a lower dosage than usual [84] and potentially reducing the negative environmental impact of synthetic pesticides. The octopamine receptors do not fit into the categories of receptors found in vertebrates. This means that octopamine receptor agonists could be a promising option for a commercial pesticide. It is crucial for the pesticides to be tailored to the specific target, have minimal impact on mammals, and employ a unique mode of action compared to other pesticides available.

2.1.3 Inhibition of voltage-gated sodium (Na^+2) channels (VGSCs)

Voltage-gated Na^+2 channels are vital transmembrane proteins that play a crucial role in transmitting electrical signals in various excitable cells, such as nerve, muscle, and neuroendocrine cell types, contributing to the establishment of resting potential as well as the initiation and spread of action potentials in neurons [85]. Pyrethrin, a secondary metabolite derived from the dried flowers of *Tanacetum cinerariifolium* (Asteraceae), contains two primary active components known as Pyrethrins I and II (cyclopropylmonoterpene esters) [24] that acts as a modulator of nerve membrane sodium current by prolonging the opening of VGSCs, resulting in over neuroexcitation, causing loss of coordinated movement control, paralysis, and death of insects. The same mode of action is followed by the other synthetic insecticides (synthetic pyrethroids derived from pyrethrins), but plant-based pyrethrins are more target-specific than synthetic one [86]. When pyrethrins are applied, they cause an immediate knock-down effect and have low toxicity to mammals, as well as very short residual activity, which prevents accumulation in food chains and ground water. Pyrethrins break down rapidly when they come into contact with air, sunlight, water, and high temperatures, which means that you might need to apply them frequently [87]. Pyrethrins are used to manage various insects and mites, such as spider mites, flies, mosquitoes, fleas, and beetles [25, 26].

Decalesides I and II, derived from *Decalepis hamiltonii* root, have a similar mode of action and are considered trisaccharides. These plant-based products are toxic to a variety of insects upon contact exposure [28]. Sabadilla, extracted from the pulverized seed of the sabadilla lily (*Schoenocaulon officinale* Schltdl. and Cham.), has been used by native American Indians as insecticide against various pests for centuries. The alkaloids produced in sabadila are collectively known as veratrine which contains two main alkaloids, namely cevadine and veratridine exist in a ratio of 2:1. The alkaloids have similar mechanism of action as pyrethrins have, however, the binding site seems to be different [30]. Sabadilla found least toxic plant-based insecticides effective either contact or ingestion and degrades rapidly when exposed to air and sunlight and has little residual toxicity. However, purified veratrine alkaloids are more toxic to mammals and effective against honeybees, caterpillars, leaf hoppers, thrips, stink bugs, and squash bugs [29, 31].

2.1.4 Inhibition of voltage-gated calcium ($Ca^{+}2$) channels (VGSCs)

Ryanodine, an alkaloid derived from the roots and woody stems of *Ryania speciosa* Vahl (Salicaceae), a plant native to South America, exhibits insecticidal properties by disrupting the $Ca^{+}2$ channels in skeletal muscle cells' sarcoplasmic reticulum. This interference leads to the release of excessive calcium ions into actin and myosin protein filaments, causing muscle contractions and eventual paralysis [88]. Ryanodine acts quickly as a poison, making it effective against insects through contact or ingestion. It has low toxicity for mammals and provides control for up to 2 weeks after the first application. When combined with piperonylbutoxide (PBO), Ryania crude extracts enhance their insecticidal activity against caterpillars, worms, potato beetles, lace bugs, aphids, and squash bugs [29, 33]. Ryania, a slow-acting stomach toxin, causes insects to cease consuming plants shortly after ingestion and exhibits greater residual activity than other plant-based insecticides, while posing less risk to humans.

2.1.5 Alteration the activity of acetylcholinesterases (AChEs) enzymes

The enzyme acetylcholinesterase (AChE) plays a crucial role in the nervous systems of most animals, including insects [89], by breaking down the neurotransmitter acetylcholine (ACh) into acetatic acid and choline [90], and enabling an activated cholinergic neuron to return to its resting state. It is important for regulating the nerve signals by preventing overstimulation of the postsynaptic neuron and also the target for most bioinsecticides and synthetic insecticides [91]. Germacrone, the primary bioactive compound derived from *Rhododendron thymifolium* L., inhibits the AchE activity by impeding cholinergic neurotransmission at synapses, leading to prolonged binding of the neurotransmitter to its postsynaptic receptor, resulting in heightened neuroexcitation, restlessness, hyper-excitability, paralysis, and ultimately death. It is found effective against post-harvest grain insects such as *Lasioderma serricorne* and *Tribolium castaneum* [92]. Similarly, coumaran, active compound extracted from *Lantana camara* L., exhibited AchE inhibition activity and found effective against houseflies and stored grain pests, in spite of its low toxicity to humans and short residual activity [93]. Recently, Li et al. [94], reported the AchE inhibition activity of new sequiterpenes; laggeranine extracted from *Laggera pterodonta* and proposed as new potential active compound for targeting AchE. The *in-vitro* evaluation of anti-AChE activity in many essential oils and their derived terpenes have been reported, but their contribution raises doubts about to insect mortality [93].

2.1.6 Activation of nicotinic acetylcholine receptors

The nicotinic acetylcholine receptors, located in the insect nervous system within the cell bodies of interneurons, motor neurons, and sensory neurons, have a crucial function in facilitating rapid excitatory neurotransmission in the central nervous system (CNS) of insects [95]. Nicotine, an alkaloid extracted from tobacco plants

(*Nicotiana tabacum* L), mostly contains phenolic compounds like nicotine and diterpene. Nicotine, nornicotine, and anabasine mimic the neurotransmitter acetylcholine and exert neurotoxic effect by binding nicotine acetyl-choline receptors (nAChRs), similar to organophosphate and carbamate insecticides [24]. Activation of nAChRs by nicotine results in a quick increase of sodium ions, leading to depolarization and ultimately generating an action potential. Generally, the synaptic action of acetylcholine is terminated by AChE; moreover, nicotine being resistant to hydrolysis by AChE leads to prolonged activation, resulting in excessive cholinergic transmission, which ultimately leads to convulsions, paralysis, and ultimately death [95]. Nicotine acts as a potent neurotoxin, specifically targeting insects and mites, making it the most toxic plant-derived compound, but also posing significant harm to humans [45]. Neonicotinoids, including imidacloprid, acetamiprid, and thiamethoxam, are a type of insecticide with a chemical structure resembling nicotine, featuring a heterocyclic ring structure with attached nitro and cyan groups. Similar to nicotine, neonicotinoids act as an activator of nicotinic acetylcholine receptors leading to over neuroexcitation and paralysis. Neonicotinoids showed to insect nAChRs and highly selectivity in toxicity, while being less harmful to non-target organisms like birds and mammals. The compounds are also easily dissolved in water, which enables them to be applied to soils and taken up by plants, ultimately enhancing their protective capabilities. These insecticides can act as contact or ingestion poisons, causing insects to stop feeding shortly after contact and leading to death, however, exhibited negative impact on pollinators like bees [96].

2.2 Inhibition of respiratory enzyme system

Cellular respiration is the conversion of nutrient compounds into cellular energy, i.e. adenosine triphosphate (ATP) at a molecular level, specifically carried out by the electron transport chain in the mitochondria, which includes important enzymes that could be targeted by insecticides. Rotenone, a type of isoflavonoid extracted from roots and stems of *Derris* (*Derris elliptica, D. involute*), *Lonchocarpus* (*Lonchocarpus utilis, L. urucu*), and *Tephrosia virginiana* [24], used as a botanical insecticide.

Rotenone works as both contact and ingestion toxin by inhibiting ubiquinone oxidoreductase (Complex I) of the electron transport chain in mitochondria. Rotenone inhibits the nicotinamide adenine dinucleotide (NADH) dehydrogenase activity and stops electrons flow from NADH to coenzyme Q, so prevents formation of ATP from NADH resulting ATP levels fall rapidly leads to cell death. By inhibiting Complex I, it also increases the levels of reactive oxygen species (ROS) and nitrogen oxide (NO), resulting in the production of free radicals and subsequent lipid peroxidation in cellular components, ultimately leading to programmed cell death (apoptosis). Rotenone has a slow effect and can be broken down by air and sunlight. It takes several days to kill insects by affecting their nerve and muscle cells, leading to them stopping feeding and eventually dying within hours to a few days of exposure. Furthermore, this plant-based insecticide has a broad spectrum activity and found effective against a wide range of insect species including caterpillars, aphids, suckers, trips, and other pests infecting fruits and vegetables [97], including the Colorado potato beetle (*Leptinotarsa decemlineata*), Plum curculio (*Conotrachelus nenuphar*), *Diabrotica,* and *Acalymma* species [98]. Rotenone is moderately harmful to mammals and extremely toxic to fish, exhibiting activity and persistence comparable to dichlorodiphenyltrichloroethane (DDT). Research conducted in the past has indicated a potential association between rotenone exposure and Parkinson's disease (PD). The rapid exposure of rats led to brain damage similar to the lesions found in humans and animals affected by Parkinson's disease [99]. Rotenoids, despite their high toxicity, show potential as a source of new complex I inhibitors and could serve as a model for developing safer and more effective pesticide derivatives.

2.3 Alteration of the hormonal system

Insect growth regulators (IGRs) encompass chemical compounds that interfere with the endocrine system of insects, essentially mimicking or inhibiting their juvenile hormones and acting as chitin synthesis inhibitors (CSIs). Insects typically release juvenile hormones to sustain their immature state, and when they attain adequate growth, the production of hormones stops, causing them to move into the adult phase. Triterpenes such as α-amyrin acetate and oleanolic acid extracted from *Catharanthus roseus* exhibited excellent growth regulator activity [54]. In addition, natural products like davanone, ipomearone, and the juvenile hormone produced by silkworm exhibit fascinating IGR activity due to their acyclic sesquiterpene composition. By consistently application of IGR to the crops, the insects will remain in their larvae state, preventing successful moulting and effectively controlling insects. The presence of Chromenes, Precocene I and II extracted from *Ageratum conyzoides* L. (Asteraceae) leads to the early transformation of the larvae and results in the production of sterile, moribund, and dwarfish adults upon exposure [100].

Azadirachtin, a complex tetranortriterpenoid limonoid, mainly extracted from the seeds of *Azadirachta indica* (Meliaceae) interacts with ecdysteroids present in insects bodies and reduces their population [101]. The neem tree, scientifically known as *Azadirachta indica,* is native to the arid regions of India and Myanmar but is now cultivated in arid, tropical, and subtropical areas across Southeast Asia, Africa, the Americas, and Australia [26, 30, 46]. Azadirachtin showed contact systemic activity which can be classified in two ways: either through direct impact on cells and tissues or indirect effects, which involve interference with the endocrine system. This compound is extremely active and functions as a deterrent for feeding and as a regulator of insect growth in a wide variety of insect groups, including Lepidoptera, Diptera, Hemiptera, Orthoptera, and Hymenoptera [46]. The growth regulatory effects of Azadirachtin involve impacting the neurosecretory system of insects and preventing the release of crucial metamorphosis hormones like prothoracicotropic hormone (PTTH) and allatostatins, which control the prothoracic glands and corpora allata. Consequently, any disruption in the functioning of these glands could result in molting defects or sterility. Azadirachtin directly affects somatic and reproductive tissues, as well as disrupts the endocrine processes, leading to impacts on feeding, developmental, and reproductive functions [46]. Neem oil, derived from cold-pressing of neem seeds and other neem-based products, are commonly utilized as non-commercial products to control insect, pest, and phytopathogens [102, 103]. Both azadirachtin and neem seed oil have been found to elevate aphid nymphal mortality rates by 80 and 77%, respectively, while also prolonging the development time of those that manage to survive to adulthood [104].

2.4 Alteration of the water balance

Certain oils derived from rice bran, cotton seed, and palm kernel, along with natural soaps (saponins), have the ability to interfere with the protective wax layer on insects, resulting in rapid water loss from the cuticle and ultimately leading to the insects' death from dehydration [105]. Furthermore, the crude oils can also disrupt insect respiration by clogging the openings known as spiracles, leading to death by suffocation, thus effectively managing various types of insects like whiteflies, mites, caterpillars, leafhoppers, and beetles [106].

2.5 Repellent

Some groups of pest control agents have unique biological properties and are closely associated with insecticidal substances, often being employed alongside insecticides in pest control tactics. They include insect repellents, deterrents, and attractants. Plant-based compounds can act as both an insecticide and repellent, depending on their concentration. The key distinction is that repellents do not eliminate insects; they instead keep them at bay by emitting strong odours or exhibiting mild toxic properties. Recent research on plants showing repellent properties has resulted in the isolation and identification of certain effective compounds. Terpenoids and sesquiterpenoid extracted from *Cymbopogon nardus* such as callicarpenal and intermedeol, respectively, significantly control the infestations by *Amblyomma cajennense* [107]. Nepetalactone (monoterpene lactone), an active compound found in catnip oil extracted from *Nepeta cataria* L. (Lamiaceae), showed repellent activity against *Aedes aegypti* L. (Culicidae) ten times more than DEET [59] and also reported as repellent against lady beetle, cockroaches, flies, termites, and mosquitoes [60, 61]. Monoterpens, the main constituent of many essential oils, showed powerful insect repellent activity. The essential oils extracted from lemongrass (*Cymbopogon flexuosus* (Poaceae)), eucalyptus (*Eucalyptus globulus (*Myrtaceae)), rosemary (*Rosmarinus officinalis* L. (Lamiaceae)), vetiver (*Vetiveria zizanioides* L. (Poaceae)), clove (*Eugenia caryophyllus* (Myrtaceae)), and thyme (*Thymus vulgaris* L. (Lamiaceae)) considered a highly effective repellent against a wide range of insects [108]. The anthraquinone tectoquinone showed repellent activity against termites [62, 63], and alstonine alkaloid has a repellent and larvicidal activity against *Anopheles gambiae* (Culicidae) [65]. Adegorite et al. [36] demonstrated that essential oil from *Chenopodium ambrosioides* L (Chenopodiaceae) can effectively repel cowpea adult insect *Callasobruchus maculatus* L. (Coleoptera: Bruchidae), leading to mortality rates that are dependent on the concentration and duration of contact. When test insects were placed in a sealed glass jar containing the leaf extracts, they experienced a 100% mortality rate due to prolonged exposure to high concentrations of the extracts.

2.6 Attractants

Plants release attractants, which are semio-chemicals or communication compounds, to draw in insects or natural predators of the insects that feed on the plant [109]. These compound influence both taste and smell receptors. For controlling insects, attractants can be utilized in three ways: monitoring infestations, luring insects to traps or baits, and disrupting their mating and feeding behaviours. As they do not directly kill insects, they are environmentally friendly and can be used to mislead insects into laying infertile eggs or choosing unfavourable oviposition sites, reducing their numbers. In an experiment, Miller [66] applied (−) and (+) limonene from white pine (*Pinus strobus* L. (Pinaceae)) to attract the white pine cone beetle, *Conophthorus coniperda* Schwarz (Curculionidae), as well as the attraction of the predator beetle, *Enoclerus nigripes* say (Cleridae), through the release of

(−)-α-pinene, as well as the sesquiterpene caryophyllene [67]. However, they should only be used as part of a comprehensive pest management strategy and not as the sole control measure.

2.7 Antifeedant

Antifeedant compounds take part in chemoreception mechanism that involves either blocking receptors that typically respond to phagostimulants or stimulating deterrent cells. These compounds discourage insects from feeding, and it is believed that plants with high levels of these compounds are less likely to be attacked by insects [110]. The antifeedant activity of azadirachtin and neem seed extracts against numerous pest is well established. Besides, numerous terpenes, terpenoids, phenolics, and flavonoids compounds also exhibited strong antifeedant activity [111, 112]. The research conducted earlier showed that phenolic compounds hinder cellular enzymes like proteases and other digestive hydrolases, as well as polyphenol oxidases (PPOs), which are involved in multiple metabolic processes [113], leading to a reduced digestibility of dietary proteins [114].

The multiple antifeedant activates of plant extract and fractions from *Calceolaria integrifolia* were attributed to inhibition of phenol oxidase, proteinase, or tyrosinase, resulting inhibition of cuticle synthesis and to molting sclerotization toxicity [115]. In a study, Qiao et al. [47] demonstrated that azadirachtin reduces the cholinergic transmission of neurons related to the suboesophageal ganglion (SOG) of *Drosophyla melanogaster*, which are strongly related to feeding behaviour. Furthermore, the ingestion of food may be reduced because of its poisonous impact following the initial consumption (secondary deterrent), encouraging astringency, unpleasant taste, or anti-digestive effect for herbivores [48, 55]. In an experiment, Chakraborty et al. [116] tested the extract from bitter gourd (*Momordica charantia* L.) against transgenic flies with impaired aversive taste sensitive neurons and demonstrated a decreased aversion when exposed to bitter gourd extract, indicating that the bitter-sensitive gustatory neurons depend on these compounds. Abdullah et al. [117] reported that 1,8-cineol, an active compound present in Galangal essential oil, showed good antifeedant and repellent activity, as well as toxicity effect against termites. Similarly, Jose and Sujatha [118] demonstrated that terpenoids, coumarin, and phenols, found in the methanol extracts of *Gliricidium sepium,* exhibited excellent antifeedant activity. The plant-derived active compounds can either inhibit larval feeding by disrupting hormonal balance or making the food unpalatable, or they can directly affect the larvae's chemosensilla, resulting in feeding deterrence. Furthermore, despite the presence of many natural plant compounds that act as antifeedants, their commercial use in crop protection is limited due to insect habituation to feeding deterrents [119].

2.8 Phototoxins

Photoactive compounds or phototoxins, a type of phytochemicals, maintain their effectiveness by being activated through two different mechanisms instead of being degraded by sunlight. In the less common mechanism, molecular oxygen from the phototoxin absorbs light energy, producing reactive oxygen species that ultimately harm essential cellular biomolecules [120]. The alternative action involves photogenotoxicity, in which plant compounds are activated by sunlight, causing DNA damage without needing oxygen. This process occurs when the phototoxin absorbs sunlight energy in its ground state, becomes excited, and interacts with ground state O2 in the target tissue, producing reactive oxygen species and exhibiting insecticidal effects. This distinctive mechanism of phototoxins differs from conventional chemical pesticides and makes it unlikely for them to develop cross-resistance [121, 122].

Various classes of plant-derived active compounds such as quinones, furanocoumarins, and thiophenes exhibited phototoxic effect. Phototoxins, 3-methyl-3-phenyl-1,4-pentadiyne found in the essential oil extracted from *Artemisia monosperma* (Asteraceae), exhibited an activity similar to DDT against the housefly and cotton leaf worm larvae under sunlight exposure. Similarly, ponticaepoxide, an acetylenic epoxide derived from *Artemisia pontica* L. (Asteraceae), inhibited mosquito larvae growth at LC_{50} of 1.47 ppm when subjected to UV light [123]. Thiophene also known as terthiophene or α-terthienyl, the main active compound found in the roots of *Tagetes minuta* L. (Asteraceae), showed high toxicity against various organisms including nematodes: red flower beetles, blood-feeding insects such as tobacco hornworm (Sphingidae) and mosquito larvae (dipteres): yellow fever mosquito (Culicidae), rockpool mosquito (Culicidae), when subjected to near-UV radiation [124]. However, till now no commercial product based on phototoxic activity has been formulated.

3. Conclusion

Sustainable agricultural development aims to enhance the quantity and quality of crop products while ensuring human health and reducing emission of toxic substances in the environment. The efficient utilization of plant-based products in plant protection is crucial to attain this goal. Standard crop protection practices lead to adverse effects, including issues related to environmental toxicity, disturbance of the natural equilibrium, and enhancing the development of resistance in insect. Therefore, botanical insecticides considered as the best alternative source of compounds having insecticidal activity and can help to protect crops from the insect and pests. Utilization of plant-based insecticides has several advantages, including insect selectivity, less toxic to humans, and no and non-bio-accumulative in the nature. Researchers and industry stakeholders are increasingly interested in exploring new bioinsecticides with different mechanisms of action and investigating the efficacy of these natural alternatives. Botanical insecticides are derived from various types of plant secondary metabolites including alkaloids, flavonoids, terpenes, essential oils, glycosides, esters, etc. These metabolites exhibited different physical/chemical characteristics and mechanisms of action and affect the insects in various ways such as insecticides, repellents, antifeedants, toxicants, growth retardants, herbicides, and attractants. Nevertheless, additional research is urgently required to investigate the impact on pest behaviour, the significance of applied methods, and the timing of application for these botanical biopesticide mixtures.

Author details

Sorabh Chaudhary[1], Sunil Kumar Yadav[2*], Pankaj Verma[3], Sushma Sagar[4] and Mehi Lal[5]

1 Department of Biology, SRM Institute of Science and Technology, Delhi-NCR Campus, Ghaziabad, Uttar Pradesh, India

2 Department of Chemistry, SRM Institute of Science and Technology, Delhi-NCR Campus, Ghaziabad, Uttar Pradesh, India

3 Department of Biotechnology, IILM University, Greater Noida, Uttar Pradesh, India

4 Division of Vegetable Sciences, ICAR-Indian Agricultural Research Institute, Pusa, New Delhi, India

5 Division of Plant Protection, ICAR-Central Potato Research Institute-Regional Station, Meerut, Uttar Pradesh, India

*Address all correspondence to: skyhbtu@gmail.com

References

[1] Gerland P, Raftery AE, Ševcíková H, Li N, Gu D, Spoorenberg T, et al. World population stabilization unlikely this century. Science. 2014;**346**:234-237

[2] United Nations. World Population Prospects: The 2012 Revision. Population Division, Department of Economic and Social Affairs; 2013. Available from: https://www.un.org/en/development/desa/publications/world-population-prospects-the-2012-revision.html

[3] Chuadhary S, Lal M, Sagar S, Tyagi H, Kumar M, Shrama S, et al. Genetic diversity studies based on morpho-pathological and molecular variability of the *Sclerotinia sclerotiorum* population infecting potato (*Solanum tuberosum* L.). World Journal of Microbiology and Biotechnology. 2020;**36**:177

[4] Rahul SN, Khilari K, Sagar S, Chaudhary S, Kumar S, Vihan N, et al. Challenges in postharvest management of fungal diseases in fruits and vegetables-a review. South Asian Journal of Food Technology and Environment. 2015;**1**:126-130. DOI: 10.46370/sajfte.2015.v01i02.04

[5] Kumar S. Biopesticides: A need for food and environmental safety. Journal of Biofertilizers and Biopesticides. 2012;**3**(4):1-3

[6] Lal M, Kumar A, Chaudhary S, Singh RK, Sharma S, Kumar M. Antagonistic and growth enhancement activites of native pseudomonas spp. against soil and tiuber-borne diseases of potato (*Solanum tuberosum* L.). Egyptian Journal of Biological Pest Control. 2022;**32**:22

[7] Cycon M, Mrozik A, Piotrowska-Seget Z. Bioaugmentation as a strategy for the remediation of pesticide-polluted soil: A review. Chemosphere. 2017;**172**:52-71

[8] Gevao B, Semple KT, Jones KC. Bound pesticide residues in soils: A review. Environmental Pollution. 2000;**108**:3-14

[9] Meite F, Alvarez-Zaldívar P, Crochet A, Wiegert C, Payraudeau S, Imfeld G. Impact of rainfall patterns and frequency on the export of pesticides and heavy-metals from agricultural soils. The Science of the Total Environment. 2018;**616-617**:500-509

[10] Salazar-Ledesma M, Prado B, Zamora O, Siebe C. Mobility of atrazine in soils of a wastewater irrigated maize field. Agriculture, Ecosystems and Environment. 2018;**255**:73-83

[11] Chowdhury A, Pradhan S, Saha M, Sanyal N. Impact of pesticides on soil microbiological parameters and possible bioremediation strategies. Indian Journal of Microbiology. 2008;**48**:114-127

[12] Hartmann M, Frey B, Mayer J, Mäder P, Widmer F. Distinct soil microbial diversity under long-term organic and conventional farming. The ISME Journal. 2015;**9**:1177-1194

[13] Komorowicz I, Gramowska H, Barałkiewicz D. Estimation of the lake water pollution by determination of 18 elements using ICP-MS method and their statistical analysis. Journal of Environmental Science and Health Part A. 2010;**45**:348-354

[14] Damin V, Trivelin P. Herbicide's effect on nitrogen cycling in agroecosystems. In: Herbicides and Environment. London, UK: InTech Open; 2011

[15] Malik Z, Ahmad M, Abassi GH, Dawood M, Hussain A, Jamil M. Agrochemicals and soil microbes: Interaction for soil health. In: Hashmi MZ, Kumar V, Varma A, editors. Xenobiotics in the Soil Environment, Soil Biology. Springer International Publishing; 2017. pp. 139-152

[16] Verma JP, Jaiswal DK, Meena VS, Kumar A, Meena RS. Issues and challenges about sustainable agriculture production for management of natural resources to sustain soil fertility and health. Journal of Cleaner Production. 2015;**107**:793-794

[17] Zhang D, Yan M, Niu Y, Liu X, van Zwieten L, Chen D, et al. Is current biochar research addressing global soil constraints for sustainable agriculture? Agriculture, Ecosystems and Environment. 2016;**226**:25-32

[18] Bissinger BW, Roe RM. Tick repellents: Past, present, and future. Pesticide Biochemistry and Physiology. 2010;**96**:63-79

[19] Fountain ED, Wratten SD. Conservation biological control and biopesticides in agricultural. In: Fath B, editor. Reference Module in Earth Systems and Environmental Sciences. 2nd ed. Vol. 1. New Zealand: Elsevier; 2013. pp. 377-381

[20] Pavela R, Benelli G. Essential oils as eco-friendly biopesticides? Challenges and constraints. Trends in Plant Science. 2016;**21**:1000-1007

[21] Copping LG, Menn JJ. Biopesticides: A review of their action, applications, and efficacy. Pest Management Science. 2000;**56**(8):651-676

[22] Chandler D, Bailey AS, Tatchell GM, Davidson G, Greaves J, Grant WP. The development, regulation, and use of biopesticides for integrated pest management. Philosophical Transactions of the Royal Society of London. Series B, Biological Sciences. 2011;**366**(1573):1987-1998

[23] Pavela R. History, presence and perspective of using plant extracts as commercial botanical insecticides and farm products for protection against insects—A review. Plant Protection Science. 2016;**52**:229-241

[24] Isman M. Botanical insecticides, deterrents, and repellents in modern agriculture and an increasingly regulated world. Annual Review of Entomology. 2006;**51**:45-66

[25] Casida J, Quistad G. Golden age of insecticide research: Past, present, or future? Annual Review of Entomology. 1998;**43**:1-16

[26] Rattan RS. Mechanism of action of insecticidal secondary metabolites of plant origin. Crop Protection. 2010;**29**:913-920

[27] Singh D, Agarwal SK. Himachalol and β-himachalene: Insecticidal principles of himalayan cedarwood oil. Journal of Chemical Ecology. 1988;**14**:1145-1151

[28] Rajashekar Y, Rao L, Shivanandappa T. Decaleside: A new class of natural insecticide targeting tarsal gustatory sites. Die Naturwissenschaften. 2012;**99**:843-852

[29] Shivanandappa T, Rajashekar Y. Mode of action of plant-derived natural insecticides. In: Singh D, editor. Advances in Plant Biopesticides. 1st ed. New Delhi, India: Springer; 2014. pp. 323-345

[30] Copping LG, Duke SO. Natural products that have been used

commercially as crop protection agents. Pest Management Science. 2007;**63**:524-554

[31] Hare J, Morse J. Toxicity, persistence, and potency of sabadilla alkaloid formulations to citrus thrips (Thysanoptera: Thripidae). Journal of Economic Entomology. 1997;**90**:326-332

[32] Ujváry I. Pest control agents from natural products. In: Krieger R, editor. Hayes Handbook of Pesticide Toxicology. 3rd ed. New York, NY, USA: Academic Press; 2010. pp. 119-229

[33] Bloomquist JR. Insecticides: Chemistries and characteristics. In: Radcliffe's IPM World Textbook. 2nd ed. St. Paul, MI, USA: University of Minnesota; 1999

[34] Rajashekar Y, Raghavendra A, Bakthavatsalam N. Acetylcholinesterase inhibition by biofumigant (Coumaran) from leaves of *Lantana camara* in stored grain and household insect pests. Biomedical Research. 2014;**2014**:187019

[35] Picollo MI, Toloza AC, Mougabure CG, Zygadlo J, Zerba E. Anticholinesterase and pediculicidal activities of monoterpenoids. Fitoterapia. 2008;**79**:271-278

[36] Adegorite SO, Aladesida AA, John IB, Kayoed-Isola TM. Bioefficacy of *Chenopodium ambrosioides* L. (Chenopodiaceae) on adult *Callosobruchus maculatus* F. (Coleoptera: Bruchidae). World Journal of Biology Pharmacy and Health Sciences. 2024;**17**(2):192-200

[37] Höld K, Sirisoma N, Ikeda T, Narahashi T, Casida J. α-Thujone (the active component of absinthe): γ—Aminobutyric acidtype a receptor modulation and metabolic detoxification. Proceedings of the National Academy of Sciences of the United States of America. 2000;**97**:3826-3831

[38] Tong F, Coats JR. Effects of monoterpenoid insecticides on [3H]-TBOB binding in house fly GABA receptor and 36Cl—Uptake in American cockroach ventral nerve cord. Pesticide Biochemistry and Physiology. 2010;**98**:317-324

[39] Kostyukovsky M, Rafaeli A, Gileadi C, Demchenko N, Shaaya E. Activation of octopaminergic receptors by essential oil constituents isolated from aromatic plants: Possible mode of action against insect pests. Pest Management Science. 2002;**58**:1101-1106

[40] Reynoso MMN, Lucia A, Zerba EN, Alzogaray RA. The octopamine receptor is a possible target for eugenol-induced hyperactivity in the blood-sucking bug *Triatoma infestans* (Hemiptera: Reduviidae). Journal of Medical Entomology. 2019;**57**:627-630

[41] Enan E. Molecular and pharmacological analysis of an Octopamine receptor from American cockroach and fruit fly in response to plant essential oils. Archives of Insect Biochemistry and Physiology. 2005;**59**:161-171

[42] Perry NSL, Houghton PJ, Theobald A, Jenner P, Perry EK. In-vitro inhibition of human erythrocyte acetylcholinesterase by *salvia lavandulaefolia* essential oil and constituent terpenes. The Journal of Pharmacy and Pharmacology. 2000;**52**:895-902

[43] Ryan MF, Byrne O. Plant-insect coevolution and inhibition of acetylcholinesterase. Journal of Chemical Ecology. 1988;**14**:1965-1975

[44] Tooley P. Crop protection. In: Food and Drugs. London, UK: John Murray Albermark Street; 1971

[45] Okwute SK. Plants as potential sources of pesticidal agents: A review. In: Soundararajan RP, editor. Pesticides—Advances in Chemical and Botanical Pesticides. 1st ed; eBook. London, UK: IntechOpen; 2012. pp. 207-232

[46] Mordue AJ, Nisbet AJ. Azadirachtin from the neem tree *Azadirachta indica*: Its action against insects. Anais da Sociedade Entomológica do Brasil. 2000;**29**:615-632

[47] Qiao J, Zou X, Lai D, Yan Y, Wang Q, Li W, et al. Azadirachtin blocks the calcium channel and modulates the cholinergic miniature synaptic current in the central nervous system of drosophila. Pest Management Science. 2014;**70**:1041-1047

[48] Mordue AJ, Blackwell A. Azadirachtin: An update. Journal of Insect Physiology. 1993;**39**:903-924

[49] Inwood M. Azadirachtin, a scientific gold mine. Bioorganic and Medicinal Chemistry. 2009;**17**:4096-4105

[50] Saleem MA, Wilkins RM. Precocene-1: An anti-juvenile hormone, a potential 4th generation insecticide against a malathion-resistant strain of *Oryzaephilus surinamensis* (L.). Pakistan Journal of Zoology. 1984;**16**:195-201

[51] Singh S, Kumar K. Anti-JH compounds and insect pest management. In: Srivastava UC, editor. Emerging Trends in Zoology; eBook. Delhi, India: Narendra Publishing House; 2011. pp. 335-350

[52] Srivastva S, Kumar K. Precocene I and II induced metamorphosis in a noctuid moth, *Spodoptera litura* Fabr. Proceedings of the National Academy of Sciences of the United States of America. 1997;**67**:213-226

[53] Khafagi W. Effects of juvenile hormone I, precocene I and precocene II on the progeny of *Microplitis rufiventris* Kok. Female when administered via its host, *Spodoptera littoralis* (Boisd.). Journal of Applied Entomology. 2004;**128**:6-10

[54] Singh D, Mehta SS, Neoliya NK, Shukla YN, Mishra M. New possible insect growth regulators from *Catharanthus roseus*. Current Science. 2003;**84**:1184-1186

[55] Lajide L, Escoubas P, Mizutani J. Termite antifeedant activity in *Xylopia aethiopica*. Phytochemistry. 1995;**40**:1105-1112

[56] González-Coloma A, Valencia F, Martín N, Hoffmann JJ, Hutter L, Marco JA, et al. Silphinene Sesquiterpenes as model insect antifeedants. Journal of Chemical Ecology. 2002;**28**:117-129

[57] Nahrstedt A. Cyanogenic compounds as protecting agent for organisms. Plant Systematics and Evolution. 1985;**150**:35-47

[58] Morimoto M, Fukumoto H, Hiratani M, Chavasiri W, Komai K. Insect antifeedants, pterocarpans and pterocarpol, in heartwood of *Pterocarpus macrocarpus* Kruz. Bioscience, Biotechnology, and Biochemistry. 2006;**70**:1864-1868

[59] Agarwal M, Walia S, Dhingra S. Pest control properties of turmeric leaf oil against Spilosomaobliqua, Dysdercuskoenigii and Triboliumcastaneum. In: Proceedings of the 2nd all India People's Technology Congress, Calcultta, India. 1999. pp. 1-7

[60] Suszkiw J. Catnip Compounds Curb Asian Lady Beetles; Agricultural Research Services (ARS) Report. Washington, DC, USA: US Department of Agriculture; 2009. Available from: https://www.ars.usda.gov/news-events/news/research-news/2009/catnip-compounds-curb-asian-lady-beetles/

[61] McElvain SM, Bright RD, Johnson PR. The constituents of the volatile oil of catnip. I. nepetalic acid, nepetalactone and related compounds. Journal of the American Chemical Society. 1941;**63**:1558-1563

[62] Rudman P, Da Costa EWB, Gay FJ, Wetherly AH. Relationship of tectoquinone to durability in tectonagrandis. Nature. 1958;**181**:721-722

[63] Wolcott GN. Organic termite repellents tested against *Cryptotermes Brevis* walker. Journal of Agriculture of the University of Puerto Rico. 1955;**39**:115-149

[64] Sandermann VW, Dietrichs HH. Investigations on termiteproof wood species. Holz als Roh- und Werkstoff. 1957;**15**:281-297

[65] Ileke K, Ogungbite O, Alstoniaboonei D. Wild oil extract in the management of mosquito (*Anopheles gambiae*), a vector of malaria disease. Journal of Coastal Life Medicine. 2015;**3**:557-563

[66] Miller DR. Limonene: Attractant kairomone for white pine cone beetles (Coleoptera: Scolytidae) in an eastern white pine seed orchard in western North Carolina. Journal of Economic Entomology. 2007;**100**:815-822

[67] Degenhardt J, Hiltpold I, Köllner TG, Frey M, Gierl A, Gershenzon J, et al. Restoring a maize root signal that attracts insect-killing nematodes to control a major pest. Proceedings of the National Academy of Sciences of the United States of America. 2009;**106**:13213-13218

[68] Curto MJM. Ecologiaquímica: A comunicaçãoquímicananatureza. In: Lobo AM, Lourenço AM, editors. Biossíntese de Produtos Naturais. 1st ed. Lisboa, Portugal: IST Press; 2008. pp. 227-240

[69] Narwal S. Allelopathy in ecological agriculture. In: Narwal SS, Hoagland RE, Dilday RH, editors. Allelopathy in Ecological Agriculture and Forestry. Dharwad, India: Kluwer Academic; 2000. pp. 11-32

[70] Zhao X, Zheng GW, Niu XM, Li WQ, Wang FS, Li SH. Terpenes from *Eupatorium adenophorum* and their allelopathic effects on Arabidopsis seeds germination (dagger). Journal of Agricultural and Food Chemistry. 2009;**57**:478-482

[71] Bertin C, Weston LA, Huang T, Jander G, Owens T, Meinwald J, et al. Grass roots chemistry: Meta-tyrosine, an herbicidal nonprotein amino acid. Proceedings of the National Academy of Sciences of the United States of America. 2007;**104**:16964

[72] Nakano H, Nakajima E, Fujii Y, Yamada K, Shigemori H, Hasegawa K. Leaching of the allelopathic substance, –tryptophan from the foliage of mesquite (*Prosopis juliflora* (Sw.) DC.) plants by water spraying. Plant Growth Regulation. 2003;**40**:49-52

[73] Duke S, Blair A, Dayan F, Johnson R, Cook D, Bajsa-Hirschel J. Is (−)-catechin a novel weapon of spotted knapweed (*Centaurea stoebe*)? Journal of Chemical Ecology. 2009;**35**:141-153

[74] Somala N, Laosinwattana C, Chotsaeng N, Teerarak M. Citronella

essential oil-based nanoemulsion as a post-emergence natural herbicide. Scientific Reports. 2023;**13**:20851

[75] Chaudhari AK, Singh VK, Kedia A, Das S, Dubey NK. Essential oils and their bioactive compounds as eco-friendly novel green pesticides for management of storage insect pests: Prospects and retrospects. Environmental Science and Pollution Research. 2021;**28**:18918-18940

[76] Devrnja N, Milutinovi'c M, Savi'c J. When scent becomes a weapon—Plant essential oils as potent bioinsecticides. Sustainability. 2022;**14**:6847

[77] Mullin CA, González-Coloma A, Gutiérrez C, Reina M, Eichenseer H, Hollister B, et al. Antifeedant effects of some novel terpenoids on Chrysomelidae beetles: Comparisons with alkaloids on an alkaloid-adapted and nonadapted species. Journal of Chemical Ecology. 1997;**23**:1851-1866

[78] Evans PD, Maqueira B. Insect Octopamine receptors: A new classification scheme based on studies of cloned drosophila G-protein coupled receptors. Invertebrate Neuroscience. 2005;**5**:111-118

[79] Roeder T. Tyramine and Octopamine: Ruling behavior and metabolism. Annual Review of Entomology. 2005;**50**:447-477

[80] Finetti L, Roeder T, Calò G, Bernacchia G. The insect type 1 tyramine receptors: From structure to behavior. Insects. 2021;**12**:315

[81] Farooqui T. Review of octopamine in insect nervous system. Open Access Insect Physiology. 2012;**4**:1

[82] Evans PD. Multiple receptor types for octopamine in the locust. The Journal of Physiology. 1981;**318**:99-122

[83] Evans PD. Molecular studies on insect octopamine receptors. EXS. 1993;**63**:286-296

[84] Jankowska M, Lapied B, Jankowski W, Stankiewicz M. The unusual action of essential oil component, menthol, in potentiating the effect of the carbamate insecticide, Bendiocarb. Pesticide Biochemistry and Physiology. 2019;**158**:101-111

[85] Wang J, Ou SW, Wang YJ. Distribution and function of voltage-gated sodium channels in the nervous system. Channels (Austin, Tex.). 2017;**11**(6):534-554

[86] Du Y, Nomura Y, Satar G, Hu Z, Nauen R, He S, et al. Molecular evidence for dual pyrethroid-receptor sites on a mosquito sodium channel. Proceedings of the National Academy of Sciences of the United States of America. 2013;**110**:11785-11790

[87] Casida JE, Quistad GB. Pyrethrum Flowers: Production, Chemistry, Toxicology, and Uses. 1st ed. Oxford, UK: Oxford University Press; 1995

[88] Nauen R. Insecticide mode of action: Return of the ryanodine receptor. Pest Management Science. 2006;**62**(690):692

[89] Kim YH, Lee SH. Which acetylcholinesterase functions as the main catalytic enzyme in the class Insecta? Insect Biochemistry and Molecular Biology. 2013;**43**:47-53

[90] López MD, Pascual-Villalobos MJ. Mode of inhibition of acetylcholinesterase by monoterpenoids and implications for pest control. Industrial Crops and Products. 2010;**31**:284-288

[91] Jankowska M, Rogalska J, Wyszkowska J, Stankiewicz M. Molecular

targets for components of essential oils in the insect nervous system—A review. Molecules. 2017;**23**:34

[92] Liang JY, Yang Y, An YY, Shao YZ, He CY, Zhang J, et al. Insecticidal and acetylcholine esterase inhibition activity of *Rhododendron thymifolium* essential oil and its main constituent against two stored product insects. Journal of Environmental Science and Health. Part. B. 2021;**56**(4):423-430

[93] Isman M, Tak JH. Inhibition of acetylcholinesterase by essential oils and monoterpenoids: A relevant mode of action for insecticidal essential oils? Biopesticides International. 2017;**13**:71-78

[94] Li J, Li F, Wu G, Gui F, Li H, Xu L, et al. Acetylcholinesterase inhibitory activity of sesquiterpenoids isolated from *Laggera pterodonta*. Frontiers in Plant Science. 2023;**14**:1074184

[95] Green B, Welch K, Panter K, Lee S. Plant toxins that affect nicotinic acetylcholine receptors: A review. Chemical Research in Toxicology. 2013;**26**:1129-1138

[96] Taillebois E, Cartereau A, Jones AK, Thany SH. Neonicotinoid insecticides mode of action on insect nicotinic acetylcholine receptors using binding studies. Pesticide Biochemistry and Physiology. 2018;**151**:59-66

[97] Tomlin CDS. The Pesticide Manual. 12th ed. Surrey, England: British Crop Protection Council; 2000. pp. 178-179

[98] Weinzierl R. Insect management for commercial vegetable crops. In: Steffey K, editor. 1998 Illinois Agricultural Pest Management Handbook. University of Illinois at Urbana-Champaign; 1998. pp. 181-208

[99] Betarbet R, Sherer TB, MacKenzie G, Garcia-Osuna M, Panov AV, Greenamyre JT. Chronic systemic pesticide exposure reproduces features of Parkinson's disease. Nature Neuroscience. 2000;**3**:1301-1306

[100] Kamboj A, Saluja A. *Ageratum conyzoides* L.: A review on its phytochemical and pharmacological profile. International Journal of Green Pharmacy. 2008;**2**:59-68

[101] Coelho CAA, De Souza NA, Feder MD, Da Silva CE, Garcia EDS, Azambuja P, et al. Effects of azadirachtin on the development and mortality of *Lutzomyia longipalpis* larvae (Diptera: Psychodidae: Phlebotominae). Journal of Medical Entomology. 2006;**43**:262-266

[102] Lal M, Chaudhary S, Rawal S, Sharma S, Kumar M, Chakrabarti SK. Evaluation of bio-agents and neem based products against late blight disease (*Phytophthora infestans*) of potato. Indian Phytopathology. 2021;**74**:181-187

[103] Kumar V, Chaudhary VP, Kumar D, Kumar A, Sagar S, Chaudhary S. Efficacy of botanicals and fungicides against *Rhizoctonia solani* inciting sheath blight disease on Rice (*Oryza sativa* L.). Journal of Applied and Natural Science. 2017;**9**(4):1916-1920

[104] Kraiss H, Cullen EM. Insect growth regulator effects of azadirachtin and neem oil on survivorship, development and fecundity of *Aphis* glycines (Homoptera: Aphididae) and its predator, *Harmonia axyridis* (Coleoptera: Coccinellidae). Pest Management Science. 2008;**64**(6):660-668

[105] Cui C, Yang Y, Zhao T, Zou K, Peng C, Cai H, et al. Insecticidal activity and insecticidal mechanism of total saponins from *Camellia oleifera*. Molecules. 2019;**24**:4518

[106] Walia S, Saha S, Rana V. Phytochemical pesticides. In: Singh D, editor. Advances in Plant Biopesticides. New Delhi, India: Springer; 2014. pp. 295-322

[107] Soares SF, Borges LMF, Braga RS, Ferreira LL, Louly CCB, Tresvenzol LMF, et al. Repellent activity of plant derived compounds against *Amblyomma cajennense* (Ascari: Ixodidae) nymphs. Veterinary Parasitology. 2010;**167**(1):67-73

[108] Tripathi AK, Prajapati V, Aggarwal KK, Kumar S, Prajapti V, Kumar S, et al. Effect of volatile oil constituents of *Mentha* species against stored grain pests, *Callosobruchus maculatus* and *Tribolium castaneum*. Journal of Medicinal and Aromatic Plant Sciences. 2000;**22**:549-556

[109] Weinzierl R, Henn T, Koehler PG, Tucker CL. Insect attractants and traps. In: IFAS Extension, Institute of Food and Agricultural Sciences. Gainesville, FL, USA: University of Florida; 1995. pp. 1-9

[110] Zapata N, Budia F, Viñuela E, Medina P. Antifeedant and growth inhibitory effects of extracts and drimanes of *Drimys winteri* stem bark against *Spodoptera littoralis* (Lep., Noctuidae). Industrial Crops and Products. 2009;**30**:119-125

[111] Koul O. Phytochemicals and insect control: An antifeedant approach. Critical Reviews in Plant Sciences. 2008;**27**:1-24

[112] Mierziak J, Kostyn K, Kulma A. Flavonoids as important molecules of plant interactions with the environment. Molecules. 2014;**19**:16240-16265

[113] Céspedes CL, Torres P, Marın JC, Arciniegas A, de Vivar AR, Pérez-Castorena AL, et al. Insect growth inhibition by tocotrienols and hydroquinones from *Roldana barba-johannis*. Phytochemistry. 2004;**65**:1963-1975

[114] Rhoades DF. Evolution of plant defense against herbivores. In: Rosenthal GA, Janzen DH, editors. Herbivores: Their Interaction with Secondary Plant Metabolites. New York, NY, USA: Academic Press; 1979. pp. 1-55

[115] Céspedes CL, Lina-Garcia L, Kubo I, Salazar JR, Ariza-Castolo A, Alarcon J, et al. *Calceolaria integrifolia* s. l. Complex, reduces feeding and growth of *Acanthoscelides obtectus*, and *Epilachna varivestis*. A new source of bioactive compounds against dry bean pests. Industrial Crops and Products. 2016;**89**:257-267

[116] Chakraborty R, Thakur TS, Chauhan R, Manzar Z, Das G, Mitra A. Exploring the neural mechanisms of aversion to bitter gourd phytochemicals in insects using drosophila. Journal of Asia-Pacific Entomology. 2021;**24**:749-758

[117] Abdullah F, Subramanian P, Ibrahim H, Abdul Malek SN, Lee GS, Hong SL. Chemical composition, antifeedant, repellent, and toxicity activities of the rhizomes of galangal, *Alpinia galanga* against Asian subterranean termites, *Coptotermes gestroi* and *Coptotermes curvignathus* (Isoptera: Rhinotermitidae). Journal of Insect Science. 2015;**15**(1):7

[118] Jose S, Sujatha K. Antifeedant activity of different solvent extracts of *Gliricidia sepium* against third in star larvae of *Helicoverpa armigera* (Hubner) (Lepidoptera: Noctuidae). International Journal of Advanced Research in Biological Sciences. 2017;**4**(4):201-204

[119] Huang XP, Renwick JAA. Cross habituation to feeding deterrents and acceptance of a marginal host plant by *Pieris rapae* larvae. Entomologia Experimentalis et Applicata. 1995;**76**:295-302

[120] Arnason T, Towers GHN, PhilogÈNe BJR, Lambert JDH. The role of natural photosensitizers in plant resistance to insects. In: Plant Resistance to Insects. Washington, DC, USA: American Chemical Society; 1983. pp. 139-151

[121] Berenbaum M. Charge of the light brigade: Phototoxicity as a defense against insects. In: Light-Activated Pesticides. Washington, DC, USA: American Chemical Society; 1987

[122] Towers G. Interactions of light with phytochemicals in some natural and novel systems. Canadian Journal of Botany. 2011;**62**:2900-2911

[123] Marchant YY, Cooper GK. Structure and function relationships in polyacetylene photoactivity. In: Light-Activated Pesticides. Washington, DC, USA: American Chemical Society; 1987. pp. 241-254

[124] Nivsarkar M, Cherian B, Padh H. Alpha-terthienyl: A plant-derived new generation insecticide. Current Science. 2001;**81**:667-672

Chapter 9

Biopesticides for Sustainable Agriculture: A Review of Their Role in Integrated Pest Management

Kahsay Tadesse Mawcha, Grace Kinyanjui, Stella W. Simiyu, Olubukola Oluranti Babalola and Dennis Obonyo Ndolo

Abstract

Biopesticides include living organisms or products derived from natural sources, like genes and metabolites or natural-identical synthetic sources, that can be used to control pests. They are valuable pest management tools in sustainable agriculture. They protect crops from a wide range of pests and diseases while exhibiting specificity against target organisms and with minimal environmental impact. Incorporation of biopesticides into integrated pest management (IPM) programs provides a more holistic approach for growers to maximize crop yields, reduce overreliance on chemical pesticides, safeguard agroecosystems, and enhance crop yield and quality. This review explores the different classes of biopesticides (including biochemical, microbial, and plant-incorporated protectants), their modes of action, and their potential to enhance IPM strategies. The prospects of integrating emerging biopesticide technologies, such as nano-biopesticides and RNA-based biopesticides, into the existing IPM programs are also discussed.

Keywords: crop protection, pest control, integrated pest management, resistance management, emerging technologies, natural sources

1. Introduction

Africa is home to more than 13% of the world's population [1]. Before 2000, sub-Saharan Africa was a net exporter of food, but now struggles with a $7 billion food deficit, which accounts for less than 3% of the global agricultural exports [1]. Agricultural export performance in sub-Saharan Africa has significantly declined in the last two decades. The region's share of global agricultural trade value decreased from 8.4% in 1965 to 2% in 2001 [2]. To feed the growing population, a significant increase in food production will be required by 2050. However, a large portion (40%) of crops in the region is currently lost annually due to pests and diseases. The most common pest control option available to farmers is applying chemical pesticides. However, misuse or overuse of chemical pesticides is an unsustainable solution with adverse human health and environmental impacts. There is therefore a need for complementary options to synthetic chemical pesticides [3]. Integrated pest management

(IPM) provides a more holistic approach for growers to manage crop pests, maximize yields, improve crop quality, and reduce the overreliance on chemical pesticides. IPM involves the combination of multiple pest control strategies including cultural, physical, biological control agents, biopesticides, and judicious chemical control methods.

Biopesticide is a generic term generally applied to a substance or mix of substances derived from nature, or are nature-identical synthetic substances, such as microorganisms, botanicals, or semiochemicals that may be formulated and applied like a conventional chemical pesticide and normally used for short-term pest control [3, 4]. Biopesticides are becoming an essential biocontrol tool, as part of IPM strategies for growers worldwide. Biopesticides generally have low to no toxicity to nontarget beneficial organisms and wildlife, reduced environmental contamination and residues, and limited to no likelihood of developing resistance to pests [5]. Biopesticides are environmentally safe, and sustainable, not associated with the release of greenhouse gases and their use mitigates the impacts of climate change [6]. Besides ecological safety and target specificity toward their target pests [7], biopesticide use promotes biodiversity by preserving natural enemies and pollinators. Additionally, their biodegradability reduces pollution concerns associated with some chemical pesticides [7, 8]. Due to their targeted action and effectiveness in small quantities, biopesticides are gaining wide application on crops. They are playing an increasingly significant role in agriculture, promoting sustainable practices and a healthier environment [9], thus complying with the climate-smart agriculture principles [10], and ultimately, the Paris Agreement, the Nagoya protocol of Article 6 (Access to Genetic Resources), and Article 7 (safeguards the rights of indigenous and local communities regarding their traditional knowledge and ensures they benefit from its use).

The global chemical pesticides market is projected to increase by a 5.42% compound annual growth rate from $70.24 billion in 2019 to $101.64 billion in 2026. On the other hand, the biopesticides market is projected to grow from $4.4 billion in 2019 to $10.6 billion by 2027 and will match the chemical pesticides market by 2050 [11]. The global biopesticides market has been experiencing steady growth in recent years (**Figure 1A**), due to new regulations and increasing demand for sustainable agricultural practices. IPM practices incorporating biopesticides align with the principles of sustainable agriculture, striving to achieve a harmonious balance between

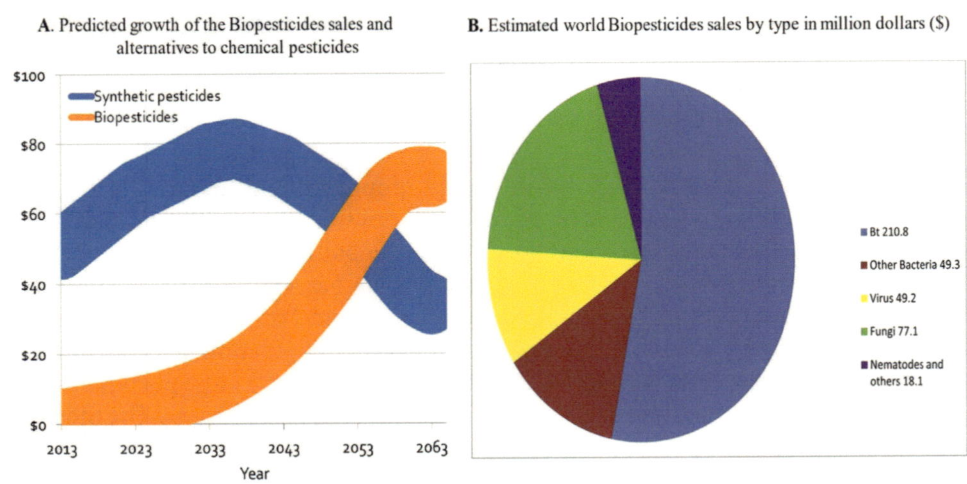

Figure 1.
Predicted growth and estimated worldwide sales of biopesticides over the next 30 years.

agricultural productivity and environmental stewardship. This approach aims to, among other things, minimize pesticide residues in our food and contribute to healthier diets, promoting long-term agricultural health by maintaining soil and crop integrity. Increasing the adoption of biopesticides will help reduce the environmental impact and minimize the risk of pesticide resistance [12, 13]. The predicted growth of biopesticide and synthetic pesticide sales (**Figure 1A** (2013–2065)), and the estimated global biopesticide sales by types in million dollars (**Figure 1B**).

2. Classes of biopesticides and their modes of action

Biopesticides are derived from biological sources such as biochemical compounds, microorganisms, plants, and other naturally derived products such as genes and metabolites [14]. They work through various mechanisms, such as (1) biochemical disruption, where they interfere with the biochemical processes of pests, affecting their growth, development, or reproduction; (2) microbial activity exhibited by several microorganisms, which infect and kill pests; (3) While biopesticides themselves are not typically natural predators or parasitoids, they can play a role in predation, parasitism, and competition in the following ways: Some biopesticides contain predatory organisms such as mites or insects that feed on pest species. For example, certain nematodes used as biopesticides actively hunt and consume pest larvae in the soil. Parasitic biopesticides, such as certain fungi, bacteria, and viruses, can infect and eventually kill pest insects. For instance, the fungus Beauveria bassiana acts as a parasite on various insect pests, leading to their death. Some biopesticides work by outcompeting pest species for resources. For example, beneficial microorganisms can outcompete harmful pathogens in the soil or on plant surfaces, thus preventing the establishment and growth of pests; and (4) repellency displayed by some biopesticides that deter pests from attacking crops [7, 15].

Biopesticides are classified into three major categories based on the type and extraction source of the active ingredient [16, 17]. (1) Microbial biopesticides comprise microorganisms such as bacteria, fungi, viruses, nematodes, or protozoans as active [18], ingredients that often target specific pest species. (2) Biochemical pesticides include compounds derived from plants, insects, and other naturally occurring bioactives, with the ability to control pests using nontoxic mechanisms. (3) Plant-incorporated protectants (PIPs) are pesticidal compounds resulting from genes transferred to plants allowing these transgenic plants to express pesticidal proteins such as transgenic *Bacillus thuringiensis* (*Bt*) toxins [14].

2.1 Microbial biopesticides

Microbial biopesticides originate from bacteria, fungi, viruses, protozoans, and nematodes. Although microbial-based pesticides control many kinds of pests, each active ingredient is relatively specific for its target pests [18]. They originate from microorganisms and they are target specific. For example, most entomopathogenic fungi target specific insects [5].

2.1.1 Bacterial biopesticides

Bacterial-based biopesticides contain bacterial species as active ingredients. *Bacillus thuringiensis* (*Bt*) is the most common species used commercially to control a diverse range of insect pest species including lepidopterans, hemipterans, and coleopterans **Figure 1B** [19, 20]. *Bacillus thuringiensis* is a spore-forming bacterium with several strains that produce various insecticidal crystal proteins during sporulation, which target different pests (**Table 1**). Bt has been used to control agriculturally important insect pests [20] For instance, *Bt* var. *kurstaki* specifically targets lepidopteran larvae, while the genes encoding crystal proteins have been transferred into crops like cotton, reducing reliance on chemical pesticides [39, 40]. The crystal proteins are toxic to specific insect pests [19, 20, 39, 40], and once ingested, they damage the gut tissues leading to paralysis of the gut. The infected insects stop feeding and eventually die from the combined effects of gut impairment and starvation. Over 100 *Bt*-based biopesticides have been developed and have received maximum attention as microbial biopesticides [20, 41, 42]. Commercially available *Bt*-based bioinsecticides include *Bt.* var. *kurstaki, Bt.* var. *israelensis, B. thuringiensis* var. *san diego,* and *Bt var. tenebrionis.* These products may be formulated into powders containing a combination of dried spores and crystal toxins, as dust or formulations in the form of concentrated liquids, oil-based flowable, wettable powders, and water-dispersible granules [43]. *Bt*-based biopesticides have a relatively narrow range of effectiveness against insect pests

Microorganism	Active ingredients and target pests	References
Bacteria (*Bt*)	*Bacillus thuringiensis Berliner*, *Bt aizawai*, and *Bt kurstaki* have high activity against lepidopteran larvae *Bt tenebrionis* shows activity against Coleopteran adults and larvae *Bt israelensis* controls gnats, flies	[15, 21–25]
Virus (Baculoviruses)	Granuloviruses (GVs) control *Plutella xylostella, Thaumatotibia leucotreta*, *Pieris rapae*, *Cydia pomonella* Nucleopolyhedroviruses (NPVs) [26] control *Helicoverpa armigera, Helicoverpa* spp. *Spodoptera exigua, S. litura* [16]	[27] [14]
	Anticarsia gemmatalis **NPV**- Velvetbean caterpillar	[28, 29]
	Buzura suppressaria	[30]
Fungi	*Beauveria bassiana* strains control aphids, thrips, whiteflies, scale insects, mealybugs, leafminers, spider mites, beetles *B. brongniartii* targets beetles, root grubs *Metarhizium anisopliae* strains target aphids, whiteflies, mealybugs, thrips, scarab beetle grubs, fruit flies *M. rileyi* controls lepidopterans *Lecanicillium lecanii* targets whiteflies, leafminers, aphids, scale insects *Paecilomyces lilacinus* controls plant parasitic nematodes *Isaria fumosorosea* targets soft-bodied insects such as whiteflies, aphids *Hirsutella thompsonii* targets spider mites *Trichoderma asperellum* against soil-borne pests	[31] [15, 20] [32] [33]
	Aschersonia aleyrodis is used to control Hemiptera	[27, 28]
	Beauveria brongniartii is used to control insects belonging to Coleoptera	[34–36]
	Nomuraea rileyi is effective in controlling Lepidoptera	[37]
Nematodes	*Steinernema* spp. and *Heterorhabditis* spp. are used on different host crops to control pests such as leafminers, thrips, weevils, beetles, cutworms, scarab grubs, lepidopterans, fungus gnats *Phasmarhabditis hermaphrodita* is used against several slug and snail species	[32, 33, 38]

Table 1.
Selected microbial pesticides used commercially to control agricultural pests.

because only a limited number of insect species belonging to the Coleoptera, Diptera, and Lepidoptera, families are susceptible to the crystal proteins. Other commercially exploited bacterial biopesticides targeting agricultural pests comprise *Burkholderia* spp., *Chromobacterium*, *Saccharopolyspora spinosa,* and *Bacillus* spp. [20, 42].

2.1.2 Fungal biopesticides

Fungal biopesticides target a wide range of insect species and include entomopathogenic fungi (EPFs) such as *Metarhizium, Beauveria, Isaria*, *Lecanicillium*, and *Hirsutella* species as active ingredients [20, 42]. Popular commercially available mycoinsecticides are often derived from species like *B. bassiana, M. anisopliae, I. fumosorosea,* and *Lecanicillium* spp. (**Table 1**). The EPF infects insects by directly penetrating the cuticle and their spores germinate and multiply in the insects' body. The fungal spores then colonize the host tissues and produce toxins that cause insect mortality. The fungal spores also emerge outside the body of the infected host and produce new conidia to continue the infection cycle. Both *M. anisopliae* and *B. bassiana* have several strains isolated from different types of host plants and are known for their broad effectiveness against insect pests like aphids, beetles, grasshoppers, and caterpillars (**Table 1**). These two ascomycetes are typically applied as conidia (spores) or mycelia, which then sporulate after application, ensuring sustained pest control [16, 43]. Biopesticides based on *Trichoderma* fungi are popularly known to control fungal soil diseases, although studies have shown that several species are also effective against insect pests [20, 31]. In addition, there is an increasing interest in the use of endophytic EPFs in the genera *Trichoderm*a and *Hypocrea* as biopesticides because, in addition to exhibiting biocontrol potential, they also act as plant growth promoters [20, 44, 45]. Other beneficial fungi such as *Cylindrobasidium laeve* for the control of *Acacia mearnsii* (black wattle) and *Colletotrichum acutatum* for the control of *Hakea sericea* (silky hakea) are some of the available mycoherbicides for integrated alien and invasive species management [11].

Naturally occurring EPFs hold immense promise as environmentally safe weapons against insect and mite pests. They possess several desirable traits for biocontrol: they selectively target pests, leaving beneficial insects like bees and predators unharmed. Additionally, they pose no threat to the growth or development of other beneficial organisms like earthworms and springtails. This makes them ideal candidates for IPM and sustainable long-term agricultural practices, promoting biodiversity and protecting the environment [11, 46]. Several fungi have been successfully mass-produced and formulated for commercial application as mycoinsecticides, paving the way for their widespread use in pest control in the agricultural, forestry, and conservation sectors [47–51].

2.1.3 Virus biopesticides

Entomopathogenic baculoviruses are well known for their utility as virus-based biopesticides targeting lepidopteran larvae and hymenopterans [26]. Baculoviruses are species-specific and similar to *Bt*-based products, they must be ingested by the target pest to initiate infection. After ingestion, they enter the body through the midgut and spread throughout the body ultimately killing the host. Viral particles are then dispersed from the liquefying dead body to the environment to continue the life cycle [52–54]. Two main groups of baculoviruses, namely *nucleopolyhedroviruses* (NPVs) and *granuloviruses* (GVs) are available as commercial isolates and infect specific target pests (**Table 1**). These groups, called occlusion bodies, differ in the structure and number of their protective protein coats. NPVs form polyhedral bodies with numerous virus particles, while GVs have smaller, granular bodies containing just one particle. Notably, each group employs a unique method to uncoat and establish infection within their host: NPVs in the nucleus and GVs within the nuclear pore complex [49–51]. The occlusion bodies are highly stable and allow the virus to survive outside the host, waiting for its next target. Baculoviruses primarily focus on lepidopteran (butterfly and moth) pests that damage crops like cotton, rice, and vegetables [55]. One example is *Heliothis zea nucleopolyhedrosis* virus (HzNPV), the first commercially successful broad-spectrum viral insecticide. Interestingly, *HzNPV* demonstrates versatility, effectively controlling pests across various crops like soybeans, sorghum, maize, tomatoes, and beans. While other entomopathogenic viruses exist, such as *tetraviruses* and *cypoviruses*, their use in crop protection remains limited compared to the widespread adoption of baculoviruses [26, 51]. A comprehensive summary of viral biopesticides and their target pests is presented by [52–54].

While viral biopesticides offer numerous advantages over conventional chemical pesticides, including specificity and environmental safety, they face certain challenges for large-scale adoption. Producing them efficiently, particularly recombinant viruses, can be costly and labor-intensive, requiring specialized equipment and lengthy procedures [38, 55–57]. Despite these hurdles, various organizations are actively supporting the development and small-scale production of these promising biopesticides. Unlike the broad spectrum of some conventional chemical insecticides, their use often remains targeted to specific pest species. In addition, successful pest control can be a slow process, as infections may take a significant amount of time to become lethal and can be easily deactivated by high temperatures and ultraviolet radiation. Understanding these vulnerabilities is crucial for choosing the pest control strategy [58] for the targeted pest and bioclimatic zone.

2.1.4 Nematode-based biopesticides

Entomopathogenic nematodes (EPNs), namely *Steinernema* spp. and *Heterorhabditis* spp. are insect-parasitic nematodes that have been exploited as biological control agents for suppressing numerous insect pests (**Table 1**). Commercial biopesticides based on EPNs contain nonfeeding, third-stage infective juveniles (IJs) as the active ingredients [21]. EPNs, in their IJ stage, barely millimeters long, use their keen senses to locate insects through carbon dioxide emissions, vibrations, and other chemical cues. The IJs enter the body of the suitable insect host through natural openings like the mouth, anus, or breathing holes, and establish in the hemocoel. These nematodes release their symbiotic bacterial components, setting off a natural pest control cascade. The EPNs in association with the released pathogenic bacteria kills the host within 48 h by septicemia and toxemia. After the host dies, infectious stages of EPNs feed on the insect's liquefied remains and develop into adults, which multiply and create a new generation of IJs that immediately start foraging for a new host and continue the cycle. Under ideal conditions, the impact of EPNs becomes visible within 5–7 days. Insects infected by *Steinernematidae* generally turn brown or tanned, while those taken down by *Heterorhabditidae* turn a distinctive red [15, 56, 57]. The attributes of IJs including rapid mortality of target insects and active host seeking have enhanced the commercialization of several EPN species as biological insecticides [23]. In addition to EPNs, slug parasitic nematodes are also used as biopesticides to control slugs and snails (**Table 1**).

2.1.5 Protozoan biopesticides

Entomopathogenic protozoans or microsporidians are intracellular parasites that can only survive by living inside other cells. They are found almost everywhere and are pathogenic to lepidopterans and orthopterans, making them useful in IPM programs. The most common microsporidian entomopathogens include *Nosema*, *Thelohania*, and *Vairimorpha* species. Microsporidia infection starts when the spores are ingested by a host insect, such as the European corn borer, *Ostrinia nubilalis.* The spores germinate and replicate in the midgut region and release the *sporoplasm*, which is then injected into the midgut target cells. The spores then spread to different tissues and organs, multiply, and cause massive infection leading to tissue breakdown and septicemia [38, 58]. Protozoans have a specific mode of action. However, despite their pest-specific nature and slow-acting properties, using protozoa as biopesticides is not as effective as other organisms, such as fungi, viruses, and bacteria [58].

2.2 Biochemical biopesticides

Biochemical pesticides have various modes of action (**Figure 2**) and include insect-derived semiochemicals, plant-derived biopesticides, and natural insect growth regulators [14]. Insect sex pheromones are the commonly used semiochemicals in pest management. These species-specific compounds produced by insects to mediate mating behavior have become essential tools for monitoring and controlling

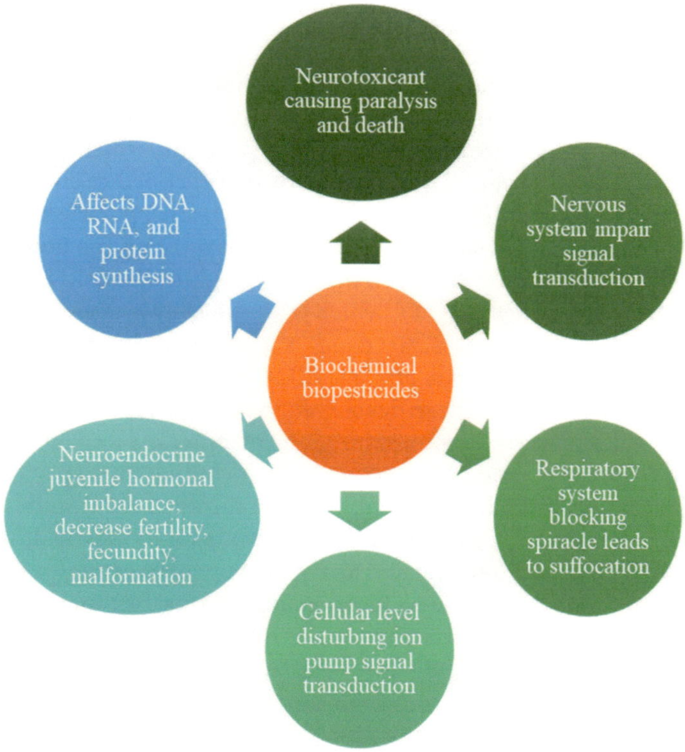

Figure 2.
Biochemical biopesticides' mode of action.

populations of agricultural pests, particularly lepidopteran species [59–61]. Monitoring allows prompt detection of pests and timely application of control measures, while strategies such as pheromone-based mass trapping and push-pull, mating disruption, and lure-and-kill/infect aim to reduce both the population levels of the target pest and crop damage [62].

Plant-derived biopesticides or botanicals are mainly composed of bioactive extracts and essential oils from plant materials [35–37]. Diverse phytochemical compounds enable botanicals to exhibit various modes of action against target pests, including feeding and oviposition deterrence, repellence, larval toxicity, ovicidal effects, growth inhibition, and other physiological disruptions [59–61]. Azadirachtin-based insecticide derived from the neem tree (*Azadiracha indica*) is the most successful botanical pesticide that has been commercialized worldwide [62, 63] and is used against diverse sucking and chewing insect pests [64]. Other plant-derived pesticides including pyrethrins, nicotine, and rotenone have been identified and play a vital role in managing a wide range of crop pests [62, 63]. Essential oils extracted from aromatic plants like cloves, thyme, rosemary, peppermint, lemongrass, cinnamon, etc. and their bioactive compounds have also demonstrated high insecticidal efficacy and are increasingly exploited in pest control programs for use as insect attractants, repellents, antifeedants, and compounds displaying ovicidal and larvicidal effects against target pests [64, 65]. Additionally, essential oils and extracts from different parts of plants have been effectively used to control plant diseases [4]. Botanicals pose fewer risks to human health compared to synthetic chemical pesticides [36, 37].

2.3 Plant-incorporated protectants (PIPs)

The most common insecticidal PIPs are the *Bt* crystal proteins expressed in genetically modified (GM) crops to reduce the abundance and damage of the targeted pests [67–70]. Once ingested, GM crops produce crystal proteins, which are toxic within the gut of the target pest. Several *Bt* crops have been developed to enhance resistance against insect pests including cotton, maize, soybeans, potatoes, tobacco, and eggplant [46]. The current research on PIPs is focused on applying double-stranded ribonucleic acid (dsRNA) to control crop insect pests [70]. The dsRNA is a key trigger molecule of RNA interference (RNAi), which is a gene-silencing mechanism targeting to silence the expression of essential genes, thereby inhibiting the development of target insects and even causing death [71, 72]. GM crops expressing dsRNA that target specific insect genes have been developed to control numerous agricultural pests including lepidopterans, hemipterans, and coleopterans [66–69]. The use of RNAi-based PIPs has several potential benefits, including high specificity and accuracy against target pests, biodegradability, and minimal environmental impact [66, 69]. Genomics involves studying the entire genome of biopesticidal organisms. Researchers use genomic sequencing and analysis to uncover genes for pesticidal compound biosynthesis, discover new biopesticidal agents, optimize existing ones, and understand resistance development in pests for designing more effective biopesticides. Molecular biology techniques are essential [73] for studying the interactions between biopesticidal compounds and their target pests at the molecular level and to identify key pathways and molecular targets [74]. **Figure 3** shows the integration of genetics, genomics, and molecular biology which play a crucial role in advancing our understanding and development of effective biocontrol agents.

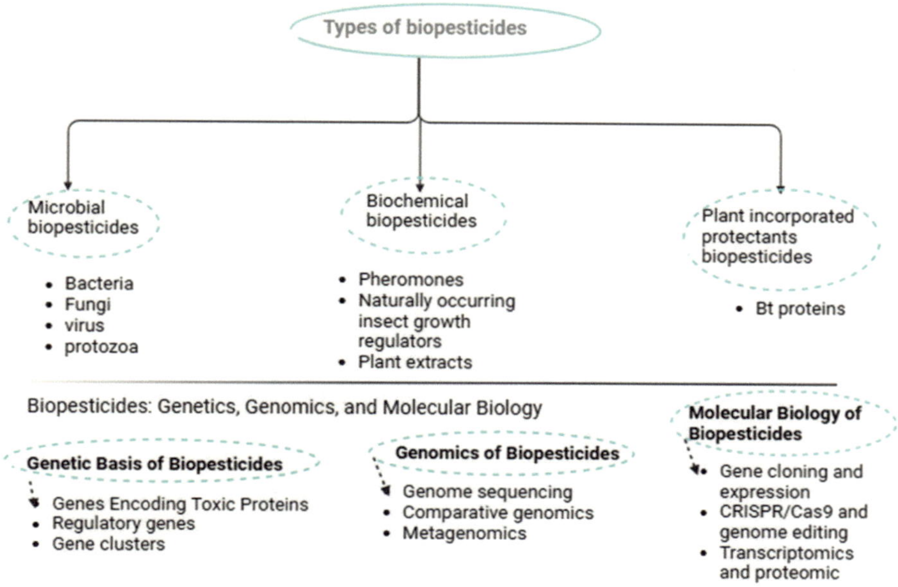

Figure 3.
Classification of biopesticides and potential advancements in biopesticide research including genetics, genomics, and molecular biology of biopesticides.

3. Incorporation of biopesticides into IPM systems

Biopesticides are a key component of IPM, a holistic and ecologically based strategy that relies on a combination of pest control tactics for effective pest management and minimal use of synthetic pesticides. The definition of IPM provided by the European Union framework stresses the growth of a healthy crop with the least possible disruption to agroecosystems and encourages natural pest control mechanisms (**Figure 4**). Pest management strategies involving biopesticides include the following: (1) biological control by using microorganisms as biopesticides, (2) application of deleterious rhizobacteria to suppress pests and competition, (3) use of natural products, and (4) nanoparticle carriers, and acknowledgment of traditional knowledge and genetic resources (Nagoya Protocol) [75].

Most biopesticides have become increasingly valuable in pest management because they are adaptable to the current IPM technologies and are highly compatible with other crop protection strategies comprised of cultural control practices, host plant resistance, mechanical and physical practices, and biological control agents such as predators and parasitoids [69, 70, 74]. The integration of biopesticides and

> 'Integrated pest management' means careful consideration of all available plant protection methods and subsequent integration of appropriate measures that discourage the development of populations of harmful organisms and keep the use of plant protection products and other forms of intervention to levels that are economically and ecologically justified and reduce or minimize risks to human health and the environment. 'Integrated pest management' emphasises the growth of a healthy crop with the least possible disruption to agro-ecosystems and encourages natural pest control mechanisms.

Figure 4.
The definition of IPM is provided by the European Union framework directive on the sustainable use of pesticides (Directive 2009/128/EC).

bioprotection agents involving both living and nonliving agents into an IPM program is illustrated in **Figure 4**. The use of insect sex pheromones is one of the focal points of IPM systems because of their practical application in detection and pest population monitoring, mass trapping, and mating disruption [34, 35, 71]. The integration of pheromone-based pest management methods into IPM programs has enhanced the management of numerous agricultural pests [75–77]. For instance, the combined use of sex pheromones and fungal-based biopesticides in the attract-and-infect strategy increases effectiveness against target pests such as fall armyworms [35, 73]. Additionally, synergism has been observed between botanical pesticides and several entomopathogenic microorganisms, namely *Bt* bacterium, fungal isolates, baculoviruses, and nematodes against different agricultural pests [78], indicating that this combination can be readily exploited in IPM. The remarkable potential of EPF is actively being harnessed for use in IPM programs. Mass production techniques and ecological approaches are constantly refined to optimize their effectiveness and ensure seamless integration into sustainable agricultural practices [43, 47]. Furthermore, promising results have been achieved by combining fungal biopesticides with traditional insecticides, often leading to significantly boosted pest mortality rates. For example, studies have shown that combining *B. bassiana* with low-dose insecticides enhances potato beetle control, while its synergy with neem oil proves effective against tobacco thrips eggs and nymphs [11, 16, 26, 31, 43–51] without any long-term negative impact on the environment. Insect-resistant transgenic crops expressing insecticidal crystal proteins such as *Bt* toxins are also compatible with other pest control tactics [74], and thus better harmonization with existing IPM strategies can achieve great success in pest management with minimal environmental impact. The integration of biopesticides with various IPM control practices and other pest control strategies is presented in **Figure 5**.

4. Future outlook

Continued research and development of biopesticides and their incorporation into IPM strategies can contribute to sustainable agriculture and reduce overreliance on chemical pesticides. However, successful biopesticide utilization in IPM requires overcoming the industry's challenges such as restrictive regulations and complex registration processes that hinder product commercialization [79]. This would ensure that farmers have access to more of these products. While the key components of an IPM program including cultural, physical, and chemical control methods and the use of biological control agents and biopesticides have been practically implemented, it is necessary to integrate emerging biopesticide technologies such as nanotechnology and RNAi to optimize the productivity and sustainability of existing IPM systems as well as increase the adoption of biopesticides in modern agriculture [80]. Biotechnology, genetic engineering, and nanotechnology offer great potential in developing novel products such as nano-biopesticides and RNAi-based PIPs [19]. Nano-biopesticides are biological protection products formulated from nanomaterials to enhance their efficacy, achieve controlled release of the active ingredient, and reduce environmental contamination [4]. The development of nano-biopesticide formulations holds promise for further improving the efficacy of microbial- and plant-based pesticides and enabling better utilization of these biopesticides in IPM systems. For instance, *Bt*-based nano-insecticides show improved insecticidal activity [74–76, 81] and the utilization of these nano-formulations can promote sustainable pest management [82]. Additionally, there is a need for more research to optimize the RNAi technology

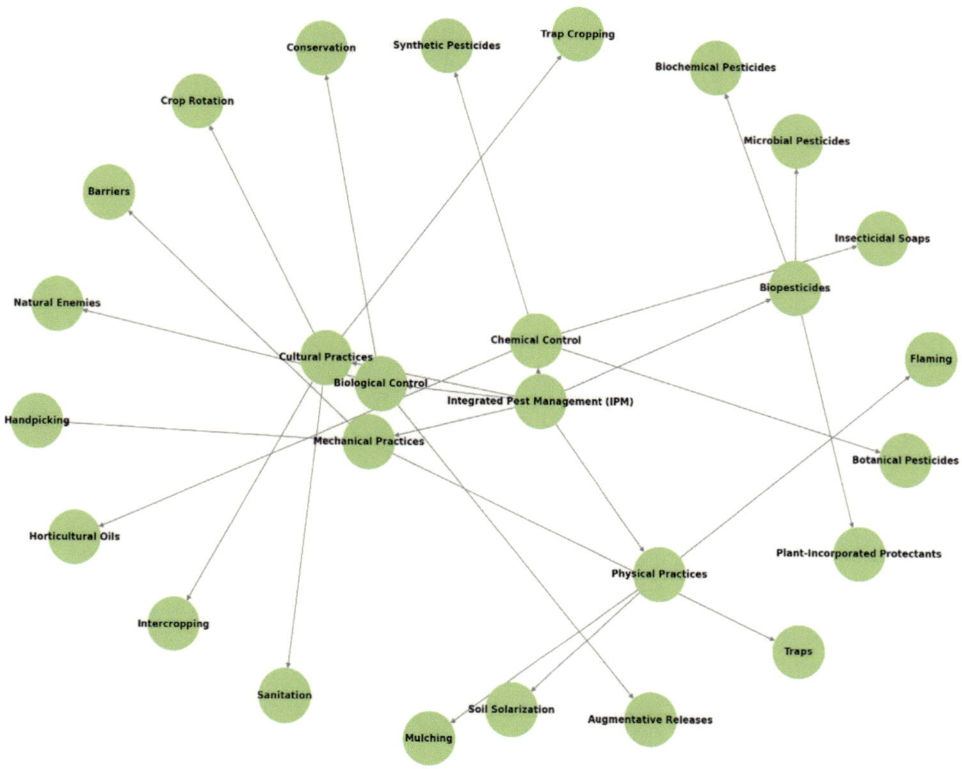

Figure 5.
The integration of biopesticides with integrated pest management (IPM) and other related pest control practices. This visual representation helps to understand the comprehensive approach of IPM and depicts IPM at the center, surrounded by different pest control practices and specifics related to biopesticides, protection, and other pest control practices.

and integrate RNA-based biopesticides into the existing IPM strategies. Future actions should also focus on lowering the production costs of novel biocontrol products including RNAi-based PIPs and other transgenic crops to promote their commercial application and integration into diversified and effective IPM programs. Additionally, the application of molecular technologies such as genomics, metagenomics, proteomics, and transcriptomics are becoming increasingly important in elucidating the relationship between pests, biocontrol agents, and host plants and increase our understanding of the underlying biocontrol mechanism to help improve the efficacy and robustness of the biological control agents and biopesticide formulations.

Acknowledgements

The authors are grateful to the International Center for Genetic Engineering and Biotechnology for financial support.

Conflict of interest

The authors declare no conflict of interest.

Author details

Kahsay Tadesse Mawcha[1,2]*, Grace Kinyanjui[3], Stella W. Simiyu[4], Olubukola Oluranti Babalola[5]* and Dennis Obonyo Ndolo[1]

1 International Center for Genetic Engineering and Biotechnology, Cape Town, South Africa

2 Department of Plant Sciences, Aksum University, Aksum, Tigray, Ethiopia

3 Department of Biological Sciences, University of Embu, Embu, Kenya

4 Faculty of Agriculture, Food and Environmental Sciences, Catholic University of the Sacred Heart, Piacenza, Italy

5 Food Security and Safety Focus Area, Faculty of Natural and Agricultural Sciences, North-West University, PMB 2740, Mmabatho, South Africa

*Address all correspondence to: kahsay.mawcha@icgeb.org and olubukola.babalola@nwu.ac.za

References

[1] Bjornlund V, Bjornlund H, van R. Andre. Why food insecurity persists in sub-Saharan Africa: A review of existing evidence. Food Security. 2022;**14**:845-864. DOI: 10.1007/s12571-022-01256-1

[2] Kandiero T, Randa J. Agricultural exports: Important issues for sub-Saharan Africa. African Development Review. 2004;**16**:1-35. DOI: 10.1111/j.1467-8268.2004.00082.x

[3] Cordeau S, Triolet M, Wayman S, Steinberg C, Jean-Philippe G. Bioherbicides: Dead in the water? A review of the existing products for integrated weed management. Crop Protection. 2016;**87**:44-49. DOI: 10.1016/j.cropro.2016.04.016

[4] Ayilara MS, Adeleke BS, Akinola SA, Fayose CA, Adeyemi UT, Gbadegesin LA, et al. Biopesticides as a promising alternative to synthetic pesticides: A case for microbial pesticides, phytopesticides, and nanobiopesticides. Frontiers in Microbiology. 2023;**14**:1040901. DOI: 10.3389/fmicb.2023.1040901

[5] EPA. What are Biopesticides? 2023. pp. 1-3

[6] Borges S, Alkassab AT, Collison E, Hinarejos S, Jones B, McVey E. Overview of the testing and assessment of effects of microbial pesticides on bees: Strengths, challenges and perspectives. Apidologie. 2021;**52**:1256-1277. DOI: 10.1007/s13592-021-00900-7

[7] Essiedu JA, Adepoju FO, Maria I. Benefits and limitations in using biopesticides: A review. In: Proceedings of the VII International Young Researchers' Conference. In Proceedings of the Physics, Technology, Innovations (PTI-2020). Ekaterinburg, Russia: AIP Publishing; 2020. p. 080002

[8] Sharma S, Malik P. Biopesticides: Types and applications. International Journal of Advance in Pharmacy, Biology and Chemistry. 2012;**1**:508-515

[9] EPA. Ingredients used in pesticide products: Pesticides. What are biopesticides? Available Online.

[10] FAO. Climate- Smart Agriculture Source Book. 2nd ed. Rome, Italy: Food and Agriculture Organization of the United Nations; 2017. ISBN 978-92-5-109988-9

[11] Lacey LA, Grzywacz D, Shapiro-Ilan DI, Frutos R, Brownbridge M, Goettel MS. Insect pathogens as biological control agents: Back to the future. Journal of Invertebrate Pathology. 2015;**132**:1-41. DOI: 10.1016/j.jip.2015.07.009

[12] Essiedu JA, Adepoju FO, Ivantsova MN. Benefits and limitations in using biopesticides: A review. In: Proceedings of the in Proceedings of the VII International Young Researchers' Conference—Physics, Technology, Innovations (PTI-2020); Ekaterinburg, Russia. 2022. pp. 18-22

[13] FAO. The 3 Pillars of Climate-Smart Agriculture and the Sustainable Development Goals Synergies & Trade-Offs. Vol. 112024. p. 20

[14] Kumar J, Ramlal A, Mallick D, Mishra V. An overview of some biopesticides and their importance in plant protection for commercial acceptance. Plants. 2021;**10**:1185

[15] Stenberg JA, Sundh I, Becher PG, Björkman C, Dubey M, Egan PA, et al. When is it biological control? A framework of definitions, mechanisms,

and classifications. Journal of Pest Science. 2021;**94**:665-676. DOI: 10.1007/s10340-021-01354-7

[16] Ruiu L. Microbial biopesticides in agroecosystems. Agronomy. 2018:235

[17] Parveen SS, Rashtrapal PS. Integrated Pest management strategies using endophytic entomopathogenic fungi. Plant Science Today. 2024:568-574. DOI: 10.14719/pst.2740

[18] Madhu L. Sharma biopesticides: An eco-friendly approach. International Journal of Pure and Applied Research. 2019;**2**:31-34

[19] Kumar P, Kamle M, Borah R, Mahato DK, Sharma B. *Bacillus thuringiensis* as microbial biopesticide: Uses and application for sustainable agriculture. Egyptian Journal of Biological Pest Control. 2021;**31**:95-102. DOI: 10.1186/s41938-021-00440-3

[20] Mazid SK. A review on the use of biopesticides in insect Pest management. International Journal of Advanced Science and Technology. 2011;**1**:169-178

[21] Bhat AH, Chaubey AK, Askary TH. Global distribution of entomopathogenic nematodes, steinernema and heterorhabditis. Egyptian Journal of Biological Pest Control. 2020;**30**:1-5. DOI: 10.1186/s41938-020-0212-y

[22] Koul O. Microbial biopesticides: Opportunities and challenges. CABI Review. 2011;**6**:1-26

[23] Koppenhöfer AM, Shapiro-Ilan DI, Hiltpold I. Entomopathogenic nematodes in sustainable food production. Frontiers in Sustainable Food Systems. 2020, 2020;**4**:125. DOI: 10.3389/fsufs.2020.00125

[24] Senthil-Nathan SA. Review of biopesticides and their mode of action against insect pests. Environmental Sustainability: Role of Green Technologies. 2014;**9**:49-63. DOI: 10.1145/2806777.2806838

[25] Rajput VS, Jhala J, Vijay SA. Biopesticides and their mode of action against insect pests: A review. International Journal of Chemical Studies. 2020;**8**:2856-2862

[26] Gelaye Y, Negash B. The role of baculoviruses in controlling insect pests: A review. Cogent Food & Agriculture. 2023;**9**:225413. DOI: 10.1080/23311932.2023.2254139

[27] Liu X, Cao A, Yan D, Ouyang C, Wang Q, Li Y. Overview of mechanisms and uses of biopesticides. International Journal of Pest Management. 2019;**67**:65-72. DOI: 10.1080/09670874.2019.1664789

[28] Yang MM, Li ML, Zhang Y, Wang YZ, Qu LJ, Wang QH, et al. Baculoviruses and insect pests control in China. African Journal of Microbiology Research. 2012;**16**:214-218. Available from: http://www.academicjournals.org/AJMR

[29] Panazzi AR. History and contemporary perspectives of the integrated pest management of soybean in Brazil. Neotropical Entomology. 2013;**42**:119-127. DOI: 10.1007/s13744-013-0111-y

[30] Moscardi F, De Souza ML, De Castro ME, Lara Moscardi M, Szewczyk B. Baculovirus pesticides: Present state and future perspectives. In: Microbes and Microbial Technology: Agricultural and Environmental Applications. New York, NY: Springer; 2011

[31] van Lenteren JC, Bolckmans K, Köhl J, Ravensberg WJ, Urbaneja A. Biological control using invertebrates and microorganisms: Plenty of new

opportunitie. BioControl. 2018;**63**:39-59. DOI: 10.1007/s10526-017-9801-4

[32] Goble TA, Costet L, Robène I, Nibouche S, Rutherford RS, Conlong DE, et al. Beauveria brongniartii on white grubs attacking sugarcane in South Africa. Journal of Invertebrate Pathology. 2012;**111**:225-236

[33] Faria M, Souza DA, Sanches MM, Schmidt FG, Oliveira CM, Benito NP, et al. Evaluation of key parameters for developing a *Metarhizium rileyi*-based biopesticide against *Spodoptera frugiperda* (Lepidoptera: Noctuidae) in maize: Laboratory, greenhouse, and field trials. Pest Management Science. 2022;**78**:146-154

[34] Lacey LA, Liu TX, Buchman JL, Munyaneza JE, Goolsby JA, Horton D. Entomopathogenic fungi (hypocreales) for control of potato psyllid, bactericera cockerelli (sulc) (hemiptera: Triozidae) in an area endemic for zebra chip disease of potato. Biological Control. 2011;**36**:271-278. DOI: 10.1016/j.biocontrol.2010.11.012

[35] McCoy CW, Samson RA, Boucias DG, Osborne LS, Pena J, Buss L. Pathogens Infecting Insects and Mites of Citrus. LLC Friends of Microbes: Winter Park, FL, USA; 2009. p. 193

[36] Townsend RJ, Nelson TL, Jackson T. Beauveria brongniartii–a potential biocontrol agent for use against manuka beetle larvae damaging dairy pastures on cape Foulwind. New Zealand Plant Protection. 2010;**63**:224-228. DOI: 10.30843/nzpp.2010.63.6572

[37] Thakre M, Thakur M, Malik N, Ganger S. Mass scale cultivation of entomopathogenic fungus nomuraea rileyi using agricultural products and agro wastes. Journal of Biopesticides. 2011;**4**:176-179

[38] Gupta SDA. Biopesticides: An ecofriendly approach for pest control. Journal of Biopesticides. 2010;**3**:186-188

[39] Tanda Y, Kaya HK. Insect Pathology. Harcourt Brace Jovanovich Publishers, San Diego: Academic Press Inc.; 1993

[40] Roy A, Moktan B, Sarkar PK. Characteristics of bacillus cereus isolates from legume-based Indian fermented foods. Food Control. 2007;**18**:1555-1564

[41] Gupta S, Dikshit AK. Biopesticides: An ecofriendly approach for pest control. Journal of Biopesticides. 2019;**2**:186-188

[42] Kachhawa D. Microorganisms as a biopesticides. Journal of Entomology and Zoology Studies. 2017;**5**:468-473

[43] Boyetchko S, Pedersen E, Punja Z, Reddy MS. Formulations of biopesticides. In: Biopesticides: Use and Delivery. 1999. pp. 487-508

[44] Akutse KS, Subramanian S, Maniania NK, Dubois T, Ekesi S. Biopesticide research and product development in Africa for sustainable agriculture and food security—Experiences from the International Centre of Insect Physiology and Ecology (Icipe). Frontiers in Sustainable Food Systems. 2020;**4**:563016. DOI: 10.3389/fsufs.2020.563016

[45] Sharma R, Sharma P. Fungal entomopathogens: A systematic review. Egyptian Journal of Biological Pest Control. 2021;**31**:1-3. DOI: 10.1186/s41938-021-00404-7

[46] Ritesh K, Pratik S, Susanth VR, Pratush B, Mohammed N. Potential and prospects of trichoderma in plant protection. Advances in Agriculture. 2023;**2023**:5573662. DOI: 10.1155/2023/5573662

[47] Poveda J, Baptista P, Soledad S, Velasco P. Editorial: Beneficial effects of fungal endophytes in major agricultural crops. Frontiers in Plant Science. 2022;**13**:1061112. DOI: 10.3389/fpls.2022.1061112

[48] Kim JJ, Goettel MS, Gillespie DR. Evaluation of *Lecanicillium* longisporum, vertalec against the cotton aphid, aphis gossypii, and cucumber powdery mildew, sphaerotheca fuliginea in a greenhouse environment. Crop Protection. 2021;**29**:540-544

[49] Kaur D. Testing native isolates of cylindrobasidium laeve for suppression of regrowth of acer Macrophyllum [Thesis Submitted in Partial Fulfillment of The Requirements for The Degree of Master of Pest Management]2004. p. 66

[50] Goettel MS, Koike M, Kim JJ, Aiuchi D, Shinya R, Brodeur J. Potential of *Lecanicillium spp*. for the management of insects, nematodes, and plant diseases. Journal of Invertebrate Pathology. 2008;**98**:256-261. DOI: 10.1016/j.jip.2008.01.009

[51] Chandler D, Bailey AS, Tatchell GM, Davidson G, Greaves J, Grant WP. The development, regulation and use of biopesticides for integrated Pest management. Philosophical Transactions of the Royal Society B Biology Science. 2011;**366**:1987-1998. DOI: 10.1098/rstb.2010.0390

[52] Singh A, Bhardwaj R, Singh IK. Biocontrol agents: Potential of biopesticides for integrated pest management. In: Giri B et al., editors. Biofertilizers for Sustainable Agriculture and Environment, Soil Biology. India: Springer International Publishing; 2019

[53] Usta C. Microorganisms in biological pest control—A review (bacterial toxin application and effect of environmental factors). Current Progress in Biological Research. 2013;**13**:287-317. DOI: 10.5772/55786

[54] Cory JS. Assessing the risks of releasing genetically modified virus insecticides: Progress to date. Crop Protection. 2000;**19**:779-785

[55] Maciel RM, Luski PG, Sutil WP, Gonçalves J, Hayashida R, de Queiroz AP, et al. The use of baculovirus spodoptera SfMNPV alone and combined with herbicides and adjuvant to control spodoptera frugiperda (Smith, 1797) (Lepidoptera: Noctuidae). Biological Control. 2024;**188**:105408

[56] Moscardi F, de Souza ML, de Castro MEB, Moscardi ML, Szewczyk B. Baculovirus pesticides: Present state and future perspectives. In: Ahmad I, Ahmad F, Pichtel J, editors. Microbes and Microbial Technology. Springer: New York; 2011. pp. 415-445

[57] Sarwar M. Biopesticides: An effective and environmental friendly insect-pests inhibitor line of action. International Journal of Engineering and Advanced Research Technology. 2015;**1**:10-15

[58] Venkatramanan V, Shah S, Prasad R. Sustainable Bioeconomy. Singapore: Springer; 2021. DOI: 10.1007/978-981-15-7321-7

[59] Tarasco E, Fanelli E, Salvemini C, El-Khoury Y, Troccoli A, Vovlas A, et al. Entomopathogenic nematodes and their symbiotic bacteria: From genes to field uses. Frontiers in Insect Science. 2023;**3**:1195254. DOI: 10.3389/finsc.2023.1195254

[60] Rae R, Sheehy L, McDonald-Howard K. Thirty years of slug control using the parasitic nematode phasmarhabditis hermaphrodita and beyond. Pest Management Science.

2023;**79**:3408-3424. DOI: 10.1002/ps.7636

[61] Hikal WM, Baeshen RS, Said AAH. Botanical insecticide as simple extractives for pest control. Cogent Biology. 2017;**3**:1404274. DOI: 10.1080/23312025.2017.1404274

[62] Witzgall P, Kirsch P, Cork A. Sex pheromones and their impact on pest management. Journal of Chemical Ecology. 2010;**36**:80-100

[63] Idris H, Suryani E, Gustia H, Ramadhan AIJ. The effect of various essential oil and solvent additives on the botanical pesticide of piper aduncum essential oil on formulation antifungal activity. Results in Engineering. 2022;**16**:100644. DOI: 10.1016/j.rineng.2022.100644

[64] Kilani-Morakchi S, Morakchi-Goudjil H, Sifi K. Azadirachtin-based insecticide: Overview, R. assessments, and future directions. Frontiers in Agronomy. 2021;**10**:676208. DOI: 10.3389/fagro.2021

[65] Lengai GMW, Muthomi JW, Mbega ER. Phytochemical activity and role of botanical pesticides in pest management for sustainable agricultural crop production, scientific African. Cogent Biology. 2017;**3**:1404274. DOI: 10.1016/j.sciaf.2019.e00239

[66] Assadpour E, Can Karaça A, Fasamanesh M, Mahdavi SA, Shariat-Alavi M, Feng J, et al. Application of essential oils as natural biopesticides; recent advances. Critical Reviews in Food Science and Nutrition. 2023;**20**:1-21

[67] Devrnja N, Milutinović M, Savic J. When scent becomes a weapon—Plant essential oils as potent bioinsecticides. Sustainability. 2022;**14**:6847. DOI: 10.3390/su14116847

[68] Malahlela M, Thibane VS, Mudau F. Nematocidal activity of fermented extracts from lantana camara plant parts against meloidogyne javanica on tomato. International Journal of Vegetable Science. 2021;**27**:20-28. DOI: 10.1080/19315260.2019.1697981

[69] Figueira A, Marques-Souza H, Camargo R, Barbosa G, Possignolo I, Peres L, et al. RNA interference as a gene silencing tool to control tuta absoluta in tomato (solanum lycopersicum). PeerJ. 2016;**10**(4):e2673. DOI: 10.7287/PEERJ.PREPRINTS.2406V1

[70] Li X, Liu X, Lu W, Yin X, An S. Application progress of plant-mediated RNAi in pest control. Frontiers in Bioengineering and Biotechnology. 2022;**10**:963026. DOI: 10.3389/fbioe.2022.963026

[71] Gressel J, Meir S, Herschkovitz Y, Al-Ahmad H, Babalola OO, Amsellam Z. Transgenic biocontrol agents to overcome evolutionary barriers. In: Integrating New Technologies for Striga Control Chapter 23, Pp Towards Ending the Witch-Hunt, Edit. Singapore. ISBN 10-981-270-708-5: World Scientific Publishing Co. Pte. Ltd.; 2007. pp. 313-324

[72] Niu J, Shen G, Christiaens O, Smagghe G, He L, Wang J. Beyond insects: Current status and achievements of RNA interference in mite pests and future perspectives. Pest Management Science. 2018;**74**:2680-2687

[73] Yalçın M. Perspective Chapter: Pheromone-Based Techniques in Sustainable Pest Management. London, UK: IntechOpen; 2023

[74] Anderson JA, Ellsworth PC, Faria JC, Head GP, Owen MD, Pilcher CD, et al. Genetically engineered crops: Importance of diversified integrated

Pest management for agricultural sustainability. Frontiers in Bioengineering and Biotechnology. 2019;**7**:24. DOI: 10.3389/fbioe.2019.00024

[75] Nagashima Y, Araki Y, Nishida K, Kuramitsu S, Wakabayashi K, Shimato S, et al. Efficacy of intraoperative irrigation with artificial cerebrospinal fluid in chronic subdural hematoma surgery: Study protocol for a multicenter randomized controlled trial. Trials. 2024;**25**:6. DOI: 10.1186/s13063-023-07889-7

[76] Suresh K, Archana S. Biopesticides for integrated crop management: Environmental and regulatory aspects. Biofertilizers Biopesticides. 2014;**5**:1-5. DOI: 10.4172/2155-6202.1000e12

[77] Chandler D, Davidson G, Grant WP, Greaves J, Tatchell M. Microbial biopesticides for integrated crop management: An assessment of environmental and regulatory sustainability. Trends in Food Science and Technology. 2008;**19**:275-283

[78] Reddy DS, Chowdary N. Botanical biopesticide combination concept—A viable option for pest management in organic farming. Egyptian Journal of Biological Pest Control. 2021;**31**:23. DOI: 10.1186/s41938-021-00366-w

[79] Rafiq S, Rashid S, Mantoo MA, Askary TH, Qurat-ul-ain b S, Nusrat ul I. Biopesticides as an opportunistic control measures against insect pests—A review. Just Agriculture. 2022;**2**:2582-8223

[80] Abdollahdokht D, Gao Y, Faramarz S, Poustforoosh A, Abbasi M, Asadikaram G, et al. Conventional agrochemicals towards nano-biopesticides: An overview on recent advances. Chemical and Biological Technologies in Agriculture. 2022;**9**:13. DOI: 10.1186/s40538-021-00281-0

[81] Rizvi SA, George J, Reddy GV, Zeng X, Guerrero A. Latest latest developments in insect sex pheromone research and its application in agricultural pest management. Insects. 2021;**12**:484. DOI: 10.3390/insects12060484

[82] Pan X, Guo X, Zhai T, Zhang D, Rao W, Cao F, et al. Nanobiopesticides in sustainable agriculture: Developments, challenges, and perspectives. Environmental Science: Nano. 2023;**10**:41-61. DOI: doinn10.1039/D2EN00605G

Chapter 10

Perspective Chapter: Biopesticides to Produce Zero-Residue Vegetables

Bibhuti Bhusan Sahoo, Amitabh Panda, Shubhasri Sahoo and Swaraj Meher

Abstract

Vegetables are a vital source of phytonutrients that are essential for human health. However, inadequate availability and phytotoxicity lead to various diseases in our daily lives. Therefore, it is crucial to produce and consume vegetables with minimal or no residues to ensure they are safe for consumption. Biological pesticides are natural products that can be consumed by humans without being harmful to them. The use of bio-pesticides is an effective way to manage different biotic stresses in vegetable production. By adopting this approach, we can confidently ensure safer and healthier vegetable production. This chapter provides a brief explanation of the different types of biopesticides, their mode of action, and their application in vegetable production toward the production of zero residues vegetables.

Keywords: biopesticides, ecofriendly, integrated pest management, vegetable crops, zero toxicity

1. Introduction

Phytonutrients in vegetables are the most essential part of our daily diet because they contribute to strengthening our immunity. In this post-pandemic era, we are more concentrated on boosting our immunity with fruits and vegetables. Our goal is not to protect against the virus, rather to prevent it from infecting us. As a result, we are striving to eliminate medicines by substituting fruits and vegetables for them. The presence of fruits and vegetables in our daily diet ensures an adequate intake of fiber, as a digestive aid. Fiber provides a feeling of fullness and reduces overeating, but the consumption of fruits and vegetables in India is lower than the WHO recommended quantity of 400 g because of limited availability. The limited availability of vegetables may be a result of losses happening at different stages of their entire lifecycle, which can happen at the producer (farmer) level as well as the wholesaler and retailer level. Furthermore, post-harvest losses happen at several points of the supply chain, including harvesting, grading and packaging, handling, shipping, storage, and marketing. Almost all these factors have an impact on the availability of vegetables in India [1]. Very often the injudicious use of undesirable agrochemicals like herbicides, insecticides, and pesticides leads to the spoilage of vegetables expressing a high magnitude of phytotoxicity. Less availability coupled with phytotoxicity is a major cause of hidden hunger. In contrast, the projected Indian population by 2050–2051 is estimated to be 182.35 crores [2]. The consumption of food products has been rising, and the projected requirements will more than double. It is estimated that fruits and vegetable imports will amount to more than 58.2 and 47.0 billion dollars in 2050 (based on 2009 US dollars), compared with negligible vegetable trade in 2009. Feeding the nation is our primary responsibility. Hence it is very clear that, though we are somehow self-sufficient in producing more fruits and vegetables, we are far from producing toxicity-free quality foods. The present day requires a toxin-free diet. Today, our goal is to produce "zero residues" foods. Unseemly application of chemical pesticides is deleterious to fruits and vegetables both in terms of quality and quantity. As a result of the residual impact of these chemicals, human bodies experience severe carcinogenic effects. Human health is adversely affected by the constant deposition of these health-hazardous agrochemicals on the earth. Biopesticides are natural substances derived from animals, plants, and microorganisms like bacteria, cyanobacteria, and microalgae. They are utilized to maintain plant health. With nano biopesticides, specific pests and diseases can be managed as well, and the toxicity of crops can be minimized. Biopesticides create a sustainable environment that is safe and toxicity-free for growing quality foods. The production and consumption of quality foods will eventually result in a sustainable environment for humans. Thus, biopesticides can also be viewed as gifts from God for the coming future.

2. Need of the hour

Global food security is threatened by various factors, including biotic stress caused by viruses and insect pests. But we have an innovative and effective solution to protect crops and maintain a healthy balance between chemical pesticides and the environment, i.e., biopesticides. There are three methods of applying biopesticide: seed treatment, soil soaking, and foliar spraying. Among these, seed priming, which involves providing helpful bacteria to seeds, is a highly effective technique. This method aids in accumulating inorganic compounds and remediating metal-contaminated soils through plants. The beneficial microbes in biopriming can also induce resistance responses against herbivores and phytopathogens by controlling three phytohormones—salicylic acid (SA), jasmonic acid (JA), and ethylene. Elicited defensive responses are classified into two types: systemic acquired resistance (SAR) and induced systemic resistance (ISR). Biopriming is a promising strategy that supports synchronized seed germination and seedling establishment, providing strong immune responses for early seedlings. This natural approach helps crops resist diseases, stay healthy, and decrease reliance on chemical inputs. This study aims to promote agricultural sustainability by using biopriming with beneficial bacteria as a biopesticide, reducing the need for chemical inputs. Microbial bio-agents are highly effective in combating plant pathogens through diverse mechanisms, such as direct antagonism, mixed interactions, and indirect antagonism. These interactions employ a range of strategies, including parasitism, predation, antibiosis, enzyme production, competition for nutrients and space, and inducing plant resistance mechanisms. Moreover, microbial bio-agents not only control plant pathogens but also significantly promote plant growth and increase stress tolerance. These bio-agents play a critical and indispensable role in safeguarding plants against the harmful effects of phytopathogens and significantly enhance overall plant health [3].

The "Green Revolution" in Indian agriculture was undoubtedly a breakthrough that led to a significant increase in crop production. However, the overuse of synthetic pesticides and fertilizers had a detrimental impact on soil quality, the environment, and human health. This resulted in several issues, such as insecticide residues, pest outbreaks, and resistance in over 500 insect species. With the global population set to reach a staggering 9 billion by 2050, the world faces a colossal 70% increase in food demand [4]. Regrettably, agriculture is still grappling with the aftermath of these negative impacts, which are further compounded by the effects of climate change. Therefore, there is an increasing awareness about the paramount importance of pesticide-free, high-quality products, leading to a shift toward alternative approaches like "Green Technologies" for plant protection.

Vegetables are an incredibly valuable crop. However, they can be vulnerable to insect pests and diseases, which can cause significant damage. Chemical pesticides have been commonly used to combat these issues, but their indiscriminate use can have harmful effects on the environment and human health. Moreover, overuse of pesticides can lead to insects developing resistance to them. Therefore, it is imperative to adopt alternative pest management techniques to reduce the misuse of chemical pesticides in vegetable production. Fortunately, biopesticides have emerged as a highly promising option for vegetable growers, despite some challenges that need to be addressed [5].

Biopesticides are a safe and effective solution for controlling pests and pathogens in agriculture. These are natural substances from animals, plants, and microorganisms like bacteria, cyanobacteria, and microalgae. As per the US Environmental Protection Agency, they are derived from natural sources, such as animals, plants, bacteria, and specific minerals [4]. These pesticides consist of genes and metabolites from these sources that help protect crops. Biopesticides are environment-friendly and target specific pests, unlike conventional chemical pesticides. They provide significant advantages in agriculture by protecting crops while being environmentally safe.

3. Importance of biopesticides

The importance of biopesticides lies in their numerous advantages for agriculture and the environment. Here are some key reasons why biopesticides are important:

i. *Environment-friendly:* Biopesticides are derived from natural sources, reducing the environmental impact associated with synthetic chemical pesticides. They are typically less harmful to non-target organisms, along with beneficial insects, birds, and mammals, and have lower toxicity levels.

ii. *Reduced chemical residues:* Biopesticides leave fewer chemical residues on crops, which is beneficial for food safety and reduces potential health risks for consumers.

iii. *Target-specific:* Many biopesticides are highly specific to the pests they target, minimizing harm to beneficial insects and other organisms. This specificity helps maintain ecological balance in agricultural ecosystems.

iv. *Resistance management:* The use of biopesticides can help delay the development of pesticide resistance in pests, as they often work through different modes of action than chemical pesticides.

v. *Sustainable agriculture:* These are an integral part of integrated pest management (IPM) strategies, promoting sustainable agriculture by reducing the reliance on synthetic chemicals and minimizing the impact on ecosystems.

vi. *Safe for farmworkers:* Biopesticides are generally safer for farmworkers who handle them compared to chemical pesticides, reducing the risk of pesticide-related health issues.

vii. *Compliance with regulatory standards:* Biopesticides often have shorter re-entry intervals and pre-harvest intervals, making it easier for farmers to comply with pesticide application regulations.

viii. *Biodiversity conservation:* By preserving non-target species and promoting biodiversity in agricultural landscapes, biopesticides contribute to the long-term health and resilience of ecosystems.

ix. *Consumer demand:* With growing consumer awareness of sustainable and organic farming practices, the use of biopesticides aligns with market preferences and can enhance the marketability of agricultural products.

x. *Reduced environmental pollution:* Biopesticides are less likely to contaminate soil, water, and air compared to chemical pesticides, helping to reduce environmental pollution.

4. Classification of biopesticides

Biopesticides can be grouped into three main categories.

a. *Microbial pesticides:* These use tiny organisms like bacteria, fungi, viruses, or protozoa to control specific pests. Each microbe targets a particular type of pest, making them quite precise in their action. For instance, some fungi control certain weeds, while others target specific insects.

b. *Biochemical pesticides:* These are natural substances that manage pests without any toxicity. They work through mechanisms like interfering with mating (e.g., insect sex pheromones) or attracting pests to traps using scented plant extracts. Unlike traditional pesticides, they do not directly kill or inactivate pests.

c. *Plant-incorporated protectants (PIPs):* These are pest-controlling stuff produced by plants using added genetic material. Scientists insert genes, such as the one for the *Bacillus thuringiensis* (*Bt*) pesticidal protein, into the plant's genetic code. The modified plant then produces the substance that can eliminate specific pests, eliminating the need for external pesticide application.

d. *Botanical insecticides:* Botanical insecticides, popularly known as natural insecticides, are highly effective and safe insecticides that are derived from plants. Organic gardeners often prefer using botanical insecticides instead of synthetic materials because they are safer, healthier, and eco-friendly. To select the right plants for botanical insecticides, it is important to consider these factors, *viz.* perennials, wide distribution, removable parts, minimal water and nourishment requirements, and active ingredients that are effective at low rates.

5. Mode of action of biopesticides

The mode of action of biopesticides can vary depending on the type of biopesticide and the target pest. Here are some common modes of action for different categories of biopesticides.

5.1 Microbial biopesticides

a. *Parasitism:* Certain microorganisms, such as nematodes or fungi, can parasitize and infect specific pests. They enter the pest's body and reproduce, ultimately killing the pest.

b. *Toxin production:* Some microbial biopesticides produce toxins that are lethal to the target pest. For example, *Bacillus thuringiensis* (*Bt*) bacteria produce proteins toxic to certain insects when ingested.

c. *Competitive exclusion:* Microbes can compete with pests for resources or space, making it difficult for the pests to establish themselves.

5.2 Botanical biopesticides

a. *Repellance:* Botanical biopesticides often contain extracts from plants that pests find unpalatable or repellant. This discourages pests from feeding on or colonizing the treated plants.

b. *Growth inhibition:* Compounds in botanical biopesticides can disrupt the growth and development of pests. For example, they may interfere with the molting process in insects, preventing them from maturing.

c. *Feeding disruption:* Some botanicals can disrupt the feeding behavior of pests, making them unable to consume plant tissues.

5.3 Biochemical biopesticides

a. *Interference with pheromones:* Biochemical biopesticides can contain substances that mimic or interfere with the sex pheromones of pests, disrupting their mating behavior.

b. *Attract and kill:* They can also include compounds that attract pests to traps or baits, where they are subsequently controlled or killed.

Biochemical biopesticides act in non-toxic ways, like disrupting pests' growth, using semiochemicals, suffocating or dehydrating pests, creating barriers or irritation, and inducing resistance in plants. These modes of action are non-toxic as they do not involve direct chemical or biological interactions that result in direct lethality to the target pest organism.

5.4 Plant-incorporated protectants (PIPs)

a. *Toxin production:* Genes introduced into the plant's genetic material (e.g., *Bt genes*) enable the plant to produce toxins that are toxic to specific pests when they feed on the plant.

b. *Systemic action:* Some PIPs act systemically, meaning that the entire plant contains the protective substance. This provides continuous protection as pests feed on different parts of the plant.

6. Significance of biopesticides in vegetable cultivation

Biopesticides in vegetable cultivation are natural or biological substances that are commonly used to manage pests and diseases in vegetable crops. Here are some of the key biopesticides used in vegetable cultivation:

Bacillus thuringiensis (Bt): It is a highly useful bacterium that produces proteins that are specifically targeted to certain groups of insects like caterpillars and beetles, making it an effective biopesticide in vegetable farming. It has been successfully used to control harmful pests such as cabbage worms, tomato hornworms, and Colorado potato beetles. Moreover, formulations using *Bt* subsp. *kurstaki* and *Bt* subsp. *aizawai* have proven highly effective in managing various lepidopteran pests. They work well independently or in combination with other biopesticides or biocontrol agents on vegetables [6].

Neem-based products: Neem oil and neem extracts are derived from the neem tree (*Azadirachta indica*). They have insecticidal properties and can be effective against a variety of vegetable pests, including aphids, whiteflies, and leafhoppers. Neem extracts are widely used for plant protection due to their potent properties. Extracts from neem seeds are effective against a wide range of pests, including soft-bodied insects and mites. Aqueous extracts of neem seeds can control mosquito larvae. Neem leaf extracts (**Figure 1**) are used as biopesticides against stored grain pests and as biofertilizers. Neem extracts can be formulated into natural and eco-friendly products for easy application on plants. Nanoformulations of neem extracts have been developed using nanotechnology to enhance their stability and effectiveness. These offer controlled release of active compounds, minimizing non-target toxic effects and improving overall efficacy. Overall, neem extracts provide a sustainable and eco-friendly solution for plant protection [7].

Fungal biopesticides: Entomopathogens play a crucial role in effectively managing insect pests in vegetable cultivation and have been proven to be highly effective. The main components, *Beauveria bassiana* and *Metarhizium anisopliae*, make up 68% of microbial pesticides based on entomopathogenic fungi. These fungi are skilled at infecting and killing a wide range of insect pests in vegetable crops, including aphids, thrips, and whiteflies. *Trichoderma* spp., a beneficial fungus, serves as a biopesticide to control soil-borne pathogens like *Fusarium*, *Rhizoctonia*, and *Pythium*. This

Figure 1.
Neem leaf extract used as biopesticides against stored grain pests.

promotes plant health and reduces the risk of root diseases in vegetables. *Trichoderma* controls soil-borne pathogens by outcompeting them for nutrients and space, producing antimicrobial compounds, inducing systemic resistance, and exhibiting mycoparasitic activity. It also promotes plant growth and root development to establish a stronger defence system. Overall, *Trichoderma* is a valuable biocontrol tool for managing soil-borne pathogens and promoting plant health [8].

Predatory insects: Various useful insects, such as ladybugs (lady beetles), lacewings, and parasitoid wasps, are used as biocontrol agents to manage pest populations in vegetable cultivation. For example, ladybugs can help control aphids, while parasitoid wasps can parasitize and control caterpillar pests.

Copper-based fungicides: Copper-based biopesticides are used to manage fungal diseases in vegetables, including blights and mildews. Copper-based products like copper sulfate, copper sulfate pentahydrate, copper hydroxide, and copper oxychloride have low environmental persistence compared to synthetic chemical fungicides.

Plant extracts: Some biopesticides are derived from plant extracts. These can have repellent or toxic effects on pests. These extracts obtained from plants are capable of serving as biostimulants and plant protection agents. Examples include garlic extract, chili pest, and neem leaf pest etc. Plant extracts (**Figure 2**) are a natural and eco-friendly alternative to synthetic chemical pesticides. They contain toxic compounds that target pests through various mechanisms, reducing their reproductive capacity and causing mortality. Plant extracts can also act as repellents, antifeedants, and ovicides, preventing pests from feeding, laying eggs, and hatching. They can interfere with the growth and development of pests, affecting their life cycle and population dynamics. Additionally, some plant extracts have antimicrobial properties that protect plants from disease-causing pathogens. Overall, plant extracts

Figure 2.
Plant extracts—an eco-friendly alternative to synthetic chemical pesticides.

provide effective pest control while minimizing the negative impacts of synthetic chemical pesticides [9].

Beneficial nematodes: Entomopathogenic nematodes, such as *Steinernema* and *Heterorhabditis* species, are used to control soil-dwelling insect pests like root-knot nematodes and cutworms in vegetable crops.

Virus-based biopesticides: When it comes to controlling insect pests that harm crops, biopesticides are a highly effective solution. These biopesticides take the form of viruses that specifically target the pests, such as whiteflies and caterpillars. Two of the most effective viruses are *Nucleopolyhedrovirus* (NPV) and *Granulovirus* (GV). Additionally, Entomopathogenic viruses like *Helicoverpa armigera* NPV, *Spodoptera litura* NPV, and *Spodoptera exigua* NPV have been found to be highly effective against various insect pests in vegetables. These viruses have already been commercialized and are widely used to control tomato fruit worm (*Helicoverpa armigera*), common armyworm (*Spodoptera litura*), and beet armyworm (*S. exigua*), respectively.

Bacterial biofungicides: Beneficial bacteria, such as *Pseudomonas* and *Bacillus* species, are used to suppress soil-borne diseases like bacterial wilt and damping-off in vegetable crops.

Biochemical pesticides: Biochemical pesticides are natural or non-toxic substances used in vegetable production to control pests and diseases. These work through non-toxic mechanisms, such as disrupting mating behavior or inhibiting pathogen growth. In vegetable production, insect sex pheromones are used to disrupt the mating behavior of pests like moths. By releasing synthetic pheromones, male pests become confused, leading to reduced mating success and fewer larvae.

7. Selection and application of biopesticides

Effective selection and application of biopesticides require careful planning and attention to detail. Integrated pest management (IPM) principles should guide the decision-making process, and compatibility with other pest control methods should be considered. Proper formulation, storage, and application techniques are essential for maximizing the efficacy of biopesticides while minimizing their environmental impact. Adherence to label instructions and safety precautions is crucial for safe and successful biopesticide use in agriculture.

7.1 Integrated pest management (IPM)

a. *Assessment:* Before selecting a biopesticide, conduct a thorough assessment of the pest problem. Identify the target pests, their population levels, and the extent of damage.

b. *Monitoring:* Implement regular monitoring and scouting to track pest populations and assess the effectiveness of biopesticide applications.

c. *Thresholds:* Establish action thresholds to determine when intervention is necessary. IPM emphasizes that treatments should only be applied when pest populations reach economically damaging levels.

7.2 Compatibility with conventional pest control methods

a. *Integration:* Biopesticides can be integrated with conventional pest control methods, such as synthetic pesticides. However, compatibility should be assessed to ensure that the combined treatments do not negatively affect the efficacy of either method.

b. *Sequential applications:* Plan the timing of biopesticide applications in coordination with other control measures to maximize effectiveness while minimizing environmental impact.

7.3 Formulation and storage

a. *Biopesticide formulation:* Different biopesticides are available in various formulations, such as wettable powders, liquid concentrates, or granules. Select a formulation suitable for your specific application method and crop.

b. *Storage conditions:* Follow manufacturer recommendations for storing biopesticides. They are often sensitive to temperature, light, and moisture. Proper storage helps maintain their efficacy and shelf life.

c. *Safety precautions:* Handle biopesticides with care, following safety guidelines and wearing appropriate protective equipment as specified on the product label.

d. *Application equipment:* Ensure that application equipment, such as sprayers or dusters, is clean and properly calibrated to deliver the biopesticide evenly and effectively.

7.4 Application techniques

a. *Timing:* Apply biopesticides at the right time in the pest's life cycle or when they are most vulnerable. This can vary depending on the specific biopesticide and target pest.

b. *Coverage:* Achieve thorough coverage of the plant or soil surface to ensure contact with the pests. Pay attention to the underside of leaves and areas where pests are likely to hide.

c. *Dilution and mixing:* Follow recommended dilution rates and mixing instructions provided by the biopesticide manufacturer.

d. *Adjuvants:* Some biopesticides may require adjuvants (e.g., surfactants or stickers) to enhance their effectiveness. Use these as directed.

e. *Environmental conditions:* Consider factors like temperature, humidity, and wind speed when applying biopesticides. Adverse weather conditions may affect their performance.

f. *Record keeping:* Maintain records of biopesticide applications, including dates, rates, and observations of pest activity and control.

7.5 Adherence to label instructions

a. *Read the label:* Always read and follow the label instructions provided by the biopesticide manufacturer. The label provides critical information on usage, safety precautions, application rates, and timing.

b. *Legal and regulatory compliance:* Ensure compliance with local, national, and international regulations governing the use of biopesticides.

8. Practical instances of biopesticide application in vegetable farming across the globe

Biopesticides have been proven to be highly effective in controlling pests in vegetable cultivation, making them the ideal choice for those looking for eco-friendly pest control solutions.

i. *Bacillus thuringiensis (Bt) in organic corn and tomato production*

Bt-based biopesticides are widely used in organic vegetable farming. For instance, organic corn and tomato growers often apply *Bt* formulations to control caterpillar pests like corn earworms and tomato hornworms. It produces proteins toxic to these pests when ingested, providing an effective and organic-friendly solution [10].

ii. *Neem oil in organic vegetable farms*

Neem oil, derived from the neem tree, is commonly used in organic vegetable production. Organic farmers apply neem oil to control aphids, whiteflies, and other insect pests on a variety of vegetables, including tomatoes, cucumbers, and peppers. Neem oil acts as both a feeding deterrent and a growth regulator for these insects.

iii. *Beauveria bassiana for whitefly management in greenhouses*

Greenhouse vegetable growers often face challenges with whitefly infestations. Beauveria bassiana, a fungal biopesticide, has been effectively used to manage whiteflies in greenhouse tomato and cucumber crops. The fungus infects and kills the whiteflies, providing sustainable control.

iv. *Trichoderma spp. in soil-borne disease management*

Trichoderma species are beneficial fungi used to control soil-borne diseases in vegetable cultivation. Farmers apply Trichoderma formulations to soil or as seed treatments to suppress diseases like damping-off and root rot in crops such as lettuce and cabbage [10].

v. *Ladybugs (lady beetles) in organic lettuce production*

Ladybugs are natural predators of aphids and other soft-bodied insects. Organic lettuce farmers release ladybugs into their fields to help control aphid populations. These beneficial insects feed on aphids, reducing the need for chemical pesticides.

vi. *Entomopathogenic nematodes for root-knot nematode control*

Root-knot nematodes can damage vegetable roots, reducing crop yield. Entomopathogenic nematodes, such as *Steinernema* and *Heterorhabditis* species, are applied to the soil to parasitize and control root-knot nematode populations in vegetable crops like tomatoes and peppers [10].

vii. *Insect sex pheromones for cabbage moth control*

Vegetable growers combat cabbage moth infestations by using insect sex pheromones. These pheromones disrupt the mating behavior of male moths, reducing the chances of successful reproduction and caterpillar damage to cabbage and broccoli crops.

9. Challenges and limitations of biopesticides in vegetable production

The use of biopesticides in vegetable crops comes with several challenges and limitations. These limitations often make them less cost-effective and may not always provide the desired outcome. Limitations of using biopesticides in vegetable farming are as follows:

9.1 Resistance management

a. *Challenge:* Pests can develop resistance to biopesticides over time, reducing their effectiveness. This is especially concerning when a specific biopesticide is relied upon extensively.

b. *Limitation:* Continuous use of the same biopesticide or its over-reliance in pest management can accelerate the development of resistance, making it necessary to rotate or combine different biopesticides or integrate them with other control methods.

9.2 Regulatory hurdles

a. *Challenge:* Biopesticides are subject to regulatory scrutiny and approval processes, which can be time-consuming and costly for manufacturers. These regulations aim to ensure safety and efficacy but can pose challenges for market access.

b. *Limitation:* Some biopesticides may face delays in approval or be restricted in their use, limiting the availability of effective options for vegetable growers.

9.3 Cost considerations

a. *Challenge:* Biopesticides can sometimes be more expensive than synthetic chemical pesticides, primarily due to production and regulatory costs.

b. *Limitation:* The higher cost of biopesticides can be a limiting factor for some growers, especially small-scale farmers with limited budgets. This cost challenge can impact the adoption of biopesticide-based pest management strategies.

10. Future trends and research in biopesticide use for growing vegetables

i. *Microbial innovations:* Scientists are working on new and improved versions of microbial biopesticides, like bacteria and fungi, to better fight a wider range of vegetable pests.

ii. *Plant-based biopesticides:* Researchers are looking into new plant-derived compounds and extracts that can act as pesticides, expanding the options for biopesticides.

iii. *Insect pheromones and behavior control:* Ongoing studies focus on insect sex pheromones and other compounds that can change pest behavior, offering more precise pest control in vegetables.

iv. *Better formulations:* New biopesticide formulas are being created to last longer, stick better to plants, and stay stable. This makes them easier for farmers to use.

v. *Precision application:* Technology is improving how biopesticides are sprayed on crops. Precision sprayers and drones are making it possible to apply them more accurately, saving time, and resources.

vi. *Biotechnology and genetic modification:* Scientists are making progress in adding biopesticide genes to crops. This means that the crops themselves can produce biopesticides, reducing the need for external applications.

vii. *Resistance management:* Scientists are studying how to deal with pests that become resistant to biopesticides. They are looking at strategies like rotating different biopesticides or using them together.

viii. *Nano-biopesticides:* Nano-formulations of biopesticides are being extensively researched to improve their efficiency, stability, and practicality for vegetable farming. Nano-biopesticides are a superior alternative to traditional pesticides due to their diminutive size, enhanced solubility, stability, mobility, and reduced toxicity. These pesticides can be tested against specific pests to gauge their effectiveness across diverse crops. Nano-biopesticides have specific activities against various pests, such as pupicidal activity, larvicidal activity, and anti-feeding activity. Different types of nano-biopesticides with varying modes of action, such as sabadilla, pyrethrum, azadirachtin, and fluoroacetate, exhibit different effects against pests. For instance, the alkaloid toxin of sabadilla causes a loss of nerve cell membrane, which kills most insects immediately after use. On the other hand, pyrethrins alter the process of sodium and potassium ion exchange in insects' nerve fibers, resulting in the inhibition of the transmission of nerve impulses. Similarly, azadirachtin has antifeedant activity and causes reduced food consumption instead of control [11].

11. Indian scenario

Biopesticides are seen as a sustainable solution for pest management in agriculture. Efforts like Sikkim Organic Mission (SOM), National Programme for Organic Production (NPOP), Strengthening and Modernization of Pest. Management Approach (SMPMA), National Mission for Sustainable Agriculture (NMSA), PKVY, and ZBNF support their use in India. While subsidies for conventional pesticides hinder biopesticide research, recent bans on toxic chemicals have boosted biopesticide development. Government policies, entrepreneurial support, and designated no-pesticide zones can promote biopesticide production. Initiatives like SOM and NPOP have led to organic farming growth. The Organic Farming Policy recognizes biopesticides as inputs. NABARD's SMPMA program offers financial aid for bio-fertilizer/bio-pesticide units. NMSA focuses on biopesticide research and commercialization. PKVY and SHM support organic farming, and the government has taken steps to promote biopesticide use. Zero budget natural farming (ZBNF) minimizes costly inputs and promotes biopesticides for organic farming.

Currently, there are 361 bio-production units, including 141 in the private sector without government grants and 38 with government aid. Additionally, 35 IPM centers have received support from the Ministry of Agriculture and Farmers Welfare since 2010. State Biocontrol Laboratories were established in various states, and institutions like the Indian Council of Agricultural Research sponsored Krishi Vigyan Kendras (KVK), State Government sponsored state biocontrol labs, and National Agricultural Co-operative Marketing Federation of India (NAFED) produce microbial pesticides. Public sectors contribute 70% of biopesticide production in India. As of January 1, 2021, India had 970 registered biopesticides under the 1968 Insecticide Act [12]. Government regulations and the adverse effects of chemical pesticides are driving the adoption of environmentally friendly measures like microbial biopesticides in agriculture, both globally and in India.

12. Conclusion

Biopesticides are becoming increasingly popular as a sustainable and effective solution for managing pests and diseases in vegetable farming. Due to their natural origin, low environmental impact, and compatibility with organic practices, they are the preferred choice for many growers. By using biopesticides, growers can effectively control pests while minimizing the use of chemicals on vegetables, thus ensuring superior food safety and quality.

Biopesticides have enormous potential in vegetable farming to create more efficient, sustainable, and environmentally friendly agricultural practices. Despite challenges such as pest resistance and regulatory complexities, careful integration into integrated pest management (IPM) strategies can overcome these obstacles. With ongoing research and the development of innovative formulations like nano-biopesticides, the future of biopesticide use in vegetable farming looks bright. Growers can confidently look forward to healthier crops, safer food, and more sustainable farming practices.

Author details

Bibhuti Bhusan Sahoo[1*], Amitabh Panda[2], Shubhasri Sahoo[2] and Swaraj Meher[2]

1 Regional Research and Technology Transfer Station (OUAT), Bhawanipatna, Odisha, India

2 Krishi Vigyan Kendra (OUAT), Bhawanipatna, Odisha, India

*Address all correspondence to: bibhutihort@rediffmail.com

References

[1] Motkuri V. Vegetable consumption in India: Supply chain and prices. In: MPRA Paper No. 101979, Posted 23 Jul 2020 02:19. Hyderabad: UTC; 2020. Available from: https://mpra.ub.uni-muenchen.de/101979/

[2] Nanda AR, Haub C. The Future Population of India—A Long-Range Demographic View. New Delhi: Population Foundation of India, Tara Crescent; 2007

[3] Zehra A, Raytekar NA, Meena M, Swapanil P. Efficiency of microbial bio-agents as elicitors in plant defence mechanism under biotic stress: A review. Current Research in Microbial Sciences. 2021;**2**:100054. DOI: 10.1016/j.crmicr.2021.100054

[4] Kumar J, Ramlal A, Mallick D, Mishra V. An overview of some biopesticides and their importance in plant protection for commercial acceptance. Plants. 2021;**10**:1185. DOI: 10.3390/plants1006185

[5] Srinivasan R, Sevgan S, Ekesi S, Tamò M. Biopesticide-based sustainable pest management for safer production of vegetable legumes and brassicas in Asia and Africa. Pest Management Science. Special Issue: Natural Products in Pest Management. Sep 2019;**75**(9):2446-2454. DOI: 10.1002/ps.5480

[6] Kumar M, Kumar S, Kumar K. Role of bio-pesticide in vegetable pest management: A review. Journal of Pharmacognosy and Phytochemistry. 2019;**8**(2):1757-1763

[7] Chaudhary S, Kanwar RK, Sehgal A, Cahill DM, Barrow CJ, Sehgal R, et al. Progress on *Azadirachta indica* based biopesticides in replacing synthetic toxic pesticides. Frontiers in Plant Science. 2017;**8**:610. DOI: 10.3389/fpls.2017.00610

[8] Pacheco-Trejo J, Aquino-Torres E, Reyes-Santamaría MI, Islas-Pelcastre M, Pérez-Ríos SR, Madariaga-Navarrete A, et al. Plant defensive responses triggered by *Trichoderma* spp. as tools to face stressful conditions. Horticulturae. 2022;**8**:1181. DOI: 10.3390/horticulturae8121181

[9] Anitha P, Vishnupriya PS. Use of botanical extracts as biopesticides. Acta Scientific Biotechnology. 2021;**2**(6):03-05

[10] Verma DK, Guzmán KNR, Mohapatra B, Talukdar D, Chávez-González ML, Kumar V, et al. Recent Trends in Plant- and Microbe-Based Biopesticide for Sustainable Crop Production and Environmental Security. Singapore: Springer; 2021. DOI: 10.1007/978-981-15-4439-2_1

[11] Hazafa A, Murad M, Masood MU, Bilal S, Khan MN, Farooq Q, et al. Nano-biopesticides as an emerging technology for pest management. In: Insecticides—Impact and Benefits of Its Use for Humanity. London, United Kingdom: Intech Open Limited; 2021. DOI: 10.5772/intechopen.101285

[12] Chakraborty N, Mitra R, Pal S, Ganguly R, Acharya K, Minkina T, et al. Biopesticide consumption in India: Insights into the current trends. Agriculture. 2023;**13**(3):557. DOI: 10.3390/agriculture13030557

Chapter 11

Exploring Behavioral Methods Manipulation for Insect Pest Management

Job Chemutai Alunga

Abstract

This chapter incorporates and reviews tactics of insect pest management using stimuli to manipulate the pest's behavior for the protection of a valued resource. The methods are broadly described by separating and defining them into two groups; methods that influence behavior over a long distance. These are also described as "finding-type behaviors and include volatile chemicals that comprise pheromones, tactile, auditory, and visual stimuli." The second category consists of those manipulating insect behavior at a short distance within a centimeter (<1 cm), also known as acceptance-type behaviors (e.g., involatile chemicals). Precise prominence is placed on approaches and innovations developed through various pest behavior studies and integrating stimuli to enhance efficacy. In this book chapter, three key components of behavioral manipulation techniques are explored and discussed: Pest behavior, how the behavior of the pest is manipulated suitably, and a technique that exploits the behavioral manipulation to protect a resource from an insect pest. Notably, most of the semiochemicals used in this approach are comparatively harmless to vertebrates and beneficial insects, due to being highly selective to the target pest species. Additionally, the methods of behavioral control work together with other control techniques, including conventional methods (IPM).

Keywords: behavioral manipulation, attractants, repellents, semiochemicals, IPM, pheromones

1. Introduction

The hazards associated with the application of insecticides in controlling the injuries inflicted by insect pests have increased the potential possibilities of managing insect pests by manipulating their behavior using external and internal signals. Behavioral method manipulation, a tool for insect pest management, uses stimuli to influence the behavior of the pest for the benefit of protecting a valued resource [1]. This approach is categorized into two distinct groups:

IntechOpen

i. A category that manipulates insect pest behavior over a long distance (finding-type behaviors), these include volatile chemicals (pheromones), auditory, and visual stimuli. Through pheromones, odor-(scent) moderated alterations in insect movements create an entire set of agile behaviors that enable insects to move closer to resources from extended distances, and accomplish local pursuits to detect and consequently assess them [2].

ii. Those that manipulate behavior at a short distance (<1 cm), (acceptance-type behaviors), for example, involatile chemicals.

In this book chapter, I emphasize methods that have been developed through studies of pest behavior and those combining stimuli to increase efficacy. To define the forms of stimuli described, particularly semiochemicals, the common terms of attractant and repellent are used for long-distance stimuli and stimulant, and deterrent for short-distance stimuli. Plant repellents refer to volatiles that cause an insect to orient their movement away from the source [3]. Semiochemical practical applications, including the use of sex pheromones, may lead to modification of pest interactions and signals leading to disruptions in mating, attraction to pint-source baits for pest surveillance, and pest control by mass trapping, push-pull, and attract-and-kill techniques in pest management.

By definition, semiochemicals refer to informative molecules primarily used in plant-insect or insect-insect relations as a substitute or complementary mechanisms to insecticide tactics in diverse integrated pest management strategies [4]. Semiochemicals include allelochemicals and pheromones. Pheromones denote intraspecific communication within members of similar species. On the other hand, allelochemicals, secreted by individuals of one species, influence the behavior of individuals of a dissimilar species; a phenomenon also referred to as an interspecific effect. Allelochemicals comprise various interaction and communication molecules, including kairomones [5], allomones, antimones, apneumones, and synomones. Semiochemicals, therefore, enable manipulation of insect behavior through upsetting the existence/or reproduction of insect pests to control their crop invasions, hence being a strong pillar and foundation of IPM approaches in agriculture. Equally, semiochemicals and plant-derived compounds have been used in health care, especially in malarial control. Plant-derived compounds have been used for controlling mosquito vectors by manipulating their behavior to replace pyrethroids in infusing bed nets [6]. This user-friendly malaria control intervention has been facilitated by an in-depth understanding of the mechanical role of mosquito olfaction, odor coding, and larval ecology that has remained critical for developing novel approaches for disrupting the malaria transmission cycle, hence saving many human lives, especially in the tropics where malaria is endemic. Likewise, manipulating host-finding and acceptance behaviors can be used to transfer much of a very mobile and discriminating insect populace to plants or traps outside our prized agricultural produce and crops. Through the use of trap crop, diversionary crops, insect-refuge crops, and mass trapping can directly decrease the population density of the valued crop below an economic threshold. Studies have shown that behavioral manipulation single-handedly may often be insufficient for plant protection. However, when combined with suitably selected antibiosis factors (resistant varieties or insecticides), a pest-management system combining a behavioral aspect can yield evolutionarily stable and positive effects, hence of great benefit to organic farmers and producers.

2. Principle elements of a behavior manipulation method

The three basic components of a behavioral manipulation technique include:

i. The pest's behavior.

ii. The technique by which the pest's behavior is influenced suitably.

iii. An approach which utilizes the behavioral manipulation to protect agricultural products against pests.

Insect behavior generally is manipulated through inputs to the behavior and, more specifically, through the stimuli that generate these inputs [7]. Contemporary studies reveal, or propose, that parasites transmitted by vectors influence the phenotype of their transmitting organisms and hosts in a way or means that enhances contact among them, thereby increasing the transmission of the parasite [8, 9]. In-depth knowledge of these parasite-vector-host mechanisms and progressions is interesting not only for scientific advancement but also for understanding their significant role in public health, especially in studies involving parasites in real-world situations, such as in medicine and epidemiology [10]. Behavioral manipulations of vectors are phenomena that have been described as so complex that one single method cannot describe or understand them. With the collaboration of multiple disciplines, it should be possible to harness more knowledge, expertise, and experience in behavioral manipulation of vectors in controlling many vector-borne diseases and epidemics like malaria, yellow fever, tick-borne diseases and zika virus fever. Markedly, vector-borne diseases are responsible for more than 17% of all communicable diseases, causing more than 700,000 mortalities per year, with malaria alone causing an estimated 219 million cases globally and resulting in more than 400,000 deaths per year. Behavioral responses that impact insect dispersal are therefore important because of their effect on disease spread.

3. The choice of stimulus to use for behavioral manipulation

The choice of stimulus to use for behavioral manipulation depends on the following attributes:

Accessibility. The stimulus must remain suitable for presentation in a form that the insect can recognize and perceive.

Controllability. An ability to regulate (control) various features of a stimulus, including concentration (intensity) and endurance (longevity), will give better regulation in behavioral manipulation.

Definability and reproducibility. The more precisely a stimulus can be defined, the more precisely it can be replicated artificially.

Specificity. The greater the extent to which a stimulus can be matched with a specific pest behavior, the greater the chance that it can be used to influence that behavior. However, an ever-present stimulus in environments will likely have no effect in influencing particular insect behaviors with the exception that its concentration or quality can affect the insect within the air.

Practicability. The cost of safeguarding an agricultural product from a pest needs to be economical in relation to the control method used, especially to the environment. For example, the chemicals used should not be long-lasting in the environment and should be able to offer protection without making a farm product toxic and non-edible by humans and animals.

4. Examples of behavioral manipulation methods

Behavioral methods for the management of insect pests have been grouped into finding-type behaviors (those that act over a long distance) and acceptance-type behaviors (those that act at a short distance). Types of stimuli, especially chemicals, can be described using two typical terminologies—attractant and repellent, attractants are used for long-distance stimuli, whereas stimulants, in addition to deterrents are used for short-distance stimuli. Earlier authors found the two terminologies to only refer to the culmination outcome of a behavioral reaction toward the inducements (stimuli), however, the two terms are used for being concise and for the reason that frequently a behavior is not well explored in adequate depth or in a sufficient perspective to define definite behavioral reactions to the stimuli applied. A few times, they do not limit themselves to real-world accomplishments but to approaches that are subjected to experimental verification. This is explained by the fact that acceptance of an insect control approach is frequently dependent on external influences, including accessibility, feasibility of other pest control approaches, as well as the incidence of other pests, other than only looking at the efficiency of the technique. Therefore, changing economic situations or prospects of inefficiency of alternative pest control techniques could render an approach that is currently unfeasible to be feasible.

4.1 Finding-type behaviors (stimuli that act over long distances) and their applications

4.1.1 Attract-annihilate/attract-and-kill (A&K) system

This technique is the most utilized behavioral manipulation method in insect pest control. It entails—enticing pestilent insects toward a location where numerous quantities of insect pests can be eradicated from the site of control [11]. A long-distance attractant is used in behavioral manipulation, a process comprising devices that attract and exterminate the trapped insect pests, usually using a gel medium loaded with a pesticide. This technique is popular for various pests with behaviors having long-distance alignment, especially flying as well as walking insects, including some cockroach species and Lepidoptera pests. Frequently utilized attractants include chemical stimuli (e.g., sex pheromones), but visual stimuli are also used (intentionally or incidentally) and auditory stimuli could also be used [7]. The technique is particularly effective when it aims to control insects with gregarious habits. In Uganda, attractants (ethanol 30%) have been used in the control of the Black Coffee Twig Borer (BCTB) (*Xylosandrus compactus)* and the Coffee Berry Borer (CBB) *(Hypothenemus hampei)* in coffee plantations (**Figures 1–3**). Water infused with surfactant (soap solution) is used to drown (kill) the BCTB beetles.

Insect trap using ethanol (30%) attractant in a coffee plantation: Attract-and-kill (A&K) System.

Chemical stimuli: Sex pheromones have been selected for managing various pest species, especially Lepidoptera. The attract-annihilate method usually uses these compounds because of their attributes [12], for example, selectivity, eliciting distant

Figure 1.
Xylosandrus compactus (BCTB).

Figure 2.
BCTB damage on coffee twigs.

Figure 3.
BCTB trap with an attractant (ethanol 30%) in vial and a basin infused with soapy water.

reactions, and prolonged existence within the environment. These chemical stimuli enable monitoring of insect pests, mass trapping, and mating disruption as well as attract and kill measures in insect management.

The inadequacy of a sex pheromone attracting males can only be overcome by mixing it with an attractant for females. A blend pheromone sex-specific compounds, that attracts males, in addition to an edible bait (a blend of eugenol, geraniol, and phenethyl propionate), that mainly draw females. It has received application in managing the Japanese beetle, *Popillia japonica*. Aggregation pheromones have been used in attracting males and females, including immatures, for example, nymphs of the German cockroach *Blatella germanica*. This capacity to selectively draw female individuals enables such chemical compounds to be better matched to the attract-annihilate approach to insect management.

Visual stimuli: Visual stimuli alone are much less commonly used in attract annihilate methods than chemical stimuli. LED (light-emitting diodes) equipment having different wavelengths can currently be mass-produced owing to current technological developments, and state-of-the-art agricultural technologies based on light are beginning to draw attention. Agricultural insects and stored-product insects are influenced by luminance intensities, exposure times, and wavelengths of LEDs. It is anticipated that advancements in this technology will help establish the next generation of pest control technologies based on light and visual stimuli. Numerous hemipterans and species within various orders get drawn to light within the green-yellow section of the spectrum. Such attractant mechanisms of insects has received wide application leading to the emergence of yellow sticky traps to enable surveillance and monitoring of various pests in commercial plantations, greenhouses, and orchards. LED light sources have also been found helpful and effective in managing agrarian and storage-produced insects [13]. Using LED technology for glasshouses and silos alongside conventional traps, decreases crop losses caused by beetles, aphids, weevils, and moths. LEDs, therefore, possess great potential in integrated pest management (IPM) especially in organic farming and the production of high-value agricultural produce where chemical residues are highly undesired.

Visual and chemical stimuli: The visual and chemical stimuli are frequently used in combination [1, 14], this enhances the efficiency and efficacy of the method other than when either stimulus is used independently, for example, in Tsetse (*Glossina* spp.) management and control.

4.1.2 Distracting insect behavior through the use of attractants in addition to repellents

Distracting insect pest behavior using a stimulus that in one or another way induces or hinders coordination is applicable and successful at distant places and involves the use of attractants and repellents [15].

Attractants: These apply and work by utilizing attractants to distract a discovery (find) behavior. The method concentrates on mate discovery, predominantly in moths, popularly known as the mating disruption technique. Large quantities of artificial sex pheromones get disseminated to inhibit males from accessing females. This technique has been applied effectively for the management and control of such herbivorous insect pests as the Oriental fruit moth, *Grapholita molesta*, on stone fruits, the pink bollworm, *Pectinophora gossypiella*, on cotton. In nearly all circumstances where the technique has been tried, disruption of breeding (mating) has been at least partly attained in the treated zone. This combination of the technique with an insecticide to kill the attracted pest is thought to increase the control of bollworm over the pheromone alone [16].

Repellents: The tactic that uses repellents has mainly been used to hinder pests from locating a treasured product. Valuable repellents could be extracted from natural derivatives of insect origin, including defense secretions, they can also be solely synthetic, like several insect killers. Mosquito repellents that are used to keep mosquitoes away from biting humans are some of common practical uses of repellent volatile compounds.

4.2 Stimuli acting at close distance (acceptance-type behaviors)

Following the arrival of a valued product, insects will normally perceive an extra stimulus. Such a stimulus could enhance a behavior, retaining the pest on the valued product; it may also deter its behavior, leading to refusal and perhaps deviation from the valued product. Chemical compounds are the prominent short-range stimuli utilized in behavioral manipulation [17].

Stimulants: Several popular stimulant-based compounds play a role in controlling feeding and oviposition, especially insect feeding. Stimulants have many advantages, including enhancing exposure to toxins that need to be ingested and applicability to a wide range of pests, for example, in the case of sucrose. Feeding enhancements are normally of particularly value when combined with poisons [18] due to being able to enhance interaction between the pests and toxins which could be repressed when the insects reacts to poisons by stopping feeding, therefore, escaping a killer dosage (for example, endo-toxins produced by Bt bacteria).

Deterrents: A deterrent is a biochemical that inhibits behavior, including feeding or oviposition when introduced to a spot where such behavior usually takes place [19]. A deterrent is applied directly to a resource to stop or lessen the special effects of a pestilential behavior for example feeding in pest management. The effectiveness of the deterrent is subject to the physiological state and behavioral responses of the pest throughout the preliminary and ensuing encounters with the deterrent. Repetitive exposure and increasingly extended phases of deprivation of feeding and ovipositional sites have been shown to reduce several insect pest responses to deterrents.

Combining deterrents and attractants/stimulants: The efficiency of a deterrent-based technique can be amplified if utilized in a blend with an alternative technique that draws an insect pest to a nonvalue resource in a stimulo-deterrent deviation (3) or push-pull approach [20]. An assortment of approaches can address those challenges using deterrents applied on or next to a valued product. A stimulo-deterrent strategy (SDD) has been developed to manage herbivore insects, however, can be applied to several pests and several resource types. The strategy has only been tried and tested against a small portion of herbivores in small fields and glasshouse experiments and has not yet been explored commercially. The "push-pull" approach, a revolutionary tool for integrated pest management programs, utilizes an integration of behavior-modifying stimuli to influence and manipulate the spread and profusion of insect pests and/or natural enemies [21]. In this tactic, insect pests get deterred outwards from the principal crop ("push") by utilizing a stimulus that is deterrent or repellent. Simultaneously, the pest gets attracted ("pull") by use of an extremely apparent stimulus that attracts the pest to alternate sites (a trap crop), here the pest is focused and aggregated, enabling and easing its control.

The insect pests are concurrently attracted (pull), using extremely apparent and attractive stimuli, to alternate sites and areas such as traps or trap crops where they are focused and concentrated, enabling and easing their control.

Vibrational signals as a tool for disrupting mating: Replays of rival signals provide an impeccable illustration of a species-specific strategy for pest management while having a non-significant environmental impact. Insect pests including stinkbugs that signal each other at intermediate (medium/short range) using vibration signals, depend on the reception and discernment of co-specific substrate-borne signals and cues for detection in addition to finding prospective mates. Through presenting disrupting cues, that imitate natural stinkbug female rivalry cues, incorporated within the host plants, mating pair formation may be influenced leading to mating disruptions [22] hence controlling the pest population. Outstandingly, vibrations in addition to semiochemicals, contribute significantly to close-range communication and host approval (acceptance) of insect pests. As proposed by numerous authors, hi-jacking insect vibrational communication and interactions make it feasible to manipulate insect behaviors, comprising of mating, permanency on hosts, oviposition, and feeding to manage the target host [23].

Internal stimuli: A unique exemption to the broad view that inner stimuli are out of reach for influencing insect behavior includes the sterile male method; here a multitude of sterile males are let out to breed with wild females. This breeding brings out numerous similar inner inputs within females like those bred to standard fertile male insects. Notably various behavioral changes are induced in females by these internal contributors. Of paramount importance is that the female becomes noncompliant to additional matings, and in species in which the female mates a single time, it remains refractory forever and henceforth never being able to be fertilized and reproductive. This tactic has a big potential in the control and management of major pests like mosquitoes, Tsetse (*Glossina* spp.), and Black Coffee Twig Borer (BCTB) *Xylosandrus compactus.*

Additionally, the use of internal stimuli has been vital in the manipulation of hosts by parasites. Its application has been published and documented in hundreds of host-parasite interactions and associations across various major living organism phyla. Modifications in simple parasite behavior patterns can influence the host to orient and move toward another microhabitat, making it more prone to predation, or ensuring that by all means the transmission of the parasite occurs. This systematic manipulation of the host by the parasite may be explained as any modification in the host phenotype, with the influence and inducement of parasite genes (parasite genome) that inherently possess fitness benefits for the success of the parasite. With the background of the distorted host behavior, this largely indicates that the infected hosts respond and act in ways that enable the spread and dispersal of the parasite, and consequently the completion of the parasite's life cycle [24]. Poulin [24] further indicates that the genome of the parasite facilitates direct or indirect modification of the infected host by modifying its phenotypic traits to facilitate the parasite transmission for the completion of its lifecycle.

5. Limitations of the development and adoption of behavior-modifying tactics in integrated pest management (IPM)

a. Technical challenges

i. High specificity: Sex-based semiochemicals (pheromones) are extremely selective (specific), therefore, limiting their use in approaches including mating disruption that may be limited and necessitate management of multiple pest species. Additionally, only one sex is attracted.

ii. Ultimate eradication of the pest is hardly ever attained. Definitive control is vital when managing a pest with zero ("0") or minimum tolerance.

iii. These approaches do not provide adequate control as stand-alone controls for some pest species.

iv. They are normally less efficient when the pest pressure is high.

They require a mass application (wide-area), for example, mating disruption interventions.

b. Socio-economic challenges

i. Expensive (input costs are high).

ii. Pesticides are normally more affordable and have wider programs.

iii. They require many farmers to implement simultaneously for their effectiveness.

iv. They necessitate intensive sensitization and training programs including on-farm field schools and demonstrations. These are needed to change farmers' perceptions of the advantages and benefits in comparison to alternative approaches.

v. High competition with pesticides. Pesticides are often cheaper and have broader programs).

c. Policy-linked challenges

i. Regulatory: Some naturally derived chemicals may not tolerate exemptions.

ii. Registration: They attract the interest of high-end, largely cultivated, and widely consumed crops. Usually, the market volumes will determine and influence the interest of the industry and sector to advocate for registration.

6. Prospects

The future of behavioral manipulation and its perceived advances are envisaged to be based on ongoing research and innovations. New developments will especially depend on comprehensive and in-depth investigations and lessons learned on behaviors and semiochemical compounds in various insect species [25]. Particular research and studies on the insect physiological processes that influence insect behaviors can help characterize additional internal insect stimuli and reveal many newer methods for utilizing them, for example, insect neuropeptides. Microorganisms could be genetically engineered to provide a more economical production of semiochemicals than synthetic chemical routes.

7. Conclusion

Ongoing preferences for sustainable ecologically friendly alternatives to broad-spectrum insecticides or approaches that minimize the use of these insect killers have never been more in demand. Advancements and increased uptake of behavioral manipulation methods can facilitate meeting this demand, given that the amount of semiochemicals that are introduced into the ecosystem is comparatively small, hence making the method more environmental friendly. Additionally, most of the pheromones (semiochemicals) used are comparatively not noxious or harmless to vertebrates, crop pollinators, beneficial insects like bees, and natural enemies. Besides, semiochemicals exhibit extreme selectivity in their operation and hence very safe for non-target insects and species. To maximize their benefits, techniques of behavioral control work with other control methods such as conventional methods (IPM) for more sustainable management of insect pests. Behavioral methods of manipulation for insect pest management not only benefit the consumers but also the ecosystem and all those involved in the productive chain of agricultural products. This environment-friendly pest control method should be explored further amidst the threats of increased pest and disease occurrences and resurgences associated with the effects of climate change.

Funding

No external funding was used in this research.

Conflicts of interest

The author declares no conflict of interest.

Author details

Job Chemutai Alunga
National Coffee Research Institute (NARO-NaCORI), Mukono, Uganda

*Address all correspondence to: chemujob@yahoo.com

References

[1] Chandana CR, Nadagouda S. Behavioural manipulation of insect pests in integrated pest management?. Biological Forum – An International Journal. 2023;**15**(10):1554-1561

[2] Wechsler SP, Bhandawat V. Behavioral algorithms and neural mechanisms underlying odor-modulated locomotion in insects. Journal of Experimental Biology. 1 Jan 2023;**226**(1):jeb200261. DOI: 10.1242/jeb.200261. Epub 2023 Jan 13. PMID: 36637433; PMCID: PMC10086387

[3] Cowles RS. Manipulation of host-finding and acceptance behaviours in insects: Importance to IPM. In: Integrated Pest Management: Potential, Constraints and Challenges. 2004. pp. 185-204. DOI: 10.1079/9780851996868.0185

[4] El-Ghany NMA. Semiochemicals for controlling insect pests. Journal of Plant Protection Research. 2019;**59**(1):1-11. DOI: 10.24425/jppr.2019.126036

[5] Murali-Baskaran RK, Sharma KC, Kaushal P, Kumar J, Parthiban P, Senthil-Nathan S, et al. Role of kairomone in biological control of crop pests-a review. Physiological and Molecular Plant Pathology. 2018;**101**:3-15. DOI: 10.1016/j.pmpp.2017.07.004

[6] Muema JM, Bargul JL, Njeru SN, Onyango JO, Imbahale SS. Prospects for malaria control through manipulation of mosquito larval habitats and olfactory-mediated behavioural responses using plant-derived compounds. Parasites and Vectors. 2017;**10**(1):1-18. DOI: 10.1186/s13071-017-2122-8

[7] Foster SP, Harris MO. Behavioral manipulation methods for insect pest-management. Annual Review of Entomology. 1997;**42**(July):123-146. DOI: 10.1146/annurev.ento.42.1.123

[8] Hughes DP, Libersat F. Parasite manipulation of host behavior. Current Biology. 2019;**29**(2):R45-R47. DOI: 10.1016/j.cub.2018.12.001

[9] Safari M, Ferrari MJ, Roossinck MJ. Manipulation of aphid behavior by a persistent plant virus. Journal of Virology. 2019;**93**:e01781-18. DOI: 10.1128/JVI.01781-18

[10] Lefèvre T, Koella JC, Renaud F, Hurd H, Biron DG, Thomas F. New prospects for research on manipulation of insect vectors by pathogens. PLoS Pathogens. 2006;**2**(7):0633-0635. DOI: 10.1371/journal.ppat.0020072

[11] Mfuti DK, Tamiru A, Kirk WDJ, Akinyemi AO, Campbell H, O'Brien M, et al. Exploiting thrips aggregation pheromones to develop a lure-and-kill strategy for the management of the bean flower thrips. Agronomy. 2021;**11**(7):1269. DOI: 10.3390/agronomy11071269

[12] Reddy GVP, Sharma A, Guerrero A. Advances in the use of semiochemicals in integrated pest management: Pheromones. In: Biopesticides for Sustainable Agriculture. Cambridge, UK: Burleigh Dodds Science Publishing; 2020. pp. 251, 251-281, 282. DOI: 10.19103/as.2020.0073.12

[13] Park JH, Lee HS. Phototactic behavioral response of agricultural insects and stored-product insects to light-emitting diodes (LEDs). Applied Biological Chemistry. 2017;**60**(2):137-144. DOI: 10.1007/s13765-017-0263-2

[14] Guignard Q, Slippers B, Allison J. Chemical and visual ecology of the symphyta. Agricultural and Forest Entomology. 2022;**24**(4):453-465. DOI: 10.1111/afe.12510

[15] Rehermann G, Spitaler U, Sahle K, Cossu CS, Donne LD, Bianchi F, et al. Behavioral manipulation of *Drosophila suzukii* for pest control: High attraction to yeast enhances insecticide efficacy when applied on leaves. Pest Management Science. 2022;**78**(3):896-904. DOI: 10.1002/ps.6699

[16] Rodriguez-Saona CR, Stelinski LL. Behaviour modifying strategies in IPM: Theory and practice. In: Peshin R, Dhawan A, editors. Innovation-Development Process. Springer Science and Business Media B.V; 2009. DOI: 10.1007/978-1-4020-8992-3_11

[17] Ramalakshmi V, Dash L, Padhy D, Rout S, Tripathy B. Role of semiochemicals in pest management. In: Rout S, editor. Agriculture and Forestry: Current Trends, Perspectives, Issues—I. Vijayawada, Andhra Pradesh; 2020. pp. 209-228

[18] Mafra-Neto A, De Lame FM, Fettig CJ, Munson AS, Perring TM, Stelinski LL, et al. Manipulation of insect behavior with specialized pheromone and lure application technology (SPLAT®). ACS Symposium Series. 2013;**1141**:31-58. DOI: 10.1021/bk-2013-1141.ch004

[19] Agarwal ML, Sunil V. Basic behavioural patterns in insects and applications of behavioural manipulation in insect pest management. Journal of Entomology and Zoology Studies. 2020;**8**(2):991-996. Available from: http://www.entomoljournal.com

[20] Cook SM, Khan ZR, Pickett JA. The use of push-pull strategies in integrated pest management. Annual Review of Entomology. 2007;**52**(February):375-400. DOI: 10.1146/annurev.ento.52.110405.091407

[21] Zhang Z, Sun X, Luo Z, Gao Y, Chen Z. The manipulation mechanism of "push–pull" habitat management strategy and advances in its application. Acta Ecologica Sinica. 2013;**33**(2):94-101. DOI: 10.1016/j.chnaes.2013.01.005

[22] Mazzoni V, Anfora G. Behavioral manipulation for pest control. Insects. 2021;**12**(4):10-11. DOI: 10.3390/insects12040287

[23] Avosani S, Nieri R, Mazzoni V, Anfora G, Hamouche Z, Zippari C, et al. Intruding into a conversation: How behavioral manipulation could support management of *Xylella fastidiosa* and its insect vectors. Journal of Pest Science. 2024;**97**(1):17-33. DOI: 10.1007/s10340-023-01631-7

[24] Poulin R. Parasite manipulation of host behavior: An update and frequently asked questions. In: Advances in the Study of Behavior. 1st ed. Vol. 41, Issue C. Cambridge, Massachusetts: Elsevier Inc.; 2010. DOI: 10.1016/S0065-3454(10)41005-0

[25] Sharma A, Sandhi RK, Reddy GVP. A review of interactions between insect biological control agents and semiochemicals. Insects. 5 Dec 2019;**10**(12):439. DOI: 10.3390/insects10120439. PMID: 31817457; PMCID: PMC6955951

Chapter 12

Artificial Intelligence (AI) and Insecticides in Pest Control

Seyed Reza Mousavi and Arman Hakemi

Abstract

The growth in global food demand has prompted the development of pest control strategies, with pesticides being widely employed as a means of controlling pests. However, the excessive use of pesticides has resulted in environmental contamination, the development of pesticide resistance, and adverse effects on human health. Integrated Pest Management (IPM) represents a more sustainable approach, but its implementation is frequently impeded by a dearth of trained personnel and the high costs associated with it. The application of artificial intelligence (AI) has the potential to transform IPM practices, enhancing pest management and reducing reliance on pesticides. This study investigates the potential of AI in IPM, specifically its use in analyzing data from a range of sources, including imaging, acoustic sensors, and weather data. AI systems have the capacity to identify pests, predict diseases, and suggest management strategies, thereby reducing the reliance on pesticides. The integration of AI into IPM can address the shortage of professional personnel and the high cost of traditional IPM methods. This novel approach has the potential to transform the pesticide industry and make food production more sustainable and environmentally friendly.

Keywords: artificial intelligence, integrated pest management, insecticides, pest control, sustainable agriculture

1. Introduction

The term "pest" is used to describe diseases that affect human resources or threaten human health. These may include insects, mites, ticks, nematodes, mollusks, vertebrates, microorganisms, and plants. However, labeling pests simply because they compete with humans is not environmentally friendly. This is affected by particle density. Pest control can be used to deter pests. Correct identification of pests is essential for pest control. Bacteria thought to be harmful can disrupt ecosystems and cause new pest problems. It is important to understand the feeding and relationships between pests and their natural enemies. Unless appropriate pest management measures are implemented, the impact of major pests on resource yield is significant. Sometimes the pests that cause damage are different from those that have the potential to become pests if changes are made. Migratory diseases often destroy crops and cause serious damage. It is important to remember that insects are harmless and beneficial to the ecosystem. This includes the use of entomopathogens, selective insecticides, insect pheromones, genetic manipulation, and computer modeling to determine what is causing the disease. Implementing integrated management systems requires collaboration among many groups, including agronomists, ecosystem management experts, and pest experts. Use IPM strategies when indicated, because it requires personnel and resources. The goal is to maintain ecosystem balance and assess financial losses before taking action. IPM considers the entire ecosystem, including human factors, and includes a deep understanding of crop biology, identification of key pests, and assessment of environmental factors that may affect pest spraying. This approach anticipates the unexpected, reflects the diversity of ecosystems, and is effective. Adhering to IPM principles enables managers to use resources effectively, efficiently, and ethically. It has the potential to transform agriculture and bridge the digital divide in rural communities, thereby contributing to the Fourth Industrial Revolution.

There are many benefits to using smart technology in agriculture, including the ability to manage and farm more efficiently. It helps with water management, crop production, pest management, nutrient management, and crop rotation strategies. By analyzing data on conditions such as rainfall, temperature, wind speed, and solar radiation, AI can improve crop management, solve problems such as weeds and pests, and increase the profitability of agribusiness. AI is also being used in the development of unmanned tractors, agricultural robots, pesticides, and crop and soil monitoring. It is constantly evolving. The Internet of Things (IoT) helps in instantaneously monitoring important factors such as humidity, temperature, and soil health, reducing resource usage. It also supports precision agriculture, smart greenhouse management, data analytics, agricultural drones (UAVs), and animal monitoring. IoT identifies patterns and relationships between pesticides and their toxicity levels. This method can be used to predict pest toxicity and assess human health risks. In other words, machine learning algorithms analyze data collected by AI against massive datasets, allowing AI to detect pests and suggest remediation plans.

2. Pest control strategies

It is estimated that current agricultural profits will fall by 60% if effective protection is not provided. For example, the loss of rice, an important part of the human diet, is directly related to the increase in world hunger. This section examines the various pest control methods currently in use and discusses the impact of technology on the success of these methods. Various types of pest control are used in insect farming (**Table 1**). The classification of these methods shows the existence of six different groups: cultural measures, physical control and control, chemical control, biotechnological control,

Pest control methods	→	Cultural practices
		Physical and mechanical control
		Chemical control
		Biological control
		Biotechnological control
		Genetic control

Table 1.
Pest control methods.

and genetic control. Among these methods, chemical control (pesticides) is the most popular pest control due to its rapid effect, low cost, high efficiency, and ease of use. Therefore, intensive use of these chemicals can pose a significant risk to nontarget organisms and humans. In addition, excessive and inappropriate use of these chemicals can cause many ecological problems and environmental damage [1].

In addition, incorrect use time and use of chemical combinations can also cause pesticide presence in products. However, carelessness in the use of pesticides, lack of knowledge, and training can cause serious damage to farmers. Given the problematic impact of pesticides on nontarget organisms, there is a strong preference for using multiple selective management strategies. They should primarily be used as part of your overall planting plan for healthy growth and pest resistance. The delivery of biotechnological tools and control of body or chemical resistance is another avenue to consider. The above methods, such as biotechnology and biological control, are used to control pests below the financial loss threshold. Significant progress has been made in pest control in recent years with the advent of biotechnological methods. These methods include disrupting relationships with pheromones, releasing pathogens, and creating plant mutations [2].

3. Integrated pest management

In the 1970s, a new pest management approach called IPM was developed. IPM is an ecological pest management system that relies on natural factors such as natural enemies and weather to control pests. The goal is to identify management strategies that affect these factors as little as possible. IPM uses pesticides only when pest management and protection indicate that they are necessary. The best pest control plans include all aspects of pest control, including passive, and consider the various management, cultural, climatic, and other influences of the pests and crops to be protected. IPM allows the use of technologies that are very different from traditional pest control methods, including natural enemies, cultural practices, crop and animal protection, microbial control, seed genetics, chemical messengers (including sex attractants) and pesticides. These techniques can be used in combination with each other, rather than together or even backwards as is often the case in pesticide control. The IPM program has six components: (1) workers: (2) producers, (3) managers; (4) information and knowledge, (5) biological and ecological knowledge required to develop the program; (6) the resources involved and their effects on the physical and biological environment vary seasonally and must be monitored and evaluated. An important point is the integration manager or consultant. It is the responsibility of the manager or consultant to continuously monitor the situation, identify pests based on experience and knowledge, and select and monitor appropriate equipment and personnel. Most importantly, pest managers always understand the ecological and biological impacts of their actions. This knowledge should be extended not only to pests but also to natural and non-pest enemies [3].

Agricultural entomologists have played an important role in the development of pest management. Taking into account the specific difficulties posed by pests, the group of entomologists first described the level of work and initiative, as well as the content of self-management.

3.1 Pest detection and monitoring systems

In the field of remote sensing, the detection of plant pests has been supported in recent years by the use of wireless sensors and some motion/vibration sensors.

1. Image acquisition
↓
2. Image pre-processing
↓
3. Image segmentation
↓
4. Feature extraction in image
↓
5. Detection and classification of plant disease

Table 2.
Basic steps for plant disease detection and classification.

Currently, acoustic sensors can record specific frequencies produced by pests. However, when these sensors are used, it is still difficult to detect large areas of pest population. At least, they can indicate that there are some faults in the area. In addition, image processing and optical equipment combined with pheromone technology can directly detect pests. Correct pest identification is critical when implementing an IPM strategy (**Table 2**). Specialized acoustic, ultrasonic, and optical sensor technologies are now available to detect, identify, and identify pests [3].

3.1.1 Acoustic sensors

Acoustic technology has been used to detect pests in soil, trees, and plants since the twentieth century. The technology has advanced to the point where it can detect defects that would otherwise go unnoticed. The advent of modern computer technology has led to the development of signal processing techniques that help separate sound fragments from background noise, spurring research into new acoustic instrument designs for detecting and tracking ground intrusion lines. Recently, acoustic technology has been used to detect insects in insect feeding areas. Insect identification is usually achieved by using low-pitched sounds (0.5–150 kHz) produced by insects during activities such as flying, feeding, and talking [4].

3.1.2 Ultrasonic sensors

Insects hiding in seeds, trees, and other plant materials can be detected by emitting ultrasonic signals while feeding. An ultrasonic testing machine consists of several components, including an ultrasonic generator, a low-band amplifier, a cold signal, and an output device. In a system that includes this product, the ultrasonic signal emitted by the sensor is returned after hitting the line and sent to a machine that can process the digital signal. Therefore, the line direction of movement and the accuracy of the field can be estimated instantly. However, this system has not yet shown consistent performance in different environments. The system works well in areas with background noise and its ability to detect multiple lines is unknown. Further research and development is needed to develop a smart, portable ultrasonic device for field pest monitoring. No new research has been conducted since the 1990s to confirm this result [4].

3.1.3 Optical sensors

This system uses sensors that use infrared light at the entrance of the device. The trap is combined with an attractant, which can be a pheromone, light, or bait, and records the time the pest enters the trap. The optical sensor emits light at the entrance of the trap, and when the target insect enters the trap where the attractant is located, the light is blocked. This movement causes a change in the voltage in the system controlling the sensor, which is the signal that the pest has entered the trap. Determining the date when the pest was first caught in the trap and its population density plays an important role in IPM. To be clear, specific preferences should be used for each species (**Figure 1**).

Optical sensors in general have the advantage of counting pests. However, this approach has two important limitations. First, it is not possible to detect the presence of a pest in a trap without the specific features for that species. Second, it does not work well when used in low light. It is clear from the studies that more research is needed to define the species more precisely [3, 4].

3.1.4 Monitoring remote climate data

Weather data such as wind direction, wind speed, temperature, humidity, precipitation, leaf humidity, solar radiation and humidity are important for these models. Since agricultural pests are cold-blooded organisms, their biological activities are based on the climate data of the environment they are in. For pests living in fruits, the development of the pests will directly affect the climate data of the environment they live in. The most important of these is technology, which is located in the human core of the system, namely, the decision makers. An integration manager or consultant is the key to this efficient operation. He/she must constantly "read" the situation, evaluate the pests based on his/her experience and knowledge, and make and select the appropriate products and agents according to the situation. Pest managers should always be aware of the ecological and biological impact of their actions not only on pests but also on natural enemies and non-pest diseases. A lack of integrated pest control personnel or consultants poses a major challenge in implementing the program. On the other hand, IPM is a business venture, but if the energy, material and/or labor required to produce the resource can be reduced then the material, power and labor will be available according to the needs of other communities [5].

Figure 1.
Design of laser scanning application.

4. Integrated pest management and artificial intelligence

Technology and techniques used in IPM help in determining the best time for remote weather monitoring, remote pheromone monitoring, and DD (Degree Day) models. Degree days (DD) can be calculated as daily average temperatures (including daily maximum and minimum temperatures). However, even for farms or large enterprises using traditional production methods, this technology requires a lot of energy and labor costs. In this case, advanced management of the system requires the use of intelligent algorithms that use the information obtained from this technology. The system should be able to identify the biological diseases of the pest (using information such as color, pattern, width, length, and drawing tools for diagnosis). The system should be able to continuously monitor malware using the same operating system. The system should be able to collect important information about the weather conditions (temperature, humidity) affecting the pests from the environment where the pests live with high accuracy. The system should be equipped with a mathematical model that can evaluate the weather data in real time. The system should be able to process the information received by processing images and sensors and evaluate the pests instantly. The system should be able to communicate with professionals or companies *via* online messaging. The system should be accessible from a smartphone or tablet with Android or iOS, depending on the application. The software should be able to recommend appropriate licensed (pesticide and/or biotechnology) products based on the latest information according to the critical time of the pest [5].

4.1 Artificial intelligence in integrated pest management

The following areas have been identified as requiring further research and analysis. Develop tools to reduce chemical residues in food and environmental pollution needs more research. Work safety among farmers needs further research. Obviously, the above information has been obtained from various studies and practical applications. Although there are many studies on IPM, processing, analyzing, and using this data require input from IPM experts. The most important challenge is the limited number of experts available to provide information to farmers. The following article discusses the integration of IPM and expertise, focusing on the application of expertise in the context of IPM. More research and development is needed on this topic. The process for entrepreneurs should be as follows:

1. Select the agricultural branch needed for the business

2. Determine the growth rate of this line (minimum: 6.4°C, maximum: 40.2°C).

3. Create a collection of information in the field

4. Create DD models of specific pests

5. Monitoring of study Software development (such as image processing algorithms)

6. Build an intelligent algorithm to interpret all this information and make the right decision

7. System Validation

8. Continue research and development activities by adding new agricultural lines to the system [1, 6, 7].

4.2 Deep learning

Three different types of learning are used in machine learning: supervised learning, semi-supervised learning, and unsupervised learning. Supervised machine learning is provided with appropriate training data, while unsupervised machine learning can learn independently without receiving training data. Semi-supervised learning systems include some, if not all, of the required training materials, and less training material is properly documented. Specialized neural network techniques are used to provide solutions in these courses [8].

A selection of the most commonly used neural network architectures in deep learning is shown below.

4.2.1 Deep neural networks

A deep neural network (DNN) is a type of specialized training. An artificial neural network is a type of ANN that has a large number of layers and uses mathematical operations to simplify the connection between the input process and the output process. The network moves from one layer to the next by calculating the probability of each event. Deep neural networks can be trained to learn about any topic and include other topics by calculating the probability of occurrence based on the network training data. This is done by conducting research on the product of interest, analyzing its features in depth, and creating assumptions that can be used to differentiate the product from many similar products. To explain this, a deep neural network trained to recognize cassava lines can be used in this way. Given an image, the network will make a comparison by comparing the image with the previously obtained features of cassava stalks. Finally, the network decides that the image was likely to be a potato stalk. A result of 1 means that the network has determined that the image is of a cassava stalk. A chance of 4 out of 5 means that the system thinks the image is closest to a potato stalk. Instead, a probability of zero provides clear evidence that the image is not actually a cassava plant. The purpose of these networks is to identify atomic-level issues. At this stage, artifacts and their connections are given partial attention. These values or weights are responsible for producing the result while the network is running.

4.2.2 Recurrent neural network

Recurrent neural networks (RNNs) represent a type of neural network design that can process data in any direction. An RNN consists of a network of neuron-like nodes; each node is connected to the next node in a unidirectional manner, and all nodes are connected to each other by successive layers. Each node has the ability to carry data and can receive data from the external network, generate data, or modify data as it is transferred from the input to the urine. Two different types of cross-connection have been identified. They are classified as infinite-pulse or finite-pulse cyclic networks. The characteristic of finite spiking recurrent neural networks is a map with no direction loops that can be expanded. In contrast, infinite loop networks have graphs with loops that cannot be expanded. Both types of recurrent neural networks have stored states that are controlled by the neural network. These stored states form the long-range memory network (LSTM) of the RNN. RNNs can process variable length input data because they can use their internal memory to process the data. Therefore, these RNNs are often used for language modeling and speech recognition. These are multilayer perceptrons where all neurons in one layer are connected to all neurons in the other layer and to all neurons in all layers. This architecture is often used in image processing. Convolutional neural networks (CNNs) can classify images without the need for experience or special human effort. The network learns the features of the images and then processes them to get the desired result. CNNs can capture all the spatial and temporal features present in the images by applying filters. A CNN consists of input and output layers and multiple hidden layers. Each input neuron is associated with a set of weights. The output is given by applying the arithmetic mean to all the weighted elements. Reduce the number of measurements as much as possible and the weights can be reused. This helps train the network to understand complex images (**Figure 2**).

In CNN, the system generates a 3D map after receiving the input image and divides the image into layers and product layer. Then, the image is divided into more layers consisting of different layers, and local connections are created from all the neuron volumes that make up the image. Finally, the input amount is converted to the output amount. There are various CNN architectures, including VGG, RESNET, DENSENET, Inception, and SQUEEZENET. One of them can measure the efficiency of the dataset and the suitability of the management platform [9].

4.3 Crop monitoring using deep learning

Plant diseases pose a major threat to global food security and have a negative impact on smallholder farmers whose livelihoods depend on the growth of all crops. In this new global context, it is estimated that smallholder farmers produce about 80% of all agricultural products. It is worth noting that this number will increase even more, especially due to the worldwide food shortage. It has been revealed that more than 50% of all income is lost each year. The same study found that pests and diseases are responsible for more than 50% of crop failures. It is clear that this statistic should concern all parties. These heavy losses not only cause great harm to farmers but also affect the limited global food supply. Therefore, a solution must be found. Agronomists have not found any drugs or treatments for these diseases to date. However, these drugs are disease-specific because they are designed to treat specific diseases. Farmers may not understand the specific diseases affecting their crops or the chemicals needed to treat them. Therefore, another solution needs to be found. Using

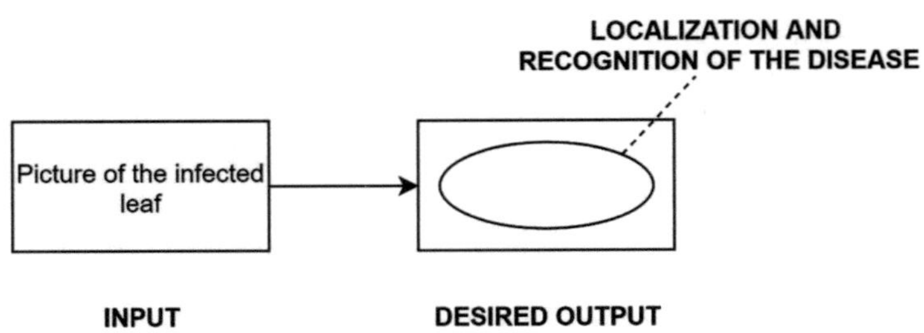

Figure 2.
The general process of the disease detection algorithm.

deep learning makes farming more efficient. The term "precision agriculture" is also used interchangeably with the term "satellite agriculture." The concept of precision agriculture involves monitoring, measuring, and responding to different types of crops through the use of decision-making systems (DSS). This approach focuses on improving agricultural resource utilization while also improving agricultural yields. Using one of the existing sensors on a GPS device, you can identify specific agricultural products and help create maps that show various conditions, including organic matter and moisture details. This is an example of precision agriculture in action. Such sensor arrays will not only produce such maps but also measure things like chlorophyll levels and the condition of water plants. These measurements, combined with satellite imagery, are useful for plants, sprays, and crops. The equipment, collectively known as Variable Rate Technology, uses data from precision agriculture to improve resource allocation and data transfer without requiring any human impact. Precision agriculture facilitates crop disease identification, drug discovery, and fruit counting. It can also directly identify crops, including determining which variety will produce the best results. Mobile cameras are often used to collect data, from deep learning to crop monitoring. This image shows a mobile camera collecting and transmitting data. This mobile camera can be an IoT camera. Also, autonomous spraying robots, autonomous painting robots, mobile cameras, and smartphones are good data collection tools. In the following sections, we will show how deep learning can be used to solve the above problems.

4.4 Plant disease and pest identification

Crops are often affected by a variety of diseases and pests, which together pose a major threat to global food security. It is not uncommon for farmers to be unaware of the presence of pests and diseases until it is too late to take control measures. Sometimes farmers may notice that some of their crops are showing unusual characteristics, such as extra green color or a bumpy appearance. They may not attribute these changes to the underlying disease, instead assuming that the plant is going through a normal growth cycle. This disease can spread throughout the farm before it is detected, reducing the benefits of the growth cycle. Farmers, especially those without advanced agricultural training, need to be equipped with the knowledge to detect diseases early and manage them effectively. In the past, support for plant disease identification was provided by agricultural associations and other organizations such as local plant clinics. Most diseases are visibly visible in the visible spectrum, so observation by a professional is an important part of disease identification. However, this approach can be ineffective due to poor quality intelligence analysis. It is important to remember that human potential is limitless. Stress, eye problems, and so forth can cause vision impairment, which can lead to inaccuracies in visual inspection.

To overcome these problems and facilitate virus identification, image-based deep learning algorithms can be used. However, this requires identifying and documenting the behaviors and characteristics associated with each disease. The following list shows the characteristics of pests and diseases that can be used for diagnosis:

4.4.1 Infection

Life cycles that can be used in the identification process. It is important to identify and correct for specific patterns found in plants associated with specific diseases. This process needs to be done at all stages to determine whether the appearance of a pattern indicates that the plant is at a particular stage of the disease. An algorithmic approach could facilitate the creation of a table containing the names of the organisms and their associated images and colors. This would allow bacteria to be identified by shape and color.

4.4.2 Zone of infection

The zone of infection is not limited to the leaf area; different diseases affect plants in different places. By locating the disease on your plants, you can more easily diagnose the disease and choose the appropriate treatment.

4.4.3 Types of fungi

An easy way to understand the different types of fungi is to identify and understand the types of fungi involved. In order to determine the type of fungi that affect plants, it is necessary to determine the characteristics of each fungus, including the color and shape they leave on the affected plant and the location of the plants they affect.

Changes can be seen on the leaves depending on the disease. If it is found that the disease leaves a strange mark on the leaves of the affected plant, it will be easy to determine whether the plant is affected by the disease. Although some of these characteristics can be determined visually, this method is unreliable and misleading. But deep learning allows the above and other interactions to be integrated into the network. The network will then receive more images for analysis. These images will include many plants, including good and bad models. These features will be loaded into the relevant system, and the network will be able to learn and process all the data. **Figure 3** shows the best way to solve the above problems, which is the identification and localization of disease on leaves. The neural network provides a map from input (image of infected plant) to output (identification of disease). Each part of the neural network represents a mathematical operation and internal connections are used to collect the input and output of the code as external connections. The most commonly

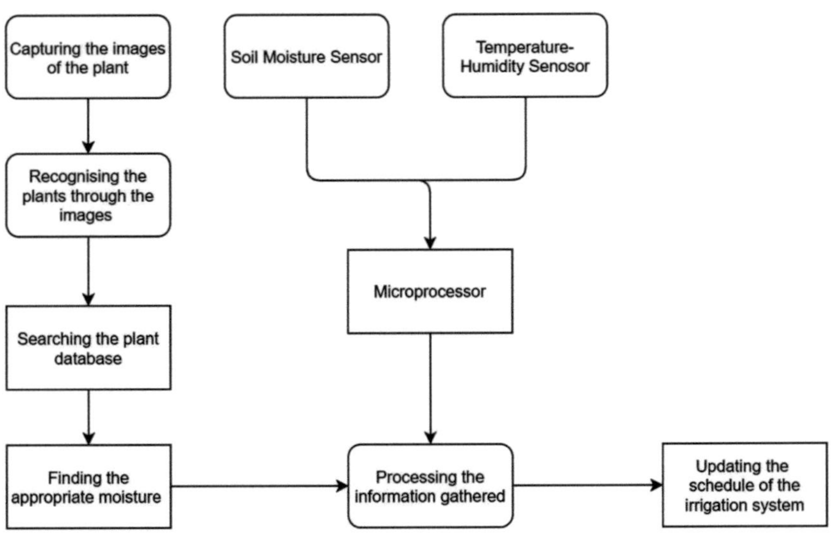

Figure 3.
The general process of smart IoT based irrigation system. Source.

Pest control methods	Cost	Environment friendly	Human health	Product quality	Labor saving	Effectiveness
Pesticides/ chemical	Low	Low	Low	Low	Low	High
Biological	Medium	High	High	Medium	Medium	Medium
Genetic	Medium	High	High	Medium	Medium	Medium
Integrated pest management system	High	Medium	Medium	High	High	High
Wireless sensor networks (IPM with AI)	Medium	High	High	High	Medium	High

Table 3.
Comparison of pest control methods.

used deep learning tools in this context are CNNs that can model similarities and identify patterns in images. The CNN architecture described above was trained and tested using different datasets containing images of healthy and diseased plants. After capturing the leaf image, the deep learning system uses the above features to analyze and identify the disease. They will also detect the disease or pest infecting the plant and provide the user with the appropriate treatment plan [10].

4.5 How to solve problems of IPM with the help of AI

The foundation of IPM is an IPM specialist who is responsible not only for the development of the system but also for routine procedures such as monitoring and selection of the appropriate procedure. Regardless of how new pesticide products are developed, decision makers still play a key role in selecting appropriate controls. Changes in plant phenology and population size, combined with climate data, are important for predicting disease growth using numerical models. The major problem is the lack of experts who can transfer information to farmers. Therefore, existing and emerging data can be processed in IPM and then analyzed using technology output technology to provide the best recommendations to agricultural producers. These methods increase management efficiency and facilitate IPM by increasing productivity and reducing costs (**Table 3**) [10].

5. Conclusion

The next section shows the application of deep learning in agriculture. Deep learning is widely used in many areas of agriculture, and this section will show how it can contribute to the success of a particular agriculture. Activities include plant disease detection, plant and pest identification, fruit counting, yield estimation, fruit harvesting, and product evaluation. Although deep learning has received much attention in many other areas such as healthcare, the analysis shows that the response to deep learning in agriculture is quite different, although still limited. This is primarily because most research results are limited to theoretical discussions rather than being translated into business applications. However, with time and the willingness of industry stakeholders to participate, these discoveries have the potential to revolutionize agriculture. In addition to deep learning, other types of image processing also have the potential to improve agriculture. However, deep learning has been shown to have better performance compared to other traditional image processing techniques. The use of this technology has the potential to reduce the number of workers required for farming, thus reducing costs. In addition, deep learning not only makes the agricultural process more efficient but also increases the overall profitability of the business. Therefore, the use of deep learning applications will help to obtain more food products to meet the food needs of the growing world population. It will also eliminate the current reduction in pests and diseases, improve resource utilization, and speed up the entire process. This will undoubtedly increase production, income, and food availability.

Conflicts of interest

The authors declare that there are no conflicts of interest regarding the research, authorship, and/or publication of this article.

Notes/thanks

We would like to extend our sincere thanks to Medical Toxicology Research Center, Mashhad University of Medical Sciences, Mashhad, Iran.

Author details

Seyed Reza Mousavi[1*] and Arman Hakemi[2]

1 Forensic Medicine and Clinical Toxicology, Medical Toxicology Research Center, Mashhad University of Medical Sciences, Mashhad, Iran

2 Clinical Toxicology, Medical Toxicology Research Center, Mashhad University of Medical Sciences, Mashhad, Iran

*Address all correspondence to: mousavir@mums.ac.ir

References

[1] Benke K, Tomkins B. Future food-production systems: Vertical farming and controlled-environment agriculture. Sustainability: Science, Practice and Policy. 2017;**13**(1):13-26

[2] Flint ML, Van den Bosch R. Introduction to Integrated Pest Management. Springer Science & Business Media; 1981, 2012. DOI: 10.1007/978-14

[3] Corallo A, Latino ME, Menegoli M. From industry 4.0 to agriculture 4.0: A framework to manage. In: 20th International Conference on Food and Environment (ICFE 2018). Rome; 2018

[4] Chokkareddy R, Thondavada N, Thakur S, Kanchi S. Recent trends in sensors for health and agricultural applications. In: Advanced Biosensors for Health Care Applications. United States: Elsevier Inc.; 2019. DOI: 10.1016/B978-0-12-815743-5.00013-5

[5] Devi DVV, Kumari GM. Real-time automation and monitoring system for modernized agriculture. International Journal of Review and Research in Applied Sciences and Engineering. 2013;**3**(1):7-12

[6] De Clercq M, Vats A, Biel A. Agriculture 4.0: The future of farming technology. In: Proceedings of the World Government Summit. Vol. 4. 2018. pp. 11-13

[7] Aubert BA, Schroeder A, Grimaudo J. IT as an enabler of sustainable farming: An empirical analysis of farmers' adoption decision of precision agriculture technology. Decision Support Systems. 2012;**54**(1):510-520. DOI: 10.1016/j.dss.2012.07.002

[8] Corallo A, Latino ME, Menegoli M. From industry 4.0 to agriculture 4.0: A framework to manage product data in Agri-food supply chain for voluntary traceability. International Journal of Nutrition and Food Engineering. 2018;**12**(5):146-150

[9] El-Sawy A, El-Bakry H, Loey M. CNN for handwritten arabic digitsrecognition based on LeNet-5. In: Proceedings of the International Conference on Advanced Intelligent Systems and Informatics 2016. Vol. 533. Cairo, Egypt: Advances in Intelligent Systems and Computing; 2016. pp. 566-575

[10] Azfar S, Nadeem A, Basit A. Pest detection and control techniques using wireless sensor network: A review. Journal of Entomology and Zoology Studies. 2015;**3**(2):92-99

Section 4

Sustainable Studies for Efficacious Pest Management

Chapter 13

Development of a Protocol to Determine and Mitigate the Evolution of Insecticide Resistance in Russian Wheat Aphid Populations in Wheat Production Areas of South Africa

Astrid Jankielsohn

Abstract

Russian wheat aphid (RWA) occurs mainly on wheat, and this insect has the potential to cause considerable damage to cultivated wheat globally. Because of the damage caused by RWA feeding, it is important to have an effective management strategy in place against RWA. Management practices used successfully against RWA in the past were chemical control and breeding for the deployment of RWA-resistant wheat. For continued success of these management practices, it is necessary to monitor the distribution of RWA populations in response to these management practices. The distribution of RWA populations was surveyed, and RWA samples were collected in the wheat production areas of South Africa during 2023. RWA clone colonies were established from field-collected samples. Insecticide resistance of RWA clone populations was determined by screening against the active ingredient, dimethoate. There was a decrease in susceptibility of RWA populations to dimethoate in areas where chemical control was used. Despite implementation of management practices, RWA populations still occur in the wheat production areas of South Africa and therefore continue to pose a threat to the wheat industry. It is important that future management practices focus on sustainable practices instead of indiscriminate use of insecticides to control RWA.

Keywords: Russian wheat aphid, resistance, insecticides, wheat, management

1. Introduction

Aphids are globally significant economic pests that can lead to substantial economic losses in a variety of crops. Specifically on wheat, wheat aphids are the most serious pests threatening wheat production and food security globally. Different wheat aphid species occur on wheat in mixed populations and feed on wheat under similar environmental conditions [1] but differ in host range and adaptability [2]. These grain aphids cause considerable economic damage to wheat and include the bird cherry oat aphid (*Rhoplosiphum padi*), the English grain aphid (*Sitobion avenae*), the greenbug (*Schizaphis graminum*), and the Russian wheat aphid (RWA) (*Diuraphis noxia*) [3]. Because of their high mobility and rapid reproduction, leading to high population pressure and crop damage, management of aphids is challenging. These aphids are parthenogenic, which enables them to significantly increase their populations in a short time [4]. The damage caused to wheat by these aphids feeding may result in significant wheat grain yield and quality losses in wheat production areas around the world [5].

The success and invasive ability of an insect pest is determined by its ability to establish in an area and the ability of this insect to spread rapidly to other areas [6]. The pressure of abiotic and biotic factors will further influence the adaptation and spread within the geographic range of establishment [7]. RWA has all these characteristics which makes it very successful at establishing and spreading to different areas [8], resulting in RWA becoming a global threat to wheat production. The first record of RWA in the wheat production areas of South Africa was in the Bethlehem area during 1978, spreading to the other wheat production areas by 1979 [9]. RWA adapted successfully to environmental changes and established quickly in these areas. The adaptability of RWA became apparent by the quick succession of different RWA biotypes in reaction to the cultivation of RWA-resistant wheat cultivars. RWA-resistant wheat cultivars were released and cultivated in South Africa during 1992 [10]. There are currently five different RWA biotypes that have been recorded in the wheat production areas of the Eastern Free State, South Africa since 2006, RWASA2 (2006); RWASA3 (2009); RWASA4 (2011); and RWASA5 (2018) [11]. RWA biotypes are mainly concentrated in the Eastern Free State dryland wheat production area, while RWASA1 (original biotype, reported in 1978) occurs mainly in the wheat irrigation areas of the Western Free State and Northern Cape and the dryland wheat production areas in the Western Cape [11].

The aim of chemical control is to keep pest numbers below damaging levels. Because it can potentially suppress wheat aphid populations in a short time, chemical insecticides are still viewed as the most important management strategy against grain aphids [4]. The most used insecticides are organophosphates and neonicotinoids and continued and widespread use of these insecticides will eventually result in insecticide resistance in aphid species. Foster et al. [12] found that widespread and frequent use of insecticides stimulates grain aphids to develop insecticide resistance. The development of resistance to insecticides resulted in a gradual increase of these grain pests, becoming a continuous problem in most wheat production regions of the world [4]. Grain aphid species not only differ in their adaptability to abiotic factors [13] but also have different adaptability to selection pressure from insecticides. Zang et al. [14] found that *Rhopalosiphum padi* and *Sitobion avenae* have developed different degrees of resistance to pesticides with various mechanisms of action, while Gong et al. [15] showed that there was a difference in susceptibilities of four grain aphid species to various insecticides. Kim et al. [16] also found species-specific susceptibility in aphids. Insecticide resistance has been confirmed in grain aphid species, *Rhopalosiphum padi*, *Sitobion avenae*, and *Schizaphis graminum* [15, 17–19]. Insecticide resistance is a major cause of failure to control aphids on wheat. Based on the adaptability and widespread occurrence of RWA in the wheat production areas of South Africa, it is necessary to develop a protocol to determine and mitigate the possible evolution of insecticide resistance in RWA populations in wheat production areas of South Africa to avoid widespread failure of insecticides and resulting damage and economic loss in the wheat industry. The preferred protocol for determining susceptibility of aphids to insecticides is the leaf dip method. Kim et al. [16] recommend the leaf dip method for testing aphids because it is easy to use, less variable, and can provide more accurate results. The leaf dip method is also the protocol recommended by the Insecticide Resistance Action Committee (IRAC).

2. Material and methods

2.1 Collection of samples of RWA populations

The distribution of RWA populations was surveyed, and RWA samples were collected in the wheat production areas of South Africa during 2023. Dryland wheat in South Africa is mainly cultivated in the winter rainfall area in the Western Cape and the summer rainfall area in the Eastern Free State. Irrigation wheat is cultivated in the Western Free State, Northern Cape, and KwaZulu Natal. Since 1992, RWA-resistant wheat cultivars were deployed in the dryland wheat production areas of the Eastern Free State as management strategy against RWA, while chemical control was the preferred management strategy against RWA in the dryland wheat production areas of the Western Cape (Swartland and Rûens wheat production areas) and the irrigation wheat production areas in KwaZulu Natal, Western Free State, and Northern Cape. Sampling sites were selected in wheat production areas, 10–20 km apart depending on the continuity of wheat fields. Leaves infested with live populations of RWA were collected and transferred to Petri dishes, sealed with parafilm, and stored in a cooler box, until deposited to glasshouse.

2.2 Creating RWA clone colonies from field collected samples

Clone colonies were established for all samples collected by transferring a single female aphid from each sample to a potted wheat cultivar. Each wheat plant with female aphid was isolated in a cage and kept in glasshouse cubicles with temperature maintained at 16/22°C (night/day). These clone populations were maintained for an average of one to two months until multiplied sufficiently to be used for insecticide-resistant screening.

2.3 Screening of RWA populations

RWA populations were screened against the active ingredient dimethoate (Dimethoate 400E EC), using the IRAC susceptibility test (Method No.: 001; version 3; June 2009). Dimethoate is an organophosphate (IRAC insecticide group 1B) with a systemic and contact mode of action. Six solutions of the active ingredient were prepared: (1) 0 ml/250 ml; (2) 0.24 ml/250 ml (25%); (3) 0.47 ml/250 ml (50%); (4) 0.66 ml/250 ml (75%); (5) 0.94 ml/250 ml (100%); and (6) 1.41 ml/250 ml (150%). Each treatment was replicated three times. Test solutions were agitated, and a fresh wheat leaf was dipped into each solution for 5 seconds. The leaves were allowed to dry and placed individually in Petri dishes with a moist filter paper. Each leaf was infested with 10 apterous adult aphids. The Petri dishes were sealed with parafilm and stored in an area where they were not exposed to direct sunlight or extremes of temperature. The maximum and minimum temperatures and RH were recorded, and mortality was assessed after 24 hours using a stereomicroscope.

2.4 Data analysis

Data were analyzed using probit analysis, and the LD 50 and susceptibility factor for each RWA population was determined.

3. Results and discussion

RWA populations were distributed throughout the wheat production areas of South Africa where insecticides were used as management practice in the dryland wheat production areas of the Western Cape and the irrigation wheat production areas of the Western Free State (**Figure 1**).

RWA population showed different levels of susceptibility to the active ingredient dimethoate (**Table 1**). There was also a decrease in the susceptibility of RWA populations to dimethoate in the Durbanville area in the Western Cape and the irrigation wheat production area at Theunissen (**Table 1**) indicated by a higher LD50 than the recommended dosage and a susceptibility factor (SF) of above 1. Variation in RWA susceptibility to chlorpyrifos was associated with a change in reproductive rates in the small grains production region of the USA [20]. It has also been shown that the evolved resistance of around 20 species in the Aphididae [21] can be linked to noticeable changes in reproductive rates of these aphids [22].

There are indications in wheat production areas in South Africa of the potential for development of resistance in RWA to the active ingredient dimethoate (**Table 1**) in these areas. The most commonly used active ingredients in the wheat production areas of South Africa are limited to two groups, organophosphates (1B) and neonicotinoids (4A) (**Table 2**). Continued and widespread use of these similar active ingredients may eventually result in insecticide resistance in RWA populations. Multiple

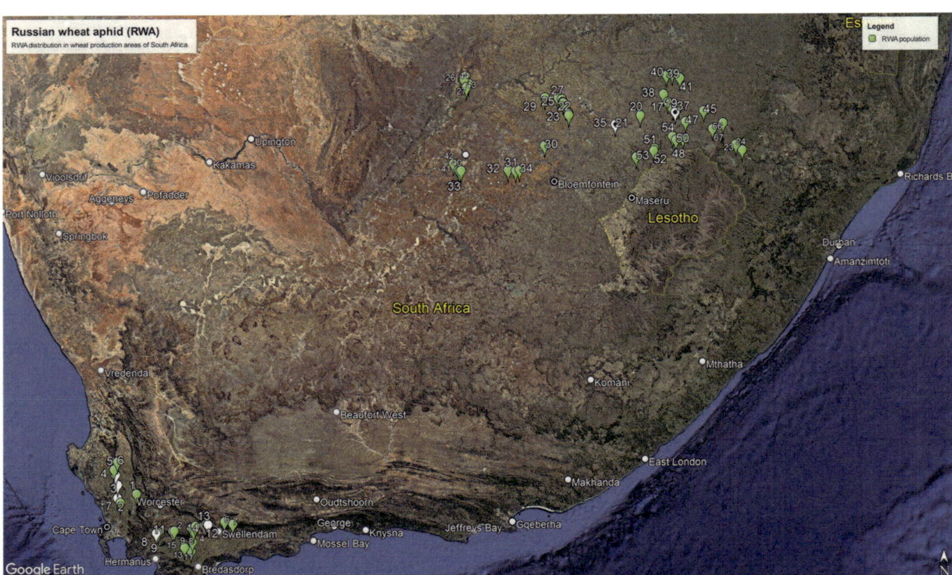

Figure 1.
Distribution of RWA populations in the wheat production areas of South Africa.

RWA Population	Slope ± SE	df	LD50 (ppm)	95% fiducial CI (ppm)	SF*	Recommended dosage (ppm)
WC1	0.950 ± 0.194	3	979	408.791–2344.027	0.26	3750
WC5	1.075 ± 0.169	3	1537.80	717.948–3293.879	0.41	3750
WC6	1.085 ± 0.172	3	969.7	446.489–2106.028	0.26	3750
WC7	0.807 ± 0.223	3	4170.50	1516.453–11,039.864	1.11	3750
WC9	2.083 ± 0.090	3	3095.58	1598.965–4931.610	0.83	3750
WC11	1.991 ± 0.094	3	2040.07	1334.645–3118.339	0.54	3750
WFS26	2.141 ± 0.091	3	4368.29	2900.646–6578.513	1.16	3750
WFS30	2.213 ± 0.087	3	1641.88	1107.132–2434.905	0.44	3750
EFS36	0.748 ± 0.252	3	458.98	147.162–1431.494	0.12	3750
EFS40	3.327 ± 0.060	3	2660.96	2027.165–3492.920	0.71	3750
EFS41	1.345 ± 0.144	3	884.05	461.984–1691.720	0.24	3750
EFS47	0.886 ± 0.206	3	1010.9	398.133–2566.844	0.27	3750

**SF: Susceptibility Factor—value below 0 indicates susceptibility; value above 0 indicates decrease in susceptibility; the higher the value above 0 the less susceptible the population*

Table 1.
Susceptibility of Russian wheat aphid (RWA) populations against active ingredient dimethoate.

Active ingredient	Group	IRAC code	Mode of action
Chlorpyrifos	Organophosphate	1B	Systemic/contact
Dimethoate	Organophosphate	1B	Systemic/contact
Imidacloprid	Neonicotinoid	4A	Systemic
Acetamiprid	Neonicotinoid	4A	Systemic
Thiamethoxam	Neonicotinoid	4A	Systemic

Table 2.
Active ingredients commonly used against grain aphids in wheat production areas of South Africa.

treatments and increased doses of insecticide have previously resulted in resistance in aphid species to several insecticides.

RWA population in wheat production areas of South Africa should be screened against all the commonly used active ingredients in these areas to determine whether there is a decrease in susceptibility of RWA to these active ingredients and consequent evolution of widespread insecticide resistance. There are several aphid species that have evolved resistance against insecticides recently [23]. Insecticide application is therefore one of the most important selection pressures causing adaptation in aphids that needs to be monitored and resolved urgently.

4. Conclusion

The results in this study show that there is a decrease in the susceptibility of some RWA populations to dimethoate. To maintain our ability to produce wheat efficiently in the future, we need continued access to effective insecticides. The development of insecticide resistance is a major limiting factor to the efficacy of insecticides. Developing and implementing an appropriate pest control strategy should preferably consider the variation of insecticide tolerance between different aphid populations. Resistance monitoring is necessary in resistance management as it provides useful information on the reaction of aphid populations to active ingredients with different modes of action. The following protocol should be followed as part of a resistance management approach:

1. Surveys to determine distribution of RWA populations during wheat growing season in all the wheat production areas of South Africa.

2. Establishment of RWA clone populations from samples collected in the field during surveys.

3. Screening of RWA clone populations against all commonly used active ingredients used in wheat production areas using the leaf dip method (IRAC susceptibility test Method No:001; version 3; June 2009)

After the susceptibility of RWA populations to active ingredients is known, implementation of resistance-delaying tactics needs to be considered. The most important of these tactics are minimizing insecticide use, applying insecticides only when needed, and rotation of insecticide classes. Organophosphates and neonicotinoids have been applied widely to manage wheat aphids in recent years, and aphids seem to become less susceptible to these insecticides. Using insecticides from different classes for a few generations and limiting insecticide use to areas only where and when it is necessary to protect yield will enable insecticide resistance management and ensure continued efficiency of insecticides.

Acknowledgements

The author wishes to acknowledge the Agricultural Research Council (ARC-Small Grain) for facilities provided, the Technology Innovation Agency (TIA) for funding the research, and Villa Crop for providing insecticides used for screening RWA.

Conflict of interest

The author declares no conflict of interest.

Author details

Astrid Jankielsohn
ARC-Small Grain, Bethlehem, South Africa

*Address all correspondence to: jankielsohna@arc.agric.za

References

[1] Sun YC, Chen FJ, Ge F. Elevated CO_2 changes interspecific competition among three species of wheat aphids: *Sitobion avenae*, *Rhopalosiphum padi*, and *Schizaphis graminum*. Environmental Entomology. 2009;**38**:26-34. DOI: 10.1603/022.038.0105

[2] Hu XS, Liu XF, Thieme T, Zhang GS, Liu TX, Zhao HY. Testing the fecundity advantage hypothesis with *Sitobion avenae, Rhopalosiphum padi*, and *Schizaphis graminum* (Hemiptera: Aphididae) feeding on ten wheat accessions. Scientific Reports. 2015;**5**:18549

[3] Crespo-Herrera LA, Singh RP, Åhman I. Field population development of bird cherry-oat aphid and greenbug (Hemiptera: Aphididae) on wheat-alien substitution and translocation lines. Euphytica. 2015;**203**:249-260. DOI: 10.1007/s10681-014-1244-8

[4] Luo K, Zhao H, Wang X, Kang Z. Prevalent pest management strategies for grain aphids: Opportunities and challenges. Frontiers in Plant Science. 2022;**12**:790919. DOI: 10.3389/fpls.2021.790919

[5] Liu XL, Yang XF, Wang CY, Wang YJ, Zhang H, Ji WQ. Molecular mapping of resistance gene to English grain aphid (*Sitobion avenae* F.) in triticum durum wheat line C273. Theoretical and Applied Genetics. 2012;**124**:287-293. DOI: 10.1007/s00122-011-1704-7

[6] Coalatti RI, MacIsaac HJ. A neutral terminology do define 'invasive' species. Diversity and Distributions. 2004;**10**:135-141

[7] Prentis PJ, Wilson JRU, Dormontt EE, Richardson DM, Lowe AJ. Adaptive evolution in invasive species. Trends in Plant Science. 2008;**13**:288-294

[8] Liu X, Marshall JL, Stary P, Edwards O, Puterka G, Dolatti L, et al. Global phylogenetics of *Diuraphis noxia* (Hemiptera: Aphididae), an invasive aphid species: Evidence for multiple invasions into North America. Journal of Economic Entomology. 2010;**103**(3):958-965. DOI: 10.1603/EC09376

[9] Walters MC, Penn F, Du Toit F, Botha TC, Aalbersberg K, Hewitt PH, et al. The Russian wheat aphid. Fmg S. Afr. 1980: Leaflet Series Wheat;**G3**:1-6

[10] Marasas C, Anandajayasekeram P, Tolmay VL, Martella P, Purchace JL, Prinsloo GJ. Socio-economic impact of Russian wheat aphid control research program. In: SACCAR/ARC Report. Gaborone, Botswana: SACCAR; 1999. 147pp. http://agris.fao.org/agris-search/search.do?recordID=QY2001000257

[11] Jankielsohn A. Russian wheat aphid distribution in wheat production areas: Consequences of management practices. In: Ansari M, editor. Current Trends in Wheat Research. IntechOpen; 2021. DOI: 10.5772/intechopen.96375

[12] Foster SP, Paul VL, Slater R, Warren A, Denholm I, Field LM, et al. A mutation (L1014F) in the voltage-gated sodium channel of the grain aphid, *Sitobion avenae*, is associated with resistance to pyrethroid insecticides. Pest Management Science. 2014;**70**:1249-1253. DOI: 10.1002/ps.3683

[13] Muneer A, Rana MA, Khalid HMS, Dilbar H, Niaz H, Muhammad I, et al. Responses of different aphid species on wheat crop with relation to abiotic factors. Pure and Applied Biology.

2023;**12**(2):991-999. DOI: 10.19045/bspab.2023.120101

[14] Zhang L, Lu H, Guo K, Yao S, Cui F. Insecticide resistance status and detoxification enzymes of wheat aphids *Sitobion avenae* and *Rhopalosiphum padi*. Science China. Life Sciences. 2017;**60**:927-930

[15] Gong P, Chen D, Wang C, Li M, Li X, Zhang Y, et al. Susceptibility of four species of aphids in wheat to seven insecticides and its relationship to detoxifying enzymes. Frontiers in Physiology. 2021;**11**:623612. DOI: 10.3389/fphys.2020.623612

[16] Kim GH, Shin WK, Ahn JW, Cho KY. Susceptibility of several insecticides on three aphids. Korean Journal of Plant Protection. 1987;**6**(2):83-88

[17] Wang K, Zhang M, Huang Y, Yang Z, Su S, Chen M. Characterisation of imidacloprid resistance in the bird cherry-oat aphid, *Rhopalosiphum padi*, a serious pest on wheat crops. Pest Management Science. 2018;**74**:1457-1465. DOI: 10.1002/ps.4834

[18] Xu T, Zhang S, Liu Y, Ma L, Li X, Zhang Y, et al. Slow resistance evolution to neonicotinoids in field populations of wheat aphids revealed by insecticide resistance monitoring in China. Pest Management Science. 2022;**78**:1428-1437. DOI: 10.1002/ps.6760

[19] Gong P, Li X, Gao H, Wang C, Li M, Zhang Y, et al. Field evolved resistance to pyrethroids, neonicotinoids, organophosphates and macrolides in *Rhopalosiphum padi* (Linnaeus) and *Sitobion avenae* (Fabricius) from China. Chemosphere. 2021;**269**:128747. DOI: 10.1016/j.chemosphere.2020.128747

[20] Brewer MJ, Kaltenbach JE. Russian wheat aphid (Homoptera: Aphididae) population variation in response to chlorpyrifos exposure. Journal of the Kansas Entomological Society. 1995;**68**(3):346-354

[21] Devonshire AL. Occurrence of insecticide-resistant aphids. In: Minks AK, Harrewijn P, editors. Aphids, Their Biology, Natural Enemies and Control, Vol. C. NY: Elsevier; 1989. pp. 123-139

[22] O'Brien PJ, Graves JB. Insecticide resistance and reproductive potential of *Aphis gossypii* glover. Southwestern Entomologist. 1992;**17**:115-121

[23] Kaleem Ullah RM, Gao F, Sikandar A, Wu H. Insights into the effects of insecticides on aphids (Hemiptera: Aphididae): Resistance mechanisms and molecular basis. International Journal of Molecular Sciences. 2023;**24**:6750. DOI: 10.3390/ijms24076750

Chapter 14

Efficiency of Essential Oil Extracted from *Azadirachta indica* on Pests of *Zea mays*

Jose Barroso

Abstract

The present study aims to evaluate the efficiency of *Azadirachta indica* essential oil against pests of the corn crop *Zea mays*. Pest populations were captured using 20 chromatrophic traps and PET bottles with attractants. To obtain the essential oil, 900 g of seeds of the plant material were placed in the oven at a temperature of 45 ± 2° C for 48 hours; then the seeds were crushed, and the solution was dried, formulating concentrations of 500, 700, 900 ppm. Mortality was verified after 1, 3, and 7 days. It was observed that *A. indica* extracts at concentrations of 500, 700, and 900 ppm reduced the populations of insect pests that affect *Zea mays* crops by an average of 75, 75, and 90%, respectively, on the 3rd day of exposure. The *Psadeletia sequax* population was completely eliminated on the 3rd day of exposure at any concentration; *Leptoglossus zonatus* resisted until the 7th day at the three concentrations tested; *Dalbulus maidis* resisted until the 7th day at a concentration of 700 ppm; and *Diatracea sacharalis, Helicoverpa armigera*, and *Euchistus heros* Fabricius resisted until the 7th day at a concentration of 500 ppm.

Keywords: insects, pests, effectiveness, *Azadiractha indica*, Formulation

1. Introduction

Organic insecticides are those produced from natural raw materials, such as essential oils and plant or bacterial extracts, which are used to control pests so as not to pose problems to the environment and public health in relation to chemical insecticides [1].

Plants with insecticidal properties have been used for a long time. The negative effects of chemical insecticides led to the discovery of other formulations, so in the first half of the last century, substances extracted from plants were widely used to control insects, due to their content in active substances [2].

The main plants from which substances with insecticidal activity are obtained belong to the genera *Nicotiana* (Solanaceae), producers of nicotine and nornicotine; genera *Derris*, *Lonchocarpus*, *Tephrosia*, and *Mundulea* (Leguminoseae), producers of rotenoids; genus *Chrysanthemum* (Asteraceae), producers of pyrethrins; and genus *Azadirachta* (Meliaceae), producers of azadirachtin [3].

Researchers around the world have been looking for better solutions against pests using products extracted from plants to constitute a renewed integrated management. In this context, the Meliaceae family has stood out, both for the number of plant species with insecticidal activity and for the efficiency of their extracts [4].

The interest in studying the species of this family is due to the important results obtained with extracts of *A. indica*, known as "neem." This is considered, among the insecticidal plants currently studied, as the most efficient [5–7].

The means of combat used against infestations in crops and even stored dry products may involve natural and synthetic products. Natural products are, in general, used at the level of small producers or family farming. These products can be of vegetable or mineral origin and include the use of sand, ash, smoke, oils, extracts, and combinations of these materials [8].

For this work, water extraction is the viable alternative, as it is the most common among farmers in developing countries, which enables them to be able to produce their own plant-based means of protection against enemies of crops and products stored. At the small farmer level, mastering simple procedures for extracting substances of plant origin is very useful, enabling the farmer to expand the nature of the extracts to be produced [9].

In Africa, and particularly in Angola, maize has great economic value, being cultivated in several regions, occupying vast areas of land for agriculture. The gains resulting from corn production have not yet reached the sustainability necessary to respond to population demand, as pest infestations in this crop are still very intense, as pest control models are not effective, causing high economic losses.

By associating these two units, the first being *Azadirachta indica* and the second *Zea mays*, we will be able to structure the most effective sustainability model. Since *A. indica* essential oil has insecticidal effects against general pests and *Zea mays* is very important economically, this study can be a contribution to achieving environmental sustainability.

Therefore, the present study aims to evaluate the efficiency of essential oils extracted from *Azadirachta indica* against pests that affect the *Zea mays* corn crop.

2. Methodology

2.1 Reference situation of adult pests in *Zea mays* crops

To understand the percentage of pests in the adult stage in *Zea mays* culture, chromatrophic traps made of cardboard and PET bottles with attractants were used. Twenty traps were used for both.

2.2 Making traps

To control pest populations, black, blue, red, and yellow cardboard measuring 50/50 centimeters was used with square spaces divided inside. For the PET bottles, the entire length was painted with oil paint, using the same colors as the chromatrophic traps. Finally, the first ones were impregnated with glue made from 500 g of rosin (petroleum resin) and half a liter of vegetable oil, and the second one was used with a composition of 30 ml of natural orange juice, 20 ml of water, and 20 ml of detergent. An entrance was opened in the center to capture insects (**Figure 1**) [10].

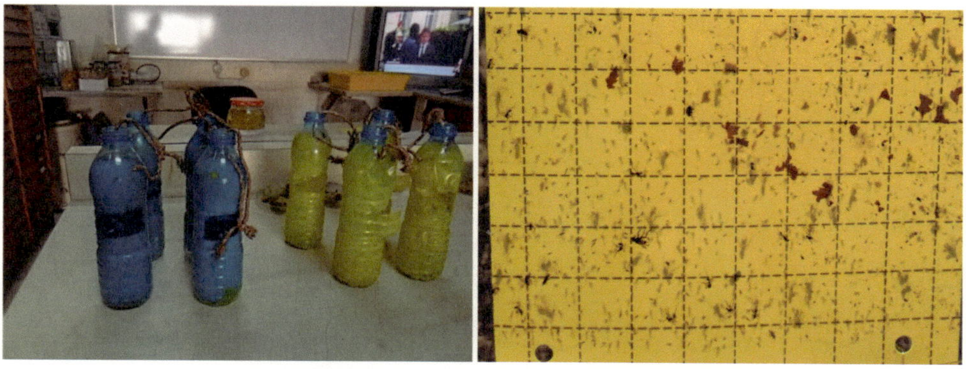

Figure 1.
Insect capture traps.

2.3 Obtaining essential oil from *Azadirachta indica* seeds

To obtain the essential oil, the cold pressing method described by Barroso [1] was used; 900 g of seeds of the plant material were placed in the oven at a temperature of 45 ± 2° C for 48 hours, subsequently ground. The seeds were placed in a Britania Diamante Black 4 blender at a speed of 900 watts.

It was pressed manually in an oil press machine. The resulting aqueous material was placed in a container and sealed with parafilm to prevent evaporation. It was left to rest in the dark at a temperature of 18 ± 2° C in an oven for 48 hours. Subsequently, a heterogeneous compound was obtained with a light phase rich in oil on the surface, an intermediate phase rich in water, and a heavy phase rich in insoluble solids. It was decanted to separate the essential oil and then filtered to remove suspended solid particles. Three concentrations of 500, 700, and 900 ppm were prepared.

2.4 Bioassays

A manual irrigator containing solutions of 500, 700, 900 ppm of *Azadirachta indica* extract was used. A *Zea mays* culture of 30–30 meters was selected and divided into three regions, each corresponding to a certain concentration.

Figure 2.
Frequency of pests on the 1st, 2nd, and 3rd day of exposure.

2.5 Mortality assessment

The mortality of insect pests was verified after 1, 3, and 7 days of exposure. It was controlled by placing traps again at the respective control time (**Figure 2**).

3. Results and discussions

3.1 Infestation level

It was found that in the culture studied, 90% of the plants suffered attacks from pests of different species. It was possible to observe damage to all plant organs, from the roots to the leaves.

3.1.1 Identified pests

Serious infestation problems were observed in the studied crops, and it was possible to capture in the traps most of the insect pests that affect the *Zea mays* crop and at their reference values to constitute problems and require intervention using a pest control method.

It was noted that the highest frequencies were observed in the species of *Leptoglossus zonatus*, *Diatracea sacharalis*, and *Euchistus heros* Fabricius, appearing on average with 13% of the registered pests. While less frequently, the species *Psedaletia sequax* and *Spodoptera frugiperda* were observed. Also, of significance were observed species such as: *Diabrotica speciosa, Dichelops* spp., *Dalbulus maidis, Agrotis ipisilon, Elasmopalpus lignosellus, Diatracea sacharalis, Helicoverpa armigera*, and *Helicoverpa zea* (**Figure 3**).

The observed differences do not imply the absence of infestation by less frequent pests. On the contrary, the fact that they are captured in traps at those concentrations allows us to identify reference values for infestation problems. Therefore, these values only reflect the reference values before the application of biological insecticides such as *S. frugiperda*.

The bibliography shows that *Spodoptera frugiperda* is a very common pest in many regions of the world, being the most important to study, as it affects all stages of

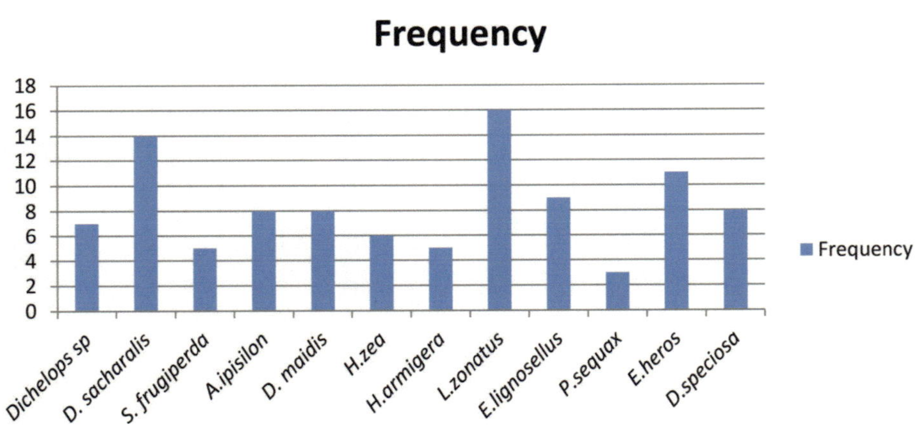

Figure 3.
Reference situation of the main insect pests identified in the Zea mays *corn crop.*

maize development and can reduce maize production by an average of up to 38%. Control of this caterpillar has been carried out with synthetic insecticides, generally expensive, with a high risk of toxicity and environmental contamination. The use of plant extracts presents a new perspective to these insecticides and can contribute to reducing crop production costs, environmental risks, and dependence on synthetic insecticides [11].

The damage caused by insect pests is not limited to crops, as seeds and cereal grains and their by-products are also subject to attack by pests, causing qualitative and quantitative losses by reducing the nutritional and commercial values of the products [12].

If we only consider the losses resulting from infestation by *S. frugiperda*, a large gap opens up regarding the losses caused by other insect pests that affect *Zea mays* crops. Because when the cumulative effect of the infestation of most of the pests that affect the *Zea mays* corn crop is observed, it is noted that losses during production are multiplied.

3.1.2 Efficiency of essential oils

In the activity of essential oils extracted from *Azadirachta indica* at a concentration of 500 ppm, it was observed that on the first day of exposure, there was a reduction of around 50% in pest populations in all species. Pest populations were further reduced on the 3rd day, achieving an average of around 25% reduction in pest populations.

On the seventh day, a reduction of around 15% was achieved. It was noted that some species were no longer found in the traps, such as *P. sequax*, which was eliminated almost in its entirety, with its average occurrence in the traps being less than 1. Meanwhile species such as *Dalbulus maidis, Diatracea sacharalis, Helicoverpa armigera, Euchistus heros* Fabricius, and *L. zonatus* were found at significant frequencies, that is, two insects per trap.

There are many other plants with insecticidal activity, such as the essential oil of *Tagetes patula* on adults of *Sitophilus zeamais*. At the highest concentrations of essential oil, 30 and 50 μL, insect mortality was 100%. At the lowest concentration used of 5 μL, mortality was 81%, proving the insecticidal effect of the *T. patula* species [13].

Mazzoneto et al. [14] described the treatment of corn leaves with the extract of the species *Chrysanthemum leucanthemum* (Asterales: Asteraceae). It was possible to observe a phage deterrent effect on the caterpillar *S. frugiperda*, that is, it reduced its feeding capacity and consequently the damage caused to plants. However, despite the author describing the larvicidal effect, using the same concentration the aqueous extract of *C. leucanthemum* did not show an ovicidal effect on the caterpillar.

Studies report that the extract of *Chrysanthemum cinerariaefolium* (Asterales: Asteraceae) had a positive effect on controlling *Tribolium castaneum* (Coleptera: Tenebrionidae) in stored rice, leading to the death of a significant number of insects [15].

Although there are many plant species with insecticidal activity and potential, not all of them have the same efficacy, which is why this study demonstrates the difference in the level of *Azadirachta indica* over other plants with a similar effect (**Figure 4**), [16].

In the activity of essential oils extracted from *Azadirachta indica* at a concentration of 700 ppm, it was observed that on the first day of exposure, it was similar to the

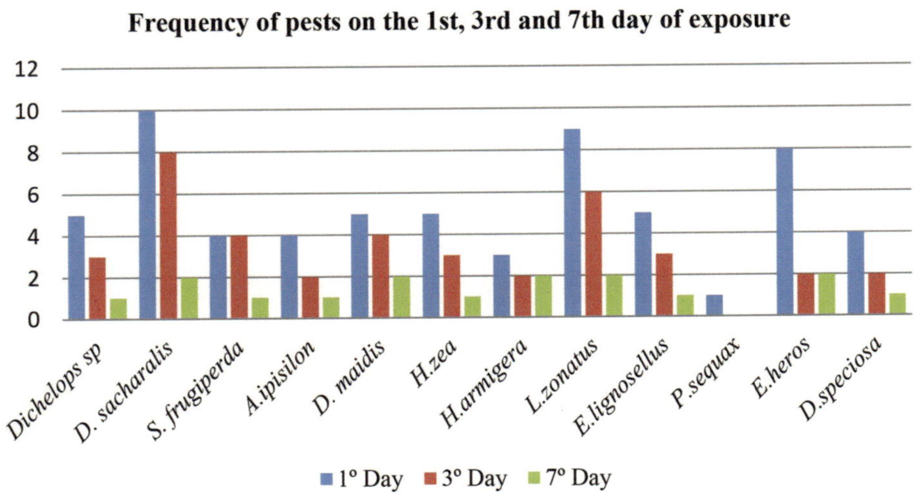

Figure 4.
Efficacy of Azadirachta indica *essential oils on* Zea mays *corn pests at concentrations of 500 ppm of the solution, during the exposure days.*

concentration used above, with an average reduction of around 50% in pest populations in all species. The same situation occurred on the 3rd day, where pest populations reduced on average by around 25%.

20% was achieved. It was noted that some species were no longer found in the traps, such as *P. sequax* and *A. ipisilon*, which were not observed in all traps, with their average occurrence being less than one insect. Meanwhile the species of *Dalbulus maidis* and *Leptoglossus zonatus* were found to have significant frequencies of two insects per trap.

Note that at this concentration, the quantity of pest insects eliminated in total increases by 50%, as in other studies carried out by the author, where the concentration factor can be a variable proportional to effectiveness.

Macagnam et al. [17] used extracts of *A. indica* in the treatment of caterpillars, with 40% mortality occurring on the fourth day of feeding of second instar caterpillars and 100% mortality on the seventh day. For fourth instar caterpillars, a mortality rate of 66.67% was observed only on the sixth day of feeding, obtaining 100% mortality on the eighth day.

Torres et al. [9] stated that the extract of *A. indica* presents insecticidal activity for caterpillars of *S. frugiperda* of the second and fourth instar, under laboratory conditions, and Barroso [1] in a study against larvae of *Anopheles* spp., also states that *A. indica* is capable of inhibiting insect feeding and chitin biosynthesis and interfering with the synthesis of ecdysone (a hormone essential to ecdysis), which can cause deformations in pupae, larvae, and adults, in addition to reducing longevity, fecundity, sterilization, and mortality.

According to Oliveira et al. [18], derivatives of *A. indica* do not kill directly but through physiological pressure and insect starvation. Viana [11] described 94.4% mortality of *S. frugiperda* using extracts of *A. indica* due to the lack of feeding (**Figure 5**).

The evaluation of the effectiveness of the essential oil extracted from *Azadirachta indica* at a concentration of 900 ppm, it was observed that on the first day of exposure there was a reduction of approximately 60% in pest populations in all species,

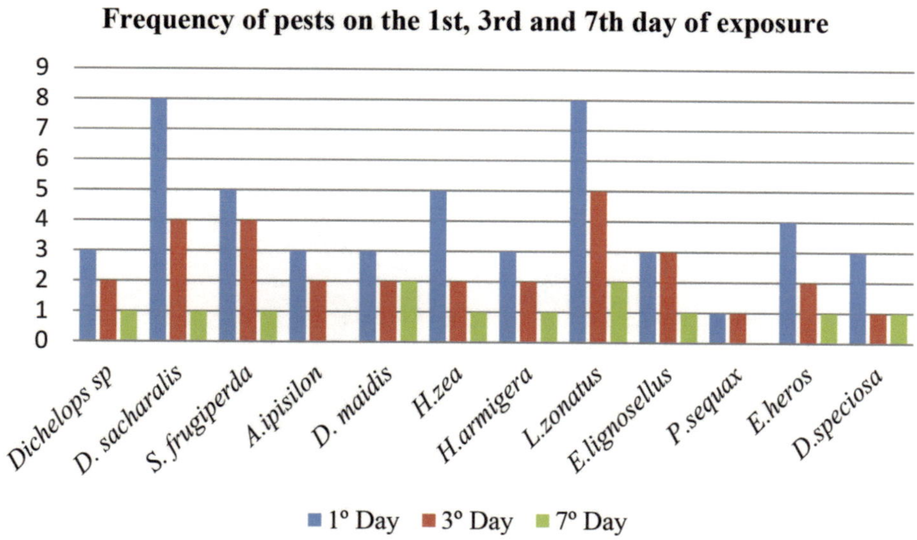

Figure 5.
Efficacy of Azadirachta indica *essential oils on* Zea mays *corn pests at concentrations of 700 ppm of the solution, during the days of exposure.*

a reduction greater than the two concentrations tested previously. Pest populations were further reduced on the 3rd day, reaching an average of around 30% of populations.

On the seventh day, an average reduction of around 5% was achieved. It was noted that some species were no longer found in the traps, such as *P. sequax*, *D. speciosa*, *A. ipisilon*, *Dichelops* sp., and *D. saccharalis*, which was not possible to capture in most of the traps, resulting in their elimination in culture. While the *Leptoglossus zonatus* population persisted, an average of two insects were observed per trap.

In this case, the effect of varying the concentration was again observed, which in this case is proportional to the effectiveness, meaning that *A. indica* extracts become more potent as their concentrations increase.

Góes et al. [19] reported that *S. frugiperda caterpillars* treated with *A. indica* extracts did not perform ecdysis, as they were unable to complete the exit from the exoskeleton, attributing this to the similarity of azadirachtin with the ecdysis hormone, which in high concentrations can disrupt transformation and prevent instar exchange, causing the death of the insect, not immediately but by reducing feeding, delaying growth, repelling adults, and reducing egg laying in the treated areas.

Lucini et al. [20], evaluating the effect of an aqueous extract of *Capsicum baccatum* on the mortality and oviposition of *Tetranychus ludeni* (Acari: Tetranychidae), concluded that the aqueous extract of *C. baccatum* did not affect the mortality of adult *T. ludeni* but reduced the oviposition of this pest as its concentration increased.

In accordance with Antonius et al. [21], it describes that the lethal effect of plant extracts on insects may differ between experiments, since the quantity and concentration of antagonistic substances may vary. The same occurs with extraction methodologies, degree of fruit maturation, and environmental conditions to which the plant was subjected during its growth, such as temperature and light (**Figure 6**).

Most of the studies covered in the literature refer to the effectiveness of *A. indica* extracts on *S. frugiperda*. However, the relevance of this study reiterates this potential

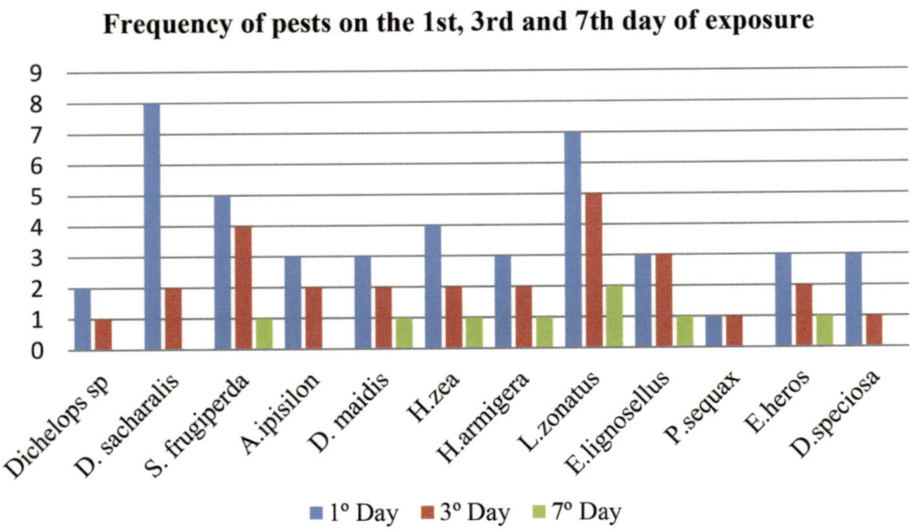

Figure 6.
Efficacy of Azadirachta indica *essential oils on* Zea mays *corn pests at concentrations of 900 ppm of the solution.*

for a group of other insect pests, such as the adult stages of *P. sequax*, *D. speciosa, A. ipisilon, Dichelops* sp., and *D. saccharalis*, where complete elimination of pests was verified on the 7th day of exposure.

4. Conclusion

A. indica essential oil at a concentration of 500 ppm reduces populations of pest insects that affect *Zea mays* by an average of 75% on the 3rd day of exposure. At the same concentration, the pest species *P. sequax* was completely eliminated on the 7th day of exposure.

The same condition was observed with a concentration of 700 ppm, where there was an average reduction of 75% of pests on the 3rd day. At the same concentration, the species of *P. sequax* and *A. ipisilon*..

The greatest effect was observed at a concentration of 900 ppm, where the pest insect population was reduced by 90% on the third day of exposure, and on the seventh day, the pest populations of *P. sequax*, *D. speciosa, A. ipisilon, Dichelops* sp., and *D. saccharalis*..

L. zonatus resisted until the 7th day, at the three concentrations tested; *D. maidis*, resisted until the 7th day at a concentration of 700 ppm, and *Diatracea sacharalis, Helicoverpa armigera*, and *Euchistus heros* Fabricius resisted until the 7th day at a concentration of 500 ppm.

DOI: http://dx.doi.org/10.5772/intechopen.1006463

Author details

Jose Barroso
Higher Institute Polytechnic Dawn of Youth, Luanda, Angola

*Address all correspondence to: barrosoandre999@gmail.com

References

[1] Barroso AJ. Assessment of the Insecticide Potential of *Azadirachta Indica* and *Ricinus communis* on Adults of *Anopheles* spp and *Aedes* spp. Luanda, Angola: REASE; 2019b

[2] Gallo D et al. Agricultural Entomology. Vol. 10. Piracicaba: Ed FEALQ; 2002. 920p

[3] Vieira PC, Fernandes JB. In: UFSC, UFRGS, editors. In Pharmacology: From Plant to Medicine. 1st ed. Florianópolis/ Porto Alegre; 1999

[4] Roel AR, Vendramim JD, Frighetto RTS, Frighetto N. Toxic activity of organic extracts of *Trichilia pallida* swartz (Meliaceae) on *Spodoptera frugiperda* (J. E. Smith). Anais da Sociedade Entomologica do Brasil. 2000;**29**:799-808

[5] Murdue AJ, Blackwell A. Azadirachtin: An update. Journal of Insect Physiology. 1993;**39**(11):903-924

[6] Schmutterer H. Potential of azadirachtin containing pesticides for integrated pest control in developing and industrialized countries. Journal of Insect Physiology. 1988;**34**(7):713-719

[7] Vendramim JD, Scampini PJ. Effect of aqueous extract of *Melia azedarach* on the development of *Spodoptera frugiperda* (J. E. Smith) in two maize genotypes. Review Agriculture. 1997;**72**:158-170

[8] Golob PE, Webley DJ. The Use of Plants and Minerals as Traditional Protectors for Stored Products. Chatham, UK: Tropical Products Institute. Agora Post pest and crop quality section. Natural Resources Institute; 1980. p. 138

[9] Torres A et al. Effect of aqueous extracts of *Azadirachtha indica*, *Mella azedarach* and aspidosperma pyrifolium on the development and oviposition of *Plutella xylostella*. Bragantia. 2009;**65**:447-457

[10] Panizzi AR, Parra J. Nutritional Ecology of Insects and Its Implications for Pest Management. São Paulo: Editora Manole; 1991

[11] Viana PA, Prates HT. Development and larval mortality of *Spodoptera frugiperda* in corn leaves treated with aqueous extract of *Azadirachta indica* leaves. Bragantia, Campinas: UL – FC; 2003;**62**:6974

[12] Anderson K et al. An economic analysis of producers decisions regarding insect control in stocked-jeain. North Central Journal of Agricultural Economics. 1990;**12**:2329

[13] Restello RM, Menegatt C, Mossi AJ. Effect of *Tagetes patula* L. (Asteraceae) essential oil on *Sitophilos zeamais* motschulsky (Coleoptera, curculionidae). Brazilian Journal of Entomology. 2009;**53**(2):304-307

[14] Mazzoneto F et al. Action of botanical insecticides on food preference and postures of *Spodoptera frugiperda* (J. E. Smith P) (Lepdoptera: Noctuidae) in corn. Entomo Brasilis. 2013;**6**(1):34-38

[15] Silva EN, Oliveira IP. Control of *Tribolium castaneum* in rice stored with different plant extracts. Revista Faculdade Montes Belos. 2015;**8**(5):34-139

[16] Barroso AJ. Evaluation of the Insecticide Potential of *Azadirachta Indica* and *Ricinus communis* on Diptera

Adults and Larvae. Luanda, Angola: REASE; 2023

[17] Macagnam R et al. Effectiveness of plant extracts to control *Spodoptera frugiperda* (J. E. Smith, 1797). Universidade do Contestado–R. Victor Sopelsa, 3000–Concórdia–SC. Brazilian Journal of Entomology. 2012;**10**(47)

[18] Oliveira MS et al. Efficiency of plant products in controlling the fall armyworm *Spodoptera frugiperda* (J. E. Smith, 1797) (Lepidoptera: Noctuidae). Science and Agrotechnology. 2007;**31**:326-330

[19] Góes GB et al. Effect of plant extracts on the control of *Spodoptera frugiperda* (J. E. Smith) (Lepdóptera: Noctuidae). Caatinga. 2003;**16**:47-49

[20] Lucini T, Scabeni C, Dedordi C, Hirose E, Shiomi HF. Effect of *Capsicum baccatum* aqueous extract on mortality and oviposition of *Tetranychus ludeni* (Acari: Tetranychidae). Scientia Agrária. 2010;**11**(4):000-000

[21] Antonius GF, Meyer JE, Snyder JC. Toxicity and repellency of hot pepper extracts to spider mite, *Tetranychus urticae* koch. Journal of Environmental Science and Health, Part B: Pesticides, Food Contaminants, and Agricultural Wastes. 2006;**41**(8):1383-1391

Chapter 15

Molecular Characterization of Bacterial Species Isolated from Tomatoes and Beans and Its Potential Ability to Degrade Carbamates Pesticides

ThankGod Onuoha

Abstract

Carbamate pesticides are synthetic chemicals broadly utilized in agribusiness to control crop pests, enhance crop quality, and amplify the life of nourishment crops. Be that as it may, its aimless utilization, inalienable properties, and the probability of having impacts on non-target life forms has made it a pollutant of concern within the environment. The research inspected the molecular characterization of bacterial species isolated from tomatoes and beans and their potential ability to degrade carbamate pesticides. Samples of tomatoes and beans were purchased randomly from Amai market in Ukwani L.G.A, Delta State. The enrichment technique was used to isolate the potential carbamate (Aldicarb) degrading bacterial species. Polymerase chain reaction (PCR) analysis was used to characterize the isolate using a universal 16SrRNA primer. *Streptococcus pyogenes* CNRZ1066, *Pseudomonas putida* strain ZR007 and *Micrococcus luteus* NCTC2665 were isolated from tomatoes.

Keywords: carbamate pesticides, tomatoes, beans, bacterial isolates, molecular characterization

1. Introduction

Carbamate pesticides are chemicals that come from a natural compound called carbamic acid. These pesticides can be separated into different groups like benzimidazole, N-methyl, N-phenyl, and thiocarbamates. The chemicals made from carbamic acid are probably insecticides that can kill a wide range of pests [1–3]. Carbamates are intensively used throughout the world and continue to be responsible for poisoning epidemics in various countries [4–6]. The carbamates are chemicals that stop AChE from working and cause the most poisonings in rural areas. Farmers are using carbamate pesticides without proper training and knowledge of how they can harm the environment. Even more concerning is that there are reports that many pesticides available to farmers are illegal and not allowed to be stored [7–9]. Buildups of pesticides, upon entering the soil, gotten to be poisonous to non-target vegetables and microbes [10–12]. A few inherent microscopic organisms, such as *Streptococcus* sp., *Pseudomonas* sp., and *Microccoccus* sp., are capable of corrupting pesticide-sullied tomatoes and beans because of persistent introduction. Be that as it may, reactions within the soil are important for cycling minerals and organic material in living systems. It changes harmful substances into carbon, which is the main food for bacteria. The breakdown of pollutants in the soil by natural microbes is an important way to clean up the soil when it has been polluted. Biodegradation is a process done by specific bacteria that break down the poison into water and carbon without needing oxygen [13–16]. This handle is imperative to the carbon substance in soil, which is the major source of carbon acclimatized by other microscopic organisms and, hence, of impressive environmental and rural significance [17]. The aim of this chapter is to determine the molecular characterization of bacterial species isolated from tomatoes and beans and their potential ability to degrade carbamate pesticides.

1.1 Microbial pesticides

Microbiological control involves the use of microorganisms that compete, fight, and control plant pests, insects, and weeds. Generally, these tiny organisms can inhibit pests by making a poison, causing a sickness, or preventing other living things from growing. Currently, some small living things are used in making bug-killing pesticides.

1.1.1 Microbes

Most of the natural pesticides available in the market are made from bacteria. The popular bacterial biopesticide is made up of tiny living things called *Bacillus* that are found in soil and can make spores. Most commercial bug sprays are made from a type of bacteria called *Bacillus thuringiensis* (Bt). This bacterium produces a toxic protein that can kill insects, and it is usually encoded by bacterial plasmids. When the parent bacterium breaks apart at the end of the sporulation cycle, it releases spores and crystal-like bodies. If insects eat these spores and crystals, they can be harmful. The protein is activated by the conditions of the insect's stomach and the movement of chemicals in its intestines [18]. The harmfulness of the protein depends upon the poison and gut receptor interaction and which decides the magnitude of pests destroyed by *B. thuringiensis* subspecies [19]. Cry proteins are turned into smaller poisons when they are eaten. These proteins attach to specific receptors in the larval midgut, creating small channels that allow more water to pass through the cell layer. When a lot of water gets into the cell, it swells up and breaks. Such bugs can die quickly from the poison or within 2–3 days from an infection in their bloodstream. Bacterial strains with a mix of 6–8 Cry proteins have been used as bug-killing pesticides. The bacterial subspecies *kurstaki* has been available for sale since the early 1960s [20]. Formulations are active against insect order Lepidoptera (moths and butterflies); Diptera (flies and mosquitoes); Coleoptera (beetles and weevils); and Hymenoptera (bee and wasps) larvae [21]. *Kurstaki* can harm gypsy moths, cabbage loopers, and other types of caterpillars. The subspecies *israelensis* is harmful to gnat flies, certain types of mosquitoes, coffee berry borers, black flies, and some types of midges. Bt *san diego* is working well at fighting bugs that harm potatoes, elm trees, and cotton plants. Bt *aizawai* is strong against wax moth and jewel back moth caterpillar. Bt *morrison* is toxic to moth and butterfly caterpillars. In Europe, *B. thuringiensis* has been used to control bugs that eat up vegetables, tomatoes, fruits, grapevines, olives, and trees [22].

In addition to Cry proteins (crystal delta endotoxins), Cyt proteins (cytolysins) have also been found as another type of bug-killing protein made by *B. thuringiensis* [22] Cytolysins stick to cell membranes like detergent [23]. The hydrophobic part of the cytolysins ties the amphipathic phospholipids; transmembrane pores are formed and cells break down by osmotic lysis [24]. The spores contain numerous proteins, which may act specifically or may act in a synergistic way [25]. With respect to their poisonous effects, Cry proteins are non-toxic to vertebrate species indeed at measurements higher than 1×10^6 kg body weight, whereas these are intensely poisonous to the caterpillars at about 1 kg body weight [26], in any, case *B. thuringiensis* can cause skin and eye irritations [27].

The acidic environment of the mammalian stomach does not favor solubilization and action of the Cry proteins. These proteins are degraded exceptionally quickly (frequently in a few seconds), from 60 kilo Dalton proteins into smaller polypeptides that are less than 2 kilo Dalton, which is similar in size to peptides consisting of 10 amino acids. The proteins are broken down quickly by enzymes in the stomach of mammals, which prevent their harmful effects. Some vertebrates do not have Cry protein binding sites on the cells of their intestines and are thus, safe from their action.

B. thuringiensis is now an important insect-killing chemical for farming. It is classified as a slightly harmful pesticide. Commercial *B. thuringiensis* products are powders made of dried spores and toxic proteins. These are applied to areas where insects feed, like leaves and roots. The commercial *B. thuringiensis* product has about 250 billion spores per gram. *B. thuringiensis* items are displayed in many places, and they make up about 1 percent of all the agricultural chemicals that are shown to people. Other types of Bacillus, including *B. sphaericus* and *B. popilliae* do not work well if kept for more than 6 months. In spite of the fact that numerous chemicals efficiently catalyze the biodegradation of pesticides, a complete understanding of the biodegradation pathway regularly requires new examinations.

2. Fate of pesticides after application to vegetables and other crops

The fate of an insecticide depends upon how it moves and acts on an individual. It could be because they are transported, distributed, transformed, or broken down. After pesticides are sprayed on the crops, they can stick to the plant's surface, be exposed to wind and sunlight, and can be washed off. The pesticide can enter the plant or stay on the surface. While still on the surface of the plant, the pesticide can evaporate and break down in sunlight, or by bacteria. All these methods can reduce the amount of pesticides in the crops, but they may also have some leftover chemicals in them. The pesticide evaporates quickly after it is sprayed on the field. Pesticides that have high vapor pressure evaporate fast, while ones that have low vapor pressure stay on the surface longer. The rate at it evaporates also depends on things like how fast the wind is blowing and how hot it is. When the wind is fast and the weather is hot, the pesticide will disappear faster. Photolysis occurs when atoms get energy from sunlight and cause pesticide breakdown. Other chemicals in the sunlight can also cause a backhanded response, and they can react with pesticides too.

Some pesticides can be broken down by bacteria and other microorganisms. Tiny organisms can use pesticides as extra food and break them down into carbon dioxide and other parts. Natural chemicals and pesticides are different because organisms cannot absorb pesticides, but they can change them at certain sites. The products that are made from the parent chemical can be more or less poisonous, in spite of the fact that the degradation of pesticides is affected by distinctive natural forms [27]. The analysts applied six pesticides (azinphos-methyl, ethion, diazinon, methidathion, phosalone, and pirimicarb) to apples and found that volatilization was the prime process followed by the solar radiations. Bacterial degradation had the least impact but for phosalone. Pirimicarb was profoundly degraded by the solar radiations. Rain wash-off can moreover be exceptionally vital when it happens rapidly after application.

3. Effects of pesticides on soil bacteria

Pesticide residues act as a source for a number of auxiliary natural toxins [28]. Such poisons include unstable organics, such as benzene, toluene, xylenes, ethyl benzenes, and trimethylbenzenes, which are discharged both as gasses and contained in leachaetes [29]. Chemicals within the environment and their impact on soil have caused expanding interest in observing intense and persistent poisonous quality and mutagenicity in soil [30]. ISO Soil Quality 190 is the major standardization body included in the development of measures for soil monitoring purposes. In any case, there is a necessity to comprehend the behavior of multispecies communities, particularly the impacts of harmful chemicals on the degrader communities.

3.1 Loss of microbial biodiversity

As numerous soil properties and forms are conferred by soil biota, it is right now broadly concurred that natural characteristics need to be observed and analyzed when checking how good or healthy the soil is [31]. Bacteria in the soil can react to the soil disturbances like pollution in many different ways. These reactions can either lead to soil stability or let the soil continue its function. Certain species may not be affected by the toxins due to resistance and thus they are not harmed by them. If the initial response is not good, but the important species can increase their population and be helpful, then the community is considered strong. If the important species are permanently affected but are replaced by other native species that can do the same job in the new conditions, we call it redundant. If all these ways to protect against contamination do not work, then we will see the lethal effects of contamination on the soil.

Some specific processes, like breaking down certain chemicals or using nitrogen with certain plants, can only be done by certain types of bacteria. The types of tiny organisms in soil are countless—up to 10 million species with a group of genes much larger than humans have been found in just a small amount of dirt. Even though most of the tiny organisms in the soil, called *archaea*, do not interact with each other and are far apart, they have a lot of different functions. This is because there are so many different species of these microbes in the soil, especially in high numbers. In simpler words, when there are fewer types of microbes in the soil, it seems that the soil is more likely to have problems with pests. Soil pollution and natural toxins often decrease the variety and distribution of soil bacteria. Mostly, when a lot of things stress the soil, like toxins and drought, it puts a lot of pressure on the tiny living things in the soil. This can have a very bad impact on them, which might be even worse when they add up together. In the case of tiny organisms in the soil, the usual rule does not seem to apply. More pressure has been linked to both an increase in sensitivity and an increase in the ability of the community to resist or adapt. During tough times, bacteria can make special resting spores to stay strong. When they are not very active and do not interact much with their surroundings, it helps them survive. This is a common strategy for soil microbes. Soil microorganisms can resist contaminants because the toxins may not be available to them in their environment for a long time. This means it is not that they are insensitive to the contaminants, but the toxins are not easily absorbed by the microorganisms over time. This is especially true for natural pollutants with high K_{ow} values [30]. Chemicals that smell good and have halogen in them are added to the soil in sludge. They tend to stick to the solid parts of the sludge, like natural fabric or small pieces of rock. If the chemicals get released because the natural fabric in the sludge breaks down, they can stick to or stay in the harder parts of the soil. This means that the choice can change or stay the same, but either way, it is really hard for natural contaminants to mix with water in soil. Microbes in the soil are often found in microprocessor inside aggregates, their spatial division from poisons can effectively decrease their presentation.

3.2 Isolation and screening of pesticide-degrading bacteria

Isolation of pesticide-degrading bacteria from tomatoes and beans samples was carried out according to the method of as discussed in Ref. [24].

The tomato sample was cut into small pieces with sanitized scissors soaked in 70% ethanol. Ten grams of a sample were weighed using a scale and then rinsed with 90 milliliters of clean water. A small amount of beans was put into a tube with 9.0 ml of clean water. One milliliter of water was serially diluted with a ratio of 1:10 in three test tubes. Then, 0.1 ml aliquots from the 10^{-2} and 10^{-3} dilutions were spread onto different media and left to grow to check for contamination of bacterial colonies at room temperature ($28 \pm 2°C$) for 48 h. The total number of colonies was counted and expressed in colony-forming unit per milliliter (cfu ml^{-1}).

3.2.1 Count of the bacterial separates

The entire heterotrophic forms (THC) were recorded employing a colony counter and means were denoted as cfu/ml. The equation for estimation is as follows: dilution of sample × no of colony shaped.

3.2.2 Identification of potential pesticide degrading bacterial isolates

Characterization of isolates was carried out as described in Refs. [2, 4, 32]. Bacterial isolates were identified on the basis of cultural, morphological, biochemical, and sugar fermentation tests.

Table 1 showed that three (3) colonies were identified to be *Streptococcus*, *Pseudomonas*, and *Micrococcus* sp. based on their gram and biochemical reaction.

3.2.2.1 Test to see if something will break down naturally

The shake-flask degradation test was used to see if bacteria could break down a pesticide in a lab experiment. We tested different types of bacteria to see if they can break down the pesticide using a specific type of solution. Nine milliliters of liquid was put into seven test tubes and made clean of germs. About 0.1 ml of aldicarb pesticide at 20 ppm was added to each test tube. This was the only thing in the test tube that the

Characteristics	1	2	3
Activities			
Shape	Spherical	Spherical	Spherical
Elevation	Convex	Low convex	Convex
Margin	Smooth	Entire	Entire
moist/not moist	moist	moist	moist
Transparency	Not transparent	Not transparent	Not transparent
Color	Creamy white	Cream	Yellow
Size	minute	Medium	Medium
Morphological			
Gram staining	+	—	+
Cell type	Cocci	bacilli	Cocci
Cell arrangement	Clusters	distinct	distinct
Biochemical			
Catalase	—	+	+
Oxidase	—	+	—
Coagulase	—	—	—
Urease	+	—	+
Indole	—	—	—
Citrate	+	+	+
Sugar fermentation			
Glucose	A	A	A
Lactose	—	—	—
Feasible bacterial	*Streptococcus* sp.	*Pseudomonas* sp.	*Micrococcus* sp.

Key; + = positive reaction, — = negative reaction.

Table 1.
Isolates of bacterial attributes in cultural medium.

bacteria could use for food and energy. After that, two drops of cell suspension were added from a previously grown isolate into each test tube. The isolate was grown in a mineral salt medium. The cell mix was made by putting a small amount of bacteria from a plate into 2 ml of mineral salt liquid. In the tubes, there was a control that was not treated. All the tubes were kept at room temperature for 7 days. These were checked to see if they looked cloudy. This would show if the isolates could use pesticides to grow.

3.2.2.2 Testing if bacteria can break down pesticides using a shaking flask

The 150 ml of mineral salt liquid was put into cone-shaped jars that can hold 250 ml. Then the test pesticide was placed in funnel-shaped sterilized containers at a concentration of 1.0%. Overnight, a small amount of each type of bacterium was put into a jar and kept warm in the research facility. Pesticide use was monitored every 2 days for 10 days by counting the growth of bacteria on nutrient agar. The thickness of the material was measured at a specific wavelength of light using a spectrophotometer. We also measured changes in the levels of ions, acidity, and temperature using specific meters.

Biodegradation rate is calculated using the formula: (OD) sample - (OD) control × 100 (19).

3.2.2.3 Statistical analysis

The information was collated. The results were communicated as optical density ± standard error (S.E). The recorded information was subjected to measurable investigation utilizing measurable bundle for social sciences (I BMSPS) Microsoft exceed expectations 2010 [26].

Researchers measured the amount of aldicarb that turned into minerals by looking at how light passes through it. They also figured out how quickly the microbes broke down aldicarb in the medium. The speed at which aldicarb breaks down was studied over different periods of time, from day 1 (0 hours) to day 10 (240 hours), with measurements taken every 2 days. The effectiveness of aldicarb got better from 0 hours (4.375 ± 0.025) to 192 hours (12.375 ± 0.025) but then decreased at 240 hours. It was noticed that after about 10 days, the amount of aldicarb decreased, as shown in **Table 2**.

So, the study found that aldicarb breaks down the most in the fluid after 8 days (192 hours) of hatching. A group of different organisms broke down waste the fastest, with some types called *Pseudomonas, Micrococcus*, and *Streptococcus* doing their best. The study shows that there was a big difference in how fast the waste broke down. We studied how three bacteria can break down chemicals in tomatoes from the Amai market that are harmful to the environment. Tomatoes got contaminated because they were around aldicarb for a long time. This caused microorganisms to change in order to survive with pesticides and to use the pesticides. Tiny living things become able to break down and get rid of man-made harmful chemicals after being exposed to them many times. This helps to clean up tomatoes, that have been contaminated, quickly and effectively.

The carbamate pesticide has been used everywhere for a long time. The first scientist to study how to break down carbamates found six bacteria that can do it. These bacteria were found in tomatoes in Austria, and they are really good at getting rid of a carbamate called Aldicarb. The first bacterium, called *Enterobacter* B-14, was able to break down Aldicarb and use the fragments for its own growth. This study found out that the bacteria *Streptococcus pyogenes, Pseudomonas putida,* and *Micrococcus leutus* can break down aldicarb. When aldicarb was the only thing the bacteria could eat and it was mixed with mineral salt water, the bacteria were able to use it and grow, making the water look cloudy. The group of isolates broke down the most waste, followed by *Pseudomonas, Micrococcus, and Streptococcus*. (See **Table 2**). The results were similar to the usual aldicarb-breaking down bacteria MTCC6362, which broke down 20% of the aldicarb in 2 days and 98% in 8 days. The same findings were also seen in studies done earlier [25, 29]. Before, a study showed that three types of bacteria from soil called *Pseudomonas, Bacillus,* and *Sphingomonas* can break down aldicarb by 69.5, 65, and 72% after being added to the soil for 10 days. The same findings were reported by [24, 30] where it was degraded from 81–95% and from 95–100%, after 8 days. Previous studies have shown that *Pseudomonas* sp., *Bacillus brevis,* and *Bacillus circulans* can break down the four forms of HCH [25, 31].

Isolates	0 h	48 h	96 h	192 h	240 h	$OD(_S)-OD(_C)$	R (%)
Micrococcus	0.025 ± .0018	0.041 ± 0.021	0.107 ± 0.001	0.068 ± 0.005	0.062 ± 0.025	0.025 ± 0.190	4.375 ± 0.018[a]
Pseudomonas	0.025 ± 0.023	0.046 ± 0.005	0.104 ± 0.035	0.091 ± 0.022	0.084 ± 0.002	0.026 ± 0.029	5.075 ± 0.020[b]
Streptococcus	0.025 ± 0.016	0.034 ± 0.045	0.058 ± 0.020	0.054 ± 0.033	0.048 ± 0.025	0.025 ± 0.007	3.175 ± 0.040[c]
Consortium	0.025 ± 0.025	0.069 ± 0.022	0.126 ± 0.013	0.162 ± 0.011	0.149 ± 0.012	0.099 ± 0.037	12.375 ± 0.011[d]
Control	0.038 ± 0.00	0.038 ± 0.00	0.038 ± 0.00	0.038 ± 0.00	0.038 ± 0.00	0.00 ± 0.00	0.00 ± 0.00[f]
p-value	0.979	0.815	0.095	0.060	0.084	0.073	0.000

Table 2.
Optical density (OD) during biodegradation test.

3.2.3 Molecular characterization of pesticide degrading bacterial isolates

The bacteria were studied in more detail by looking at their DNA. The DNA was taken out, and specific genes were amplified and sequenced.

3.2.4 Bacteria genomic DNA extraction

One group of bacteria grown on a plate was transferred to a small amount of liquid, and then shaken for 48 hours at a specific temperature. After this time, the cultures were spun at a fast speed for 5 minutes. The top layer and bottom layer were mixed with 50 microliters of TE buffer (a solution of Tris–HCl, EDTA, and water). We added 15 tiny drops of 20% SDS and three tiny drops of Proteinase K (20 mg/ml). The mixture was kept warm for 1 hour at 37 degrees Celsius. Then, 100 microliters of 5 M sodium chloride and 80 microliters of a 10% CTAB solution in 0.7 M sodium chloride were added and mixed. The mixture was heated for 10 minutes at 65 degrees Celsius and then cooled on ice for 15 minutes. Chloroform and isoamyl alcohol mix were added and then kept on ice for 5 minutes before spinning at high speed for 20 minutes. The liquid phase was changed to a new tube, isopropanol (1:0.6) was added, and the DNA was kept at −20 degrees Celsius for 16 hours to speed up the process. The DNA was separated using a machine and then washed with a solution. After that, it was left to dry for about 3 hours and finally dissolved in a buffer.

3.2.5 Polymerase chain reaction

The PCR had a total volume of 50 microliters. It had 8 microliters of DNA and 42 microliters of a cocktail that included 5× GoTaq green reaction, 10 mM of each dNTPs, 10 pmol each of 27F and 1525R specific for a ∼ 1800 bp region of the 16S rDNA gene of bacteria, and 0.3 units of Taq DNA polymerase. PCR was done using a specific heating and cooling process. The process involved heating the sample to 94 degrees Celsius for 1 minute, then repeating this cycle 29 times. Each cycle included heating the sample to 94 degrees Celsius for 30 seconds, then cooling it to 50 degrees Celsius for 1 minute, and finally heating it to 72 degrees Celsius for 1.5 minutes. The experiment ended with a final heating step to 72 degrees Celsius for 5 minutes.

3.2.6 Agarose gel electrophoresis

About 1.5 g of agarose powder was measured. While 150 milliliters of a 0.5× TAE solution was microwaved and allowed to bubbled. The mixture was allowed to cool down to about 60 degrees Celsius. We added 10 ml of ethidium bromide and mixed it gently. Then we poured it into the electrophoretic tank with the comb in place to make a gel that was about 4–5-mm thick, being careful not to make any bubbles. We poured 1× TAE buffer into the tank to make sure the gel is covered. The comb was taken away at that moment. Ten microliters of the test was mixed with 1 microliter of the 10 times concentrated loading dye [33, 34]. The samples were placed carefully into the holes made by the combs. The negative terminal of the cathodes was connected to the control pack at the end where the test was placed. The electrophoresis was allowed to run at a low electrical force until the loaded dye had moved most of the way to the negative end. Terminals are shut down and taken apart. The gel was observed using a UV light transilluminator [30].

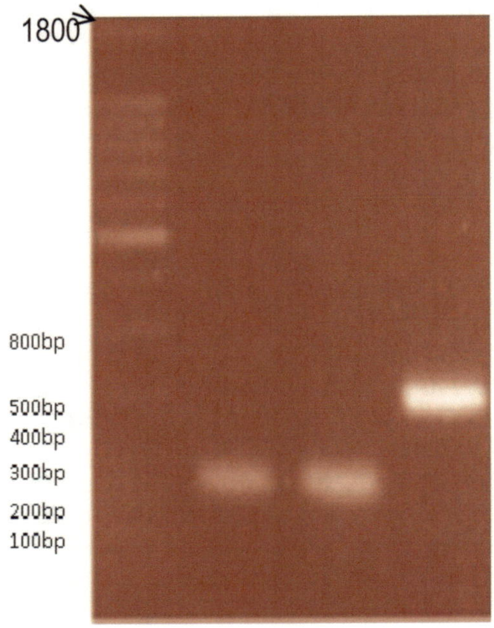

Figure 1.
Identification of Pseudomonas, Streptococcus, *and* Micrococcus *sp. through gel electrophoresis; Lanes (Left to Right); Lane 1 = Marker lane; Lane 2 = DNA of* Pseudomonas *sp.; Lane 3 = DNA of* Micrococcus *sp.; Lane 4 = DNA of* Streptococcuss *sp.*

The agarose gel electrophoresis results in **Figure 1** revealed that DNA of *Streptococcus* sp. showed band at500 bp (Lane 4) while DNA of *Pseudomonas* and *Micrococcus* sp., showed bands at 300 bp (Lanes 2 and 3).

3.2.7 Sequencing of the 16S rDNA qualities

The DNA tests were processed at the Bioscience Research facility, I. ITA, using a machine called ABI 3730×. We used specific tools called 27F and 1492R to analyze the DNA sequence. The CLC bio program was used for grouping, gathering, and arranging data. Then, the Genebank DNA database was searched for similarities using the Basic Local Alignment Search Tool (BLAST) program of the CLC bio program.

BLAST results of the sequences obtained in this study showed an identity query coverage length of 1184, 1144, and 721. The data in **Table 3** showed that the isolates from fruit vegetable (Tomatoes) with code B1, B2, and B3 belong to *Streptococcus*, *Pseudomonas*, and *Micrococcus* genera. One isolate had 89% similarity with *S. pyogenes* CNRZ1066 with accession number (NC_006449.1), second isolate had 88% similarity with *P. putida* strain ZR007 with accession number (NC_017191.1) while the third isolate had 81% similarity with *M. luteus* NCTC 2665 with accession number (NC_012803.1).

4. Discussion

The information indicates that the isolates from tomatoes with codes B1, B2, and B3 are from *Streptococcus, Pseudomonas,* and *Micrococcus* groups. One of the isolates was 89% similar to *S. pyogenes* CNRZ1066 has an accession number of NC_006449.

Isolate Code	Sequence blast	Ascension no.	Sequence identity	Query coverage length	Score bits
B1	GGATTTATTGGGCGTAAAGCGA GCGCAGGCGGTTTGATAAGTCT GAAGTTAAAGGCTGTGGCTCAA CCATAGTTCGCTTTGGAAACT GTCAAACTTGAGTGCAG AAGGGGAGAGTGGAATT CCATGTGTAGCGGTGAAAT GCGTAGATATATGGAGGAACAC CGGTGGCGAAAGCGGCTCTCTGGTCT GTAACTGACGCTGAGGCTCGAA AGCGTGGGGAGCGAACAGG ATTAGATACCCTGGTAGTCCACGCCG TAAACGATGAGTGCTAGGTGTT GGATCCTTTCCGGGATTCAGTG CCGCAGCTAACGCATTAAG CACTCCGCCTGGGGAGTACGACCG CAAGGTTGAAACTCAAAGGAATTG ACGGGGGCCCGCACAAGCGGTGGAG CATGTGGTTTAATTCGA AGCAACGCGAAGAACCTTACCAGG TCTTGACATCCCGATGCTATTTC TAGAGATAGAAAGTTACTTC GGTACATCGGTGACAGGTGG TGCATGGTTGTCGTCAGCTCGTGTC GTGAGATGTTGGGTTAAG TCCCGCAACGAGCGCAACCCCTAT TGTTAGTTGCCATCATTCAGTTGGGC ACTCTAGCGAGACTGCCGGTAA TAAACCGGAGGAAGGTGGGG ATGACGTCAAATCATCATGCCCC TTATGACCTGGGCTACACAC GTGCTACAATGGTTGGTACAACGA GTTGCGAGTCGGTGACGACGA GCTAATCTCTTAAAGCCAATCTCA GTTCGGATTGTAGGCTGCAA CTCGCCTACATGAAGTCGGAATCG CTAGTAATCGCGGATCAGCACG CCGCGGTGAATACGTTCCCGGGC CTTGTACACACCGCCCGTCAC ACCACGAGAGTTTGTAACACCCGAA GTCGGTGAGGTAACCTTTTGGA GCCAGCCGCCTAAGGTGGGA CAGATGATTGGGGTGA	NC_006449.1	*Streptococcus pyogenes* CNRZ1066	1184	839/939 (89%)
B2	GAATTATTGGGCGTAAAGGGC TCGCAGGCGGTTTCTTAAGTCTGA TGTGAAAGCCCCCGGCTCAACCGGGGA GGGTCATTGGAAACTGGGGAACTTGAGTGCAGAA GAGGAGAGTGGAATTCCACGTGTA GCGGTGAAATGCGTAGAGATGTGGA GGAACACCAGTGGCGAAGGCGACTCTCT GGTCTGTAACTGACGCTGAGGAGCGAAAG CGTGGGGAGCGAACAGGATTAGATAC CCTGGTAGTCCACGCCGTAAACGA TGAGTGCTAAGTGTTAGGGGGTTTCC GCCCCTTAGTGCTGCAGCTAACGCATTAAG CACTCCGCCTGGGGAGTACGGTCGCA AGACTGAAACTCAAAGGAATTGACGGGGGCC	NC_017191.1	*Pseudomonas putida* strain ZR007	1144	834/943 (88%)

Isolate Code	Sequence blast	Ascension no.	Sequence identity	Query coverage length	Score bits
	CGCACAAGCGGTGGAGCATGTGGTTTA ATTCGAAGCAACGCGAAGAACCTTACC AGGTCTTGACATCCTCTGACAATCC TAGAGATAGGACGTCCCCTTCGGGG GCAGAGTGACAGGTGGTGCATGGTT GTCGTCAGCTCGTGTCGTGAGA TGTTGGGTTAAGTCCCGCAA CGAGCGCAACCCTTGATCTTAGT TGCCAGCATTTAGTTGGGCACT CTAAGGTGACTGCCGGTGACAAA CCGGAGGAAGGTGGGGATGAC GTCAAATCATCATGCCCCTTATGACCTGGG CTACACACGTGCTACAATGGGCAGAACAAAG GGCAGCGAAACCGCGAGGTTAAGCCAATCCCA CAAATCTGTTCTCAGTTCGGATCGCAGTCT GCAACTCGACTGCGTGAAGCTGGAATCGC TAGTAATCGCGGATCAGCATG CCGCGGTGAATACGTTCCCGGGCCTTGT ACACACCGCCCGTCACACCACGA GAGTTTGTAACACCCGAAGTCGGTGAGGTAACC TTTTTGGAGCCAGCCGCCGAAGGTGGG ACAGATGATTGGGGTGAAGTC				
B3	GGAATTATTGGGCGTAAAGAG CTCGTAGGCGGTTTGTCGCGTCT GTCGTGAAAGTCCGGGGCTTAACCCC GGATCTGCGGTGGGTACGGGCAGACTAGAGTGCA GTAGGGGAGACTGGAATTCCTGGTG TAGCGGTGGAATGCGC AGATATCAGGAGGAACACCGATGGCGAA GGCAGGTCTCTGGGCTGTAACTGACG CTGAGGAGCGAAAGCATGGGGAGCGA ACAGGATTAGATACCCTGGTAGTCCATGCCG TAAACGTTGGGCACTAGGTGTGGGGACCA TTCCACGGTTTCCGCGCCGCAGCTAACG CATTAAGTGCCCCGCCTGGGGAGTACGGCCGC AAGGCTAAAACTCAAAGGAATTGACGGGGG CCCGCACAAGCGGCGGAGCATGCGGATT AATTCGATGCAACGCGAAGAACCTTACC AAGGCTTGACATGTTCCCGATCGCCGT AGAGATACGATTTCCCCTTTGGGGCGGG TTCACAGGTGGTGCATGGTTGTCGTCAGCTCG TGTCGTGAGATGTTGGGTTAAGTCCCGCAAC GAGCGCAACCCTCGTTCCATGTTGCCAG CACGTAATGGTGGGGACTCATGGGAGACTGCCGG GGTCAACTCGGAGGAAGGTGAG GACGACGTCAAATCATCATGCCCCTTATGTCT TGGGCTTCACGCATGCTACAATGGCCGGT ACAATGGGTTGCGATACTGTGAGGTGGAGCT AATCCCAAAAAGCCGGTCTCAGTTCGGATTG GGGTCTGCAACTCGACCCCATGAAGTCG GAGTCGCTAGTAATCGCAGATCAGCAA CGCTGCGGTGAATACGTTCCCGGGCCTTGTACACA CCGCCCGTCAAGTCACGAAAGTCGGTAACAC CCGAAGCCGGTGGCCTAACC	NC_012803.1	*Micrococcus luteus* NCTC 2665	721	738/906 (81%)

Table 3.
Identification of bacterial isolates from tomatoes.

Another isolate had 88% similarity to it. The bacteria called *P. putida* strain ZR007 has a code (NC_017191. 1) and is similar to another bacteria called *M. luteus* NCTC 2665 with a code (NC_012803. 1). The DNA sequences from this study are stored in Genbank with these codes. The results from this study are similar to the findings of [27].

Three bacteria found in tomatoes contaminated with carbamate at the Amai market were tested to see if they could break down the carbamate. Tomatoes got dirty from being around aldicarb for a long time. This caused microorganisms to change the way they break down chemicals in order to survive in places with pesticides, and to make use of them. Tiny living things, when exposed many times to man-made harmful chemicals, can become able to resist them and can develop the ability to break down these chemicals quickly, which helps to clean up contaminated tomatoes. The carbamate pesticide has been used all over the world for a long time. The first scientist to study how bacteria break down carbamates found six types of carbamate-degrading bacteria in tomatoes in Austria, and one of them, Enterobacter B-14, was able to break down aldicarb into a compound called 3,4,5,6-tetrachlorocyclohexane (7-TCCH) and use it for growth. This study found that the pesticide aldicarb can be broken down by certain bacteria like *Streptococcus pyogenes*, *Pseudomonas putida*, and *Micrococcus leutus*. When aldicarb was used alone as food and mixed with mineral salt medium (MSM) broth, the microorganisms used it and their numbers grew, making the liquid look cloudy.

The degree of mineralization of aldicarb was decided by measuring the optical thickness (OD) and calculating the rate of biodegradation of aldicarb within the medium by person microbes. The rate of aldicarb biodegradation was observed for diverse hatching period from day 1 (0 h) to day 10 (240 h) at interim of 2 days. The aldicarb debasing proficiency expanded from 0 h (4.375 ± 0.025) to 192 h (12.375 ± 0.025), but at last diminished at 240 h period. It was watched in this analysis after 10 days of brooding, the corruption of aldicarb diminished as appeared in (**Table 2**).

Hence, 192 h (8 days) of hatching period was found to be the productive degradation period of aldicarb within the fluid medium. The consortium of the separates had the most noteworthy rate of biodegradation taken after by *Pseudomonas*, *Micrococcus*, and *Streptococcus*, respectively (**Table 2**). The factual investigation uncovers that there is a tall noteworthy distinction for the rate of biodegradation l ($p < 0.05$). Biodegradation potential of the three isolated bacteria from carbamate contaminated tomatoe samples sold in some markets in Amai market has been studied. Tomatoes became contaminated due to continuous exposure to carbamate (Aldicarb) over a long period of time. This led to the evolution of metabolic determined by measuring the optical density (OD) and calculating the percentage rate of biodegradation of aldicarb in the medium by individual bacteria. The percentage rate of aldicarb biodegradation was monitored for different incubation period from day 1 (0 h) to day 10 (240 h) at intervals of 2 days. The aldicarb degrading efficiency increased from 0 h (4.375 ± 0.025) to 192 h (12.375 ± 0.025) but finally decreased at 240 h period. It was observed in this study that after 10 days of incubation, the degradation of aldicarb decreased due to long treatment of incubation. Therefore, 192 h (8 days) of incubation period was found to be the efficient degradation period of aldicarb in liquid medium [26–28]. The consortium of the isolates had the highest percentage rate of biodegradation followed by *Pseudomonas, Micrococcus*, and *Streptococcus*, respectively (**Table 2**). Microorganisms find new ways to survive in places with pesticides and using them. Tiny living things, when they come into contact with man-made harmful chemicals too many times, can become immune to them and learn how to break them down quickly to clean up contaminated tomatoes. The carbamate pesticide has been used all over the world for a long time. The first scientist to study the breakdown of carbamates found six bacteria in tomatoes in Austria that can break down carbamate (specifically aldicarb) better. The first bacterium, called Enterobacter B-14, was able to convert aldicarb into 3,4,5,6-tetrachlorocyclohexane(7-TCCH) and use it for growth. This study found that *Streptococcus pyogenes*, *Pseudomonas putida,* and *Micrococcus leutus* can break down aldicarb (a type of carbamate). When only aldicarb was used as food for the bacteria, and mixed with mineral salt medium (MSM) broth, the bacteria used it and their growth made the broth look cloudy. The amount of aldicarb that turned into minerals was mineralized. The results were similar to the usual strain that breaks down aldicarb. The usual strain can break down 20% of the aldicarb in 2 days and 98% in 8 days [25, 29] found the same results. In an earlier study, it was found that three kinds of bacteria called *Pseudomonas, Bacillus*, and *Sphingomonas* can break down 69.5, 65, and 72% of aldicarb in soil after 10 days. The same findings were reported by [24, 30], showing that the isomers α and γ degrade the fastest from 81–95% and from 95–100%, respectively, after 8 days. It has been reported in earlier studies that *Pseudomonas* sp., *Bacillus brevis and Bacillus circulans* also degrade the four isomers of HCH [25, 31].

5. Conclusion

The research discovered how to separate bacteria at a molecular level using a primer, and also how these bacteria can break down a pesticide called aldicarb. The study reveals that we need to keep checking on farms that grow food because if the food is constantly exposed to pesticides, it could make people sick when they eat it.

Author details

ThankGod Onuoha
Department of Biological Sciences, Novena University, Ogume, Nigeria

*Address all correspondence to: onuohakelechi6@gmail.com

References

[1] Adeyemi D, Anyakora C, Ukpo G, Adedayo A, Darko G. Evaluation of the levels of organochlorine pesticide residues in water samples of Lagos lagoon using solid phase extraction method. Journal of Environmental Chemistry and Ecotoxicology. 2011;**3**(6):160-166

[2] Afful S, Arim AK, Sertor-Armah Y. Spectrum of organochlorine pesticide residues in fish samples from dense basin. Research Journal of Environmental Earth Sciences. 2010;**2**(3):133-138

[3] Atuanya EI, Onuoha T, Morka E. Eco-toxicological effects of pesticide residues on soil biological sentinels. Journal of Applied Science and Environmental Management. 2018;**22**(10):1573-1577

[4] Atuanya EI, Onuoha T. Level of organochlorine pesticide residues in selected consumable vegetables commonly sold in Benin City markets. Journal of Applied Science and Environmental Management. 2018;**22**(10):1625-1630

[5] Atuanya EI, Onuoha T, Morka E, Adeola MO. Level of organochlorine pesticide residues in selected potable water commonly sold in Benin City markets. International Journal of Bioprocess and Biotechnological Advancements. 2018;**4**(1):114-119

[6] Andersson A, Pålsheden H. Comparison of the efficiency of different GLC multi-residue methods on crops containing pesticide residues. Fresenius Journal of Analytical Chemistry. 1991;**339**:365-367

[7] Agency of Toxic Substances and Disease Registry (ATSDR). Toxicological Profile for Aldrin and 9. Dieldrin. Atlanta. GA: US department of Health and Human Services Public Health Service; 2002. p. 116

[8] Boon PE, Van der Voet H, Van Raaij MTM, Van Klaveren JD. Cumulative risk assessment of the exposure to organophosphorus and carbamates insecticides in the Dutch diet. Food and Chemical Toxicology. 2008;**46**(9):3090-3098

[9] Borg J, Wanntrop H, Erne K, Hanko E. Alkylmercury poisoning in terrestrial Swedish wildlife. Vitrevy. 1969;**6**:299-379

[10] Brethour C, Weersink A. An economic evaluation of the environmental benefits from pesticide reduction. Agriculture Economy. 2001;**25**:219-226

[11] Celik S, Kunc S, Asan T. Degradation of some pesticides in the field and effect of processing. Analyst. 1995;**120**:1739-1743

[12] Chapman PM. Integrating toxicology and ecology: Putting the "eco" into ecotoxicology. Marine Pollution Bulletin. 2002;**44**:7-15

[13] Cheesbrough M. Microbiological Test: District Laboratory Practice in Tropical Countries. Part 2. New York, USA: Cambridge University Press; 2000, 447 pp

[14] Cook J, Beckett MP, Reliford B, Hammock W, Engel M. Multi-residue analysis of pesticides in fresh fruits and vegetables using procedures developed by the Florida department of agriculture and consumer services. Journal of Association of Analytical Communities International. 1999;**82**:1419-1435

[15] Daam MA, Van der Brink PJ. Implications of differences between temperate and tropical freshwater ecosystems for the ecological risk assessment of pesticides. Review. Ecotoxicology. 2010;**19**:24-37

[16] DeWitt JB. Pesticide toxicity, chronic toxicity to quail and pheasants of some chlorinated pesticides. Journa l of Agricultural Food Chemistry. 1956;**4**: 863-866

[17] Dominguez I, Agra AR, Monaghan K, Soares AM, Norueira AJ. Cholinesterase and glutathione-S-transferase activities in freshwater invertebrates as biomarkers to assess pesticide contamination. Review. Environmental Toxicology Chemistry. 2010;**29**:5-18

[18] Ecobichon DJ. Pesticide use in developing countries. Toxicology. 2001; **160**:27-33

[19] European Food Safety Authority (EFSA). General principles for the collection of national food consumption data in the view of a pan-Europe dietary survey. Journal of European Food Safety Authority. 2009;7(12):1435

[20] European Food Safety Authority (EFSA). 2008 annual report on pesticide residues according to article 32 of regulation (EC) No 396/2005. Journal of European Food Safety Authority. 2010;**8** (6):1646

[21] Elias R, Vieth A, Riva A, Horsfield B, Wilkes H. Improved assessment of biodegradation extent and prediction of petroleum quality. Organic Geochemistry. 2007;**38**:2111-2130

[22] Ennacer S, Gandaoura N, Driss R. Distribution of polychlorinated biphenyls and organochlorine pesticides in human breast milk from various locations in Tunisia. Levels of contamination, influencing factor and infant risk assessment. Environmental Research. 2008;**108**:86-93

[23] Ibiene AA, Okpokwasili GSC. Comparative toxicities of three agro-insecticide formulations on nitrifying bacteria. Report and Opinion. 2011;**3** (12):14-17

[24] Mineau PA. Review and analysis of study endpoints relevant to the assessment of "long term" pesticide toxicity in avian and mammalian wildlife. Ecotoxicology. 2005; **14**:775-799

[25] Newton I, Wyllie I, Asher A. Mortality from pesticides aldrin and dieldrin in British sparrow hawks and kestrels. Ecotoxicology. 1992;**1**:31-44

[26] Nizzetto L, Pastore C, Liu X, Camporini P, Stroppiana D, Herbert B, et al. Accumulation parameters and seasonal trends for PCBs in temperate and boreal forest plant species. Environmental Science and Technology. 2008;**42**:5911-5916

[27] Ogbeibu AE. Biostatistics: A Practical Approach to Research and Data Handling. Edo State, Nigeria: Mindex Publishing company limited; 2005. p. 265

[28] Onuoha T, Adeola M. Isolation and screening of potential pesticide degrading from selected consumable vegetable commonly sold in Benin City markets. Academia Journal of Medicinal Plants. 2019;7(3):082-085

[29] Opere BO, Ojo JO. Plasmid extraction from *Escherichia coli* and *Pseudomonas* sp. associated with pesticide residues in fruit vegetables. Journal of Xenobiotics. 2013;**2**:34-56

[30] Valavanidis A, Vlachogianni T. Integrated biomarkers in aquatic organisms as a tool for biomonitoring environmental pollution and improved ecological risk assessment, Science advances Issues: 2010. Available from: www.chem-tox-ecotox.org [Accessed: November 10, 2016].

[31] Van den Brink PJ, Bareco JM, Verboom J, Heimbach F. An individual based approach to model spatial population dynamics of invertebrates in aquatic ecosystems after pesticide contamination. Environmental Toxicology and Chemistry. 2007;**26**: 2226-2236

[32] Atuanya EI, Ekanem NO. Degradative potentials of indigenous bacteria from rubber processing factory effluent. Journal of Advance Medicine and Pharmaceutical Sciences. 2008;**2**(2): 87-92

[33] Akinnifesi TA, Asubiojo OI, Amusan AA. Effects of fungicide residues on the physicochemical characteristics of soils of a major cocoa-producing area of Nigeria. Science and Total Environment. 2006;**366**(2–3):876-879

[34] Anastassiades M, Lehotay SJ, Štajnbaher D, Schenck FJ. Fast and easy multi residue method employing acetonitrile extraction/partitioning and “dispersive solidphase extraction” for the determination of pesticide residues in produce. Journal of Association of Analytical Communities International. 2003;**86**:412-431

Chapter 16

Pest Management in Sub-Saharan Africa: Mitigating the Challenges of Insecticides Use

Ebenezer Asamani, Jerry Asalma Nboyine, Peter Quandahor, Philip Agrengsore and Joseph Adomako

Abstract

Intensification of agricultural production practices has led to monocultures of crop varieties over large areas of land. This has elevated many insects to economic pests' status. In these systems, farmers typically rely on insecticides to mitigate the damaging effects of insect pests. Although all insecticides have negative human and environmental health consequences when applied, the severity of the impact differs depending on the toxicity of their technical compound and on its formulation. Thus, the World Health Organization (WHO) recognizes five insecticide classes: (Ia) extremely hazardous, (Ib) highly hazardous, (II) moderately hazardous, (III) slightly hazardous and (U) unlikely to present an acute hazard. In African agriculture, farmers' choice of insecticides to mitigate pest damage is based on efficacy and cost. This often leads to the purchase of cheap insecticides, the majority of which are classified as extremely or highly hazardous. This chapter therefore discusses the different systems of pesticide classification, why pesticides are needed in agriculture and assesses the justification for insecticide use in selected major crops. It further examines the effects of insecticides on natural enemies of pest species and on human health. Finally, the chapter discusses sustainable methods to mitigate the negative effects of insecticide use in smallholder crop production systems.

Keywords: crop yields, yield losses, WHO classification, pest thresholds, non-target effects

1. Introduction

Pesticide is a broad term that comprises nematicides, fungicides and bactericides (including seed treatments), herbicides, insecticides, plant growth regulators, mineral oils, rodenticides and others [1]. Pesticides have been in use since the 1600s when Sulfur was used to control pests [2]. This industry emerged during the second world war, when pesticides such as dichlorodiphenyltrichloroethane (DDT) and Dichlorophenoxyacetic acid (2,4-D) were discovered and widely used around the world. However, due to the adverse effects of DDT on farm produce, human health and the environment, it was banned worldwide. Consequently, their use was now subject to strict regulations, with many pesticides no longer being accepted. The effects of pesticides on humans and the environment are known; thus, the manufacturer's active ingredient and inert substances are tested for safety before application. This activity is carried out by state-mandated pesticide registration authorities. Those that do not meet the standards established by regulatory bodies are rejected. This notwithstanding, insecticide use has increased in recent years due to population growth necessitating the production of sufficient quantities of food to meet global demand [3]. Environmental pollution has been associated with the increase in the quantity of insecticides used globally.

According to Ansari et al. [4], the agriculture sector uses an estimated 2.9 million tons of pesticides annually, accounting for approximately 85% of the total pesticide used. As reported by Perry et al. [5], pesticide consumption in the world is made of 43% herbicides, 32% insecticides, 19% fungicides and the remaining 6% is shared between growth regulators and miscellaneous agrochemicals. Insecticide use for pest control is now a global practice, with over US$6.75 billion worth of pesticides being used globally each year [2]. Reports by Zhang [6] indicated that the use of carbamates increased proportionally from 1990 to 2007 but steadily declined after 2007. Specifically, neonicotinoids, which is a group of insecticides, is currently the world's fastest-growing group of insecticides worldwide, representing about 25% of total insecticides in the global market. Currently, neonicotinoids represent the majority of insecticides used to control insect pests in the world [7]. The increase in neonicotinoid usage in the world has resulted in the decrease of organophosphates and carbamates, which used to dominate the insecticides market with a share of 59 and 18%, respectively [8]. In Africa, pesticides are imported for agricultural use, with Europe and, recently, China being the major source of pesticides [9]. Out of the 2 million tons of pesticides used globally, 25% are used in developing countries, with Africa accounting for 4% of global pesticide use. Insecticides alone account for over 25% of total pesticides in Africa [10]. There has been a 26% increase in the quantity of pesticides used in Africa from 1990 to 2017 [11]. This is because crop production is in high demand and increasing at a faster rate annually. The need to mitigate food insecurity and increase farmers' income are the main drivers. Thus, a wide range of synthetic insecticides are regularly used to manage these pests. This increases food contamination significantly, insecticide resistance in pest species and endangers consumer health in Africa. Therefore, identifying potential long-term pest management strategies for reducing chemical insecticide application is paramount for sustainable crop production in sub-Saharan Africa. This chapter is therefore aimed at reviewing the benefits, hazards and methods of harnessing insecticides while mitigating their hazards.

2. Classification of pesticides

Pesticides can be classified based on the type of activity or target organism for which the pesticide is used:

i. Insecticides (including acaricides, molluscicides and nematicides) are Substances used to kill or repel insects. It includes acaricides, molluscicides and nematicides, which are used to kill mites, snails, slugs and nematodes, respectively.

	LD_{50} for the rat (mg/kg body weight)			
	Oral		Dermal	
Class	**Solids**	**Liquids**	**Solids**	**Liquids**
Ia Extremely Hazardous	5 or less	20 or less	10 or less	40 or less
Ib Highly Hazardous	5–50	20–200	10–100	40–400
II Moderately Hazardous	50–500	200–2000	100–1000	400–4000
III Slightly Hazardous	Over 500	Over 2000	Over 1000	Over 4000

Source: WHO [13].
NB: The LD_{50} value is a statistical estimate of the number of mg of toxicant per kg of bodyweight required to kill 50% of a large population of test animals. The terms "solids" and "liquids" refer to the physical state of the product or formulation being classified.

Table 1.
Classification of pesticides based on toxicity of technical compound and on its formulation.

ii. Herbicides (including defoliants and desiccants) are substances used to destroy or inhibit the growth of plants, such as weeds. It includes defoliants, which are used to remove plant foliage and desiccants, which are used to remove water from living plant tissue.

iii. Fungicides and Bactericides are substances that destroy or inhibit the growth of fungi and substances that kill bacteria, respectively.

iv. Mineral Oils are petroleum-derived oils that are used to manage pests and diseases of plants. Petroleum oils may be referred to by many names, including horticultural oil, spray oil, dormant oil, summer oil, supreme oil, superior oil, Volck oil or white mineral oil.

v. Plant Growth Regulators are natural or synthetic substances used for controlling or modifying plant growth processes without reducing nutritive value or causing severe phytotoxicity.

vi. Rodenticides are substances that kill, repel or control rodents.

The chemical composition of pesticides can also be used to classify them into organochlorines, organophosphorus, carbamates and pyrethrin and pyrethroids [12]. Pesticides can also be grouped into organic and inorganic pesticides based on their ingredients. The activity of pesticides can also be used to classify pesticides into destroying, repelling and mitigating agents.

The World Health Organization (WHO) also classifies pesticides based on their toxicity level into extremely hazardous, highly hazardous, moderately hazardous and slightly hazardous (see **Table 1** for details) [13].

3. Why pesticides are necessary in agriculture

Since the onset of crop cultivation, pests such as weeds and pathogens have always negatively impacted the growth and productivity of agricultural crops. Their yield-reducing effects result in farmers consistently searching for mitigation strategies.

This is because severe losses (quality and quantity of produce, income and food safety hazards) occur if farmers fail to control or manage these pests [14]. Globally, approximately 20–40% of agricultural produce is lost annually due to pests and pathogens infestations [15]. To reduce these losses, pesticides are commonly used in crop production. An average of 4.6 million tons of pesticides are released into the environment annually. For insecticides, their application significantly accounts for the increase in productivity of food and fiber crops as well as prevent the incidence of vector-borne diseases [4]. In apple cultivation, Cooper and Dobson [16] reported that production would not have been possible without insecticide use to control apple worms. This has led to the percentage of marketable apples being as high as 90%. Hence, pesticides offer a solution to these losses in crops and are known to be extremely useful. It is reported that due to the increase in human population and a need for food production to be increased to meet the increasing human population, pesticide use is projected to increase in 2050 by 2.7-fold as compared to its use in the year 2000 [17]. An increase in agricultural produce yields due to pesticide usage will ultimately increase the income of farmers.

4. Insecticide resistance in Africa

Resistance development by pest species has become a major constraint to insecticide use. Major field insect pests have developed resistance to insecticides that previously controlled them effectively. In Africa, *Anopheles gambiae* Giles resistance to deltamethrin, bendiocarb, permethrin and DDT in major cotton growing areas of Burkina Faso and Cote d'Ivoire have been reported by Diabate et al. [18] and Fodjo et al. [19]. Further reports by Reid et al. [20] indicated a strong correlation between agricultural insecticide use and resistance in mosquitoes to insecticides. There is also a strong relation between the type of crop cultivated and the level of resistance in mosquito populations, with urban farms having a higher contributing factor to insecticide resistance in mosquitoes compared to rural farms. This observation is a result of the high volumes of insecticides used in urban farms compared to rural farms. Resistance to both organophosphates and pyrethroids by whitefly (*Bemisia tabaci* Gennadius) populations in Sudan cotton fields has been reported. Additionally, reports of whitefly resistance to organophosphates and deltamethrin have been reported in Burkina Faso, with evidence depicting that the pattern of resistance to these chemicals is linked to the pattern of use of insecticides in cotton fields [7]. Reports by Sene et al. [21] indicated that tomato fruit worm (*Helicoverpa armigera* Hübner) infesting tomatoes in Senegal have developed resistance to deltamethrin, while there is further information indicating that field population of tomato fruit worm is developing resistance to abamectin and profenofos.

5. Pesticide uses in selected crops

5.1 Maize (*Zea mays* L.)

Insects such as fall armyworm (*Spodoptera frugiperda* J. E. Smith), African stem borer (*Busseola fusca* Fuller), maize stalk borer (*Chilo Partellus* Swinhoe), leaf hopper (*Balclutha mbila* Nalide), cob borer (*Helicoverpa armigera* Hubner), termite (*Macrotermes* spp.) and shoot fly (*Atherigona soccata* Rondani), are major biotic constraints to increased yield of maize [22]. These insect pests attack maize from the seedling stage until harvest, and also in storage. In Sub-Sahara Africa (SSA), stem borers and fall armyworms are the major pest of maize. Many species of stem borers exist, and they account for 20–40% of yield losses in maize. In Kenya, De Groote [23] reported that about 43% of maize yield is lost due to stem borer infestation. De Groote [23] also reported that the second generation of maize stem borers can lead to 100% yield loss in Ethiopian maize farms, while the first instar causes yield loss up to 22.6%. Stem borer is controlled by applying cypermethrin 25 EC at 0.05% or deltamethrin 2.8 EC at 0.02% [22].

The fall armyworm (*S. frugiperda*) is another key pest of the crop, causing yield losses from 15 to 73% [24, 25]. In monetary terms, with the major maize-producing countries combined, yield losses estimated of USD 2.5 billion to USD 6.2 billion have been reported in SSA [26]. In general, losses recorded from fall armyworm-managed farms are lower than in those where fall armyworm is not controlled [27]. Management of fall armyworms has generally been achieved through the application of synthetic insecticides. In 2017 alone, over 72,774 liters of liquid pesticides and 4320 milligrams of powdered pesticides were used to control FAW-infesting maize farms in Ghana [28]. However, fall armyworm is known to be highly adaptable and able to develop resistance against a lot of synthetic chemicals.

5.2 Rice (*Oryza sativa* L.)

Most of the rice produced worldwide are cultivated in irrigated fields or lowland rainfed fields. This environment is suitable for the development of insect pests. Also, the continuous cultivation of rice in monocultures in irrigated fields leads to pest buildup. About 800 insect pests have been reported to be associated with rice. Of this number, 100 species of insects are reported to be pests of the crop, with only 20 of them being pest of economic importance [29]. These economic pests include green leaf hopper (*Nephotettix cincticeps* Uhler), brown plant hopper (*Nilaparvata lugens* Stål), white-backed plant hopper (*Sogatella furcifera* Horvath), stem borer (*Chilo suppressalis* Walker), rice mealy bug (*Brevennia rehi* Lindinger), rice aphid (*Hysteroneura setariae* Thomas), rice leaf folder (*Cnaphalocrocis medinalis* Guenee), rice swarming caterpillar (*Spodoptera mauritia* Boisduval) and gall midge (*Orseolia oryzae* Wood Mason) [30, 31]. Among these insects, stem borers are regarded as the most economically important pest of rice globally [32]. Prasad et al. [31] reported that stem borers alone account for about 70% of loses in rice production.

Estimates of paddy rice losses due to pest infestation range from 21 to 51% [33]. Insecticide use is the major recourse to pest management in rice. Typically, farmers apply 2–4 rounds of insecticide spray per season, and this can increase to 6–12 rounds per year in situations where the crop is cultivated all year round under irrigation [34].

5.3 Cowpea (*Vigna unguiculata* (L). Walp)

Cowpea is an important crop in the lives of relatively poor people, especially in developing countries. The crop is a host for over 100 insects on the field and in storage. Insect pests such as leaf miners, whiteflies (*Bemisia tabaci*), leafhoppers (*Empoasca* sp.), mites (*Tetranychus* spp.), thrips (*Megalurothrips sjostedti* Trybom), *Ootheca* sp., *Clavigralla* sp., *Maruca* sp. and aphids (*Aphis craccivora* Koch) attack the crop. Significant yield reduction is caused by insects that attack cowpea during the reproductive stage. These are the flower and pod-feeding insects such as flower thrips (*M. sjostedti*), legume pod borer (*Maruca vitrata* Fabricius), pod-sucking insects (*Clavigralla tomentosicollis* Stål and *Anoplocnemis curvipes* Fabricius) and phloem-feeding cowpea aphids (*A. craccivora*) [35, 36] among all the factors that impact the yield of cowpea, insect pest infestation is regarded as the paramount one having significant negative impact [36]. Yield loss in cowpeas due to insect pests alone can be as high as 6–100%.

Synthetic insecticides are the most widely used insect pest control method in cowpea production [36, 37]. Oparaeke et al. [38] reported a significant increase in the grain yield of cowpeas when insecticide treatment was applied compared to unsprayed plots. Generally, insecticide application results in a significant increase in the yield of cowpeas cultivated in Africa.

5.4 Cotton (*Gossypium hirsutum* L.)

Cotton is one of the most important industrial crops in agriculture, usually referred to as "white gold" because of its economic value. Approximately 37 African countries produce cotton, of which 30 are exporters, accounting for 10.5% of agricultural exports in SSA. Cotton from the African continent is one of the best in the world because it is handpicked [39].

More than a thousand species of insects are associated with cotton [40]. Major insect pests of cotton can be grouped into bollworms (*Helicoverpa armigera* Hübner; *Pectinophora gossypiella* Saunders), cotton Leafworm (*Alabama argillacea* Hübner), leafhoppers (*Jacobiella facialis*), aphids (*Aphis gossypii*), thrips (*Thrips tabaci*), red spider mite (*Tetranychus urticae*), cotton stainers (*Dysdercus* spp.) and whitefly (*Bemisia tabaci*) [41, 42]. The extent of damage by insect pests in cotton is dependent on location and farming practices, with reported damage ranging from 25 to 60% of potential yield [41].

Control of insect pests of cotton in SSA is solely dependent on the use of synthetic insecticides with a developed calendar system of insecticide application that allows 5–6 insecticide applications at 14-day intervals [43]. Major insecticides used in controlling whitefly in cotton fields in Africa are bifenthrin 100 g L^{-1} EC, dimethoate 400 g L^{-1} EC, chlorpyrifos 480 g L^{-1} EC, endosulfan 350 g L^{-1} EC, acetamiprid 200 g L^{-1} SL, deltamethrin 25 g L^{-1} EC, thiamethoxam 240 g L^{-1} SC and pymetrozine 500 g kg^{-1} WG.

5.5 Vegetables

A wide range of vegetables are cultivated in Africa as they are important components of daily diets. Tomato (*Solanum lycopersicum* L), dry onion (*Allium cepa* L.), watermelon (*Citrullus citrullus* (L.) Karst), cabbages (*Brassica oleracea*) and other Brassicaceae. Most of this production is done in the open field where insect pests are abundant and are regarded as the most serious problem in vegetable production. Numerous insect pests affect different types of vegetables on the continent, with some causing significant yield loss if not controlled. Insect pests attacking vegetable crops in Africa are aphids, whiteflies, stem borers, termites, diamondback moths and spider mites [44]. Insect pests do not only cause direct harm to vegetables but also serve as vectors for diseases. Example is *torrado crini* transmitted by whitefly [45].

Controlling insect pest on vegetables have been centered on the use of insecticides, both synthetic and botanical. Ngowi et al. [46] reported that 59% of pesticides used on vegetable in northern Tanzania are insecticides whiles 65% of pesticides used in Togo are insecticides [47]. In Ghana, Horna et al. [48] reported that 85% of vegetable growers use insecticides to control insect pests. Insecticides mostly used in vegetable farms include pyrethroids, organophosphates and carbamates. Vegetable farmers especially smallholder farmers mix two or more insecticides in a tank before application with the application sometimes being as high as five times in a planting season. Synthetic insecticides are mostly used by vegetable farmers with just a few using botanical insecticides [47]. In Ghana, insecticide application is done every 7 days during the main season and every 14 days during the dry season because insect pest abundance is low during the dry spell [49].

Application of insecticides in vegetable production is done purposely to control insect pests and increase yield. Amoabeng et al. [50] reported that vegetable growers that apply insecticides have higher significant income or revenue compared to vegetable growers/producers who do not apply insecticides to control insect pests during the cropping season. The revenue generated usually exceeds the cost of the protection regime used.

6. Effects of insecticides on natural enemies

The use of insecticides to manage insect pests is the most widely used control option across the world [51]. Of all pesticides used, insecticides are the most toxic [2]. Even though insecticides are known to be very effective in the control of insects, their effects on non-target species cannot be overlooked; they disrupt their activities and population [52]. The effects of insecticides on natural enemies depend on their age, sex, type of natural enemy and life stage exposed to insecticide. For example, residues of insecticides after foliar applications may hinder the emergence of adult parasitoids [53]. Frequent application of sublethal insecticides will eventually affect beneficial insects, causing them to be less effective biological control agents and resulting in low performance in parasitizing and preying on host species. Systematic insecticides will affect the population of natural enemies if their prey or host records high mortality after the application of insecticides. Stapel et al. [54] reported that parasitic wasps that fed on nectar from cotton plants treated with insecticides recorded high mortality even after 10 days of insecticide application. In Nicaraguan cabbage (*Brassica* spp.), it was observed that the application of insecticides in farms reduced parasitoid and predatory wasp populations and had a negative effect on spiders [55]. Larval parasitoids are exposed to insecticide during their developmental stages (eggs to larvae) and die along with their host when killed by the application of insecticides. Almost all larval parasitoids cannot emerge from a host that is killed by insecticides, even if the insecticides have no direct effect on the parasitoids [17].

The use of insecticides may cause the outbreak of non-target pests that benefit from the mortality of natural enemies or other competitors, a phenomenon known as secondary pests [56–58]. Other indirect effects of insecticide application on natural enemies involve the reduction in activities of macrophytes in ponds and wetlands, leading to changes in predators' environments [59]. The most widely used insecticide, neonicotinoids, affects non-target pests such as honey bees, which is a very important pollinators [17]. There is already evidence of the decrease in the population of the majority of insects in areas where neonicotinoids are constantly applied. Bringing the ecosystem back to its natural balance after disruption by insecticide application is sometimes not possible, as persistent chemicals may prevent the re-establishment of the original community structure [60].

7. Effects of insecticides on human health and the environment

Indiscriminate application of insecticides causes more harm to the environment and can lead to serious health problems. The level of impact of insecticides on human health and the environment is dependent on the exposure and toxicity level of the insecticide, as well as timing of application and landscape structure [61].

In general, insecticides take a longer time to degrade in the environment and are known to be among the most toxic class of compounds in the world [4]. The half-life of insecticides such as the neonicotinoids in soils can be as long as 19 years, especially for clothianidin [7]. Residues of insecticides accumulate in the soil, water and air, causing major issues in the environment [5]. Only about 1% of applied insecticides for pest control actually work on target pests; the rest are lost, consequently affecting some species and the ecosystem as a whole, including humans [62]. There is evidence of serious negative environmental impact due to the use of neonicotinoids to control insect pests in the agricultural landscape [63]. During aerial application of insecticides, the small particles sometimes drift to non-target areas by wind, leading to air pollution. The extent of exposure is dependent on the type of sprayer used, wind velocity, concentration level, chemical properties and temperature [4].

Insecticide residue is left on plant and earth surfaces, and some leach into the soil. The insecticide could be carried into water bodies via runoff water [64] and is known to be harmful to non-target aquatic organisms. This affects the growth, survival and reproduction of these aquatic organisms [65]. Insecticides also decrease oxygen levels in water and are able to cause behavioral changes in fish [2]. Some fish are under serious threat due to an increase in pesticide residual levels in various water bodies. Even though tap water is treated for consumption, insecticide residues have been found in it due to its solubility [50].

In terms of human health, insecticides negatively affect applicators through direct and indirect exposure, especially when safety protocols are not followed. There are four major points for insecticides to enter the human body: oral, dermal, respiratory pathways and eyes. However, the main route of pesticide entry into the human body is through ingested food containing pesticide residues [66]. Over 26 million people suffer from pesticide poisoning, and the number of deaths totals approximately 220,000 annually [67]. Bertrand [9] stated that the circumstances surrounding pesticide poisoning vary: 27% occurred during spraying, 20% by ingestion, 13% at home, 7% in the kitchen and 3% during fishing. The type of insecticide, duration of exposure and the health status of the handler or user determine the impact of insecticides on the health of humans.

The effects of insecticides on humans can be categorized into acute (headache, skin irritation, diarrhea, abdominal pain, nausea, vomiting, etc.) and chronic (damage to the immune system, stillbirth, damage to liver, lung, kidney, etc.). Although the use of protective clothing and following safe handling measures could reduce the negative impact of pesticides on users, El-Nahhal et al. [68] reported that congenital malformation among children in the areas where pesticides are used is on the rise and this indicates an indirect health risk from pesticides. Respiratory disorders, neurological disorders, reproductive (sexual/genital) syndromes, oxidative stress and cancer are some of the health-related issues due to exposure to pesticides [67, 69, 70].

8. Sustainable strategies to mitigate excessive use of insecticides

Due to the adverse effect of pesticides, environmentally friendly and low-cost effective management approaches have been advocated for in the agricultural landscapes. Integrated Pest Management (IPM) is considered a long-term solution to pest challenges that usually involves minimum to zero use of insecticides [71]. IPM is known to be very effective in pest management while being environmentally friendly [72]. It was specifically developed to minimize the effect of pesticides on natural enemies, the environment and humans. Razaq et al. [73] reported that the implementation of IPM in developed countries is more successful compared to that of developing countries mainly due to lack of knowledge among end users and fewer research efforts by scientists in developing countries. IPM basically relies on agronomic, plant resistance, physical and biological methods to keep pest populations below economic injury level [74], thus providing pest control options that protect air, water and soil resources/organisms while achieving production objectives. Sometimes, it involves the use of selective pesticides in extreme cases. However, IPM use is site-specific; thus, it depends on the cropping pattern, size of farm, floral diversity, pest pressure and the landscape surrounding the farm. It can also be used over a long period of time without adverse effects on the environment while providing good and reliable yields with low production costs. Insect pests developing resistance to insecticides in agriculture can be effectively mitigated by adopting IPM.

The natural enemy of pest is promoted in IPM systems because they form their basis. These natural enemies are conserved by reducing the number and/or quantity of insecticides applied. With diverse cropping patterns or crop rotation and proper management of areas around fields, these biological control agents are increased and are very effective in controlling insect pests.

9. Conclusion

In conclusion, this chapter showed that there are different classification systems for pesticides. For example, they can be classified based on the type of activity, chemical composition or the WHO classification system. It also showed that food production to meet the increasing human population is impossible without the use of insecticides. This was demonstrated by assessing the pest situation in some key crops and illustrating how pesticides have contributed to mitigating damage in those selected crops, thereby increasing yields. The direct and indirect harmful effects of insecticides on natural enemies of pest species were also discussed. Further, the impact of insecticides on human health and the environment was identified to be dependent on the exposure and toxicity level as well as the time of application. Finally, sustainable pest management approaches such as the use of IPM strategies that hinge on zero to minimal insecticide use were identified as the way forward in mitigating the harmful effects of insecticides in agricultural landscapes.

Acknowledgements

The authors gratefully acknowledge the management and staff of CSIR-SARI for their diverse support during the preparation of this book chapter.

Conflict of interest

The authors declare no conflict of interest in the publication of this book chapter.

Author details

Ebenezer Asamani[1], Jerry Asalma Nboyine[1*], Peter Quandahor[1], Philip Agrengsore[1] and Joseph Adomako[2]

1 CSIR-Savanna Agricultural Research Institute, Tamale, Ghana

2 CSIR-Crops Research Institute, Fumesua-Kumasi, Ghana

*Address all correspondence to: nboyinejerry@gmail.com

References

[1] FAO. International Code of Conduct on the Distribution and Use of Pesticides. Rome: Food and Agriculture Organization of the United Nations; 1986. 28 pp

[2] Mahmood I, Imadi SR, Shazadi K, Gul A, Hakeem KR. Effects of pesticides on environment. In: Plant, Soil and Microbes: Volume 1: Implications in Crop Science. 2016. pp. 253-269. DOI: 10.1007/978-3-319-27455-3_13

[3] Campos EV, Proença PL, Oliveira JL, Bakshi M, Abhilash PC, Fraceto LF. Use of botanical insecticides for sustainable agriculture: Future perspectives. Ecological Indicators. 2019;**105**:483-495

[4] Ansari MS, Moraiet MA, Ahmad S. Insecticides: Impact on the environment and human health. In: Environmental Deterioration and Human Health: Natural and Anthropogenic Determinants. 2014. pp. 99-123. DOI: 10.1007/978-94-007-7890-0_6

[5] Perry AS, Yamamoto I, Ishaaya I, Perry RY. Insecticides in Agriculture and Environment: Retrospects and Prospects. Springer Science and Business Media. Vol. 261. Berlin: Springer-Verlag; 2013. DOI: 10.1007/978-3-662-03656-3

[6] Zhang W. Global pesticide use: Profile, trend, cost/benefit and more. Proceedings of the International Academy of Ecology and Environmental Sciences. 2018;**8**(1):1

[7] Houndété TA, Kétoh GK, Hema OS, Brévault T, Glitho IA, Martin T. Insecticide resistance in field populations of Bemisia tabaci (Hemiptera: Aleyrodidae) in West Africa. Pest Management Science. 2010;**66**(11):1181-1185. DOI: 10.1002/ps.2008

[8] Stehle S, Bub S, Schulz R. Compilation and analysis of global surface water concentrations for individual insecticide compounds. Science of the Total Environment. 2018;**639**:516-525. DOI: 10.1016/j.scitotenv.2018.05.158

[9] Bertrand PG. Uses and misuses of agricultural pesticides in africa: neglected public health threats for workers and population. In: Pesticides-Use and Misuse and Their Impact in the Environment; Larramendy M, Soloneski S, editors. London, UK: IntechOpen; 2019. Available from: https://www.intechopen.com/chapters/65752

[10] Manyilizu BW. Pesticides, anthropogenic activities, history and the health of our environment: Lessons from Africa. In: Pesticides-Use and Misuse and Their Impact in the Environment. London, UK: IntechOpen; 2019. pp. 111-120

[11] Fuhrimann S, Klánová J, Přibylová P, Kohoutek J, Dalvie MA, Röösli M, et al. Qualitative assessment of 27 current-use pesticides in air at 20 sampling sites across Africa. Chemosphere. 2020;**258**:127333. DOI: 10.1016/j.chemosphere.2020.127333

[12] Yadav IC, Devi NL. Pesticides classification and its impact on human and environment. Environmental Science and Engineering. 2017;**6**:140-158

[13] WHO. The WHO Recommended Classification of Pesticides by Hazard and Guidelines to Classification. Chemical Safety and Health Unit (CHE). 2020. 98 pp. ISBN: 9789240005662. Available from: https://www.who.int/publications/i/item/9789240005662 [Accessed: May 21, 2024]

[14] Early R. Pesticides in agriculture. Food Science and Technology. 2019;**33**(1):52-55. DOI: 10.1002/fsat.3301_14.x

[15] FAO. The Future of Food and Agriculture—Trends and Challenges. 2017. Available from: http://www.fao.org/3/a-i6583e.pdf

[16] Cooper J, Dobson H. The benefits of pesticides to mankind and the environment. Crop Protection. 2007;**26**(9):1337-1348. DOI: 10.1016/j.cropro.2007.03.022

[17] Oberemok VV, Laikova KV, Gninenko YI, Zaitsev AS, Nyadar PM, Adeyemi TA. A short history of insecticides. Journal of Plant Protection Research. 2015;**55**(3)

[18] Diabate A, Baldet T, Chandre F, Akoobeto M, Guiguemde TR, Darriet F, et al. The role of agricultural use of insecticides in resistance to pyrethroids in *Anopheles gambiae* sl in Burkina Faso. The American Journal of Tropical Medicine and Hygiene. 2002;**67**(6):617-622

[19] Fodjo BK, Koudou BG, Tia E, Saric J, N'dri PB, Zoh MG, et al. Insecticides resistance status of An. gambiae in areas of varying agrochemical use in Côte D'Ivoire. BioMed Research International. 2018;**2018**(1):2874160. DOI: 10.1155/2018/2874160

[20] Reid MC, McKenzie FE. The contribution of agricultural insecticide use to increasing insecticide resistance in African malaria vectors. Malaria Journal. 2016;**15**:1-8. DOI: 10.1186/s12936-016-1162-4

[21] Sene SO, Tendeng E, Diatte M, Sylla S, Labou B, Diallo AW, et al. Insecticide resistance in field populations of the tomato fruitworm, *Helicoverpa armigera*, from Senegal. International Journal of Biological and Chemical Sciences. 2020;**14**(1):181-191. DOI: 10.4314/ijbcs.v14i1.15

[22] Dhillon MK, Kalia VK, Gujar GT. Insect-pests and their management: Current status and future need of research in quality maize. In: Maize: Nutrition Dynamics and Novel Uses. New Delhi: Springer India; 2013. pp. 95-103. DOI: 10.1007/978-81-322-1623-0_8

[23] De Groote H. Maize yield losses from stemborers in Kenya. International Journal of Tropical Insect Science. 2002;**22**(2):89-96. DOI: 10.1017/S1742758400015162

[24] Assefa F, Ayalew D. Status and control measures of fall armyworm (*Spodoptera frugiperda*) infestations in maize fields in Ethiopia: A review. Cogent Food and Agriculture. 2019;**5**(1):1641902. DOI: 10.1080/23311932.2019.1641902

[25] Chimweta M, Nyakudya IW, Jimu L, Bray Mashingaidze A. Fall armyworm [*Spodoptera frugiperda* (JE Smith)] damage in maize: Management options for flood-recession cropping smallholder farmers. International Journal of Pest Management. 2020;**66**(2):142-154. DOI: 10.1080/09670874.2019.1577514

[26] Kasoma C, Shimelis H, Laing MD. Fall armyworm invasion in Africa: Implications for maize production and breeding. Journal of Crop Improvement. 2021;**35**(1):111-146. DOI: 10.1080/15427528.2020.1802800

[27] Overton K, Maino JL, Day R, Umina PA, Bett B, Carnovale D, et al. Global crop impacts, yield losses and action thresholds for fall armyworm (*Spodoptera frugiperda*): A review. Crop Protection. 2021;**145**:105641. DOI: 10.1016/j.cropro.2021.105641

[28] MoFA. The FAW an Enemy to Food Security in Ghana. 2017. Available from: http://mofa.gov.gh/

[29] Sherif MR, Hendawy AS, El-HabashyMM.Managementofriceinsect pests. Egyptian Journal of Agricultural Research. 2005;**83**(5A):111-130

[30] Ane NU, Hussain M. Diversity of insect pests in major rice growing areas of the world. Journal of Entomology and Zoology Studies. 2016;**4**(1):36-41

[31] Prasad R, Shivay YS, Kumar D. Current status, challenges, and opportunities in rice production. In: Rice Production Worldwide. 2017. pp. 1-32. DOI: 10.1007/978-3-319-47516-5_1

[32] Abro GH, Syed TS, Shah AH, Cui J, Sattar M, Awan MS. Efficacy and economics of different insecticides against stem borers, *Scirpophaga incertulas* (Walker) in rice crop. Pakistan Journal of Zoology. 2013;**45**(4):929-933

[33] Bhatt N, Joshi S, Tiwari SN. Pests of rice. In: Pests and their Management. 2018. pp. 9-50. DOI: 10.1007/978-981-10-8687-8_2

[34] Ali M, Nessa B, Khatun M, Salam M, Kabir M. A way forward to combat insect pest in rice. Bangladesh Rice Journal. 2021;**25**(1):1-22. DOI: doi.org/10.3329/brj.v25i1.55176

[35] Agunbiade TA, Sun W, Coates BS, Djouaka R, Tamò M, Ba MN, et al. Development of reference transcriptomes for the major field insect pests of cowpea: A toolbox for insect pest management approaches in West Africa. PLoS One. 2013;**8**(11):e79929. DOI: 10.1371/journal.pone.0079929

[36] Togola A, Boukar O, Belko N, Chamarthi SK, Fatokun C, Tamo M, et al. Host plant resistance to insect pests of cowpea (*Vigna unguiculata* L. Walp.): Achievements and future prospects. Euphytica. 2017;**213**:1-16. DOI: 10.1007/s10681-017-2030-1

[37] Asante SK, Tamo M, Jackai LEN. Integrated management of cowpea insect pests using elite cultivars, date of planting and minimum insecticide application. African Crop Science Journal. 2001;**9**(4):655-665. DOI: 10.4314/acsj.v9i4.27587

[38] Oparaeke AM, Dike MC, Amatobi CI. Botanical pesticide mixtures for insect pest management on cowpea, *Vigna unguiculata* (L.) Walp plants–the legume flower bud thrips, Megalurothrips sjostedti Trybom. Journal of Sustainable Agriculture. 2006;**29**(1):5-13

[39] Amanet K, Chiamaka EO, Quansah GW, Mubeen M, Farid HU, Akram R, et al. Cotton production in Africa. In: Cotton Production. 2019. pp. 359-369. DOI: 10.1002/9781119385523.ch17

[40] Deguine JP, Ferron P, Russell D. Sustainable pest management for cotton production: A review. In: Sustainable Agriculture. 2009. pp. 411-442. DOI: 10.1007/978-90-481-2666-8_27

[41] Silvie P, Deguine JP, Nibouche S, Michel B, Vaissayre M. Potential of threshold-based interventions for cotton pest control by small farmers in West Africa. Crop Protection. 2001;**20**(4):297-301. DOI: 10.1016/S0261-2194(00)00146-0

[42] Dhawan AK. Integrated pest management in cotton. Integrated Pest Management in the Tropics. 2016;**499575**(20):16

[43] Silvie P. Plant-based extracts for cotton pest management in sub-saharan Africa: A review. Botany

Letters. 2023;**170**(1):28-41. DOI: 10.1080/23818107.2022.2122556

[44] De Bon H, Huat J, Parrot L, Sinzogan A, Martin T, Malezieux E, et al. Pesticide risks from fruit and vegetable pest management by small farmers in sub-Saharan Africa. A Review. Agronomy for Sustainable Development. 2014;**34**:723-736. DOI: 10.1007/s13593-014-0216-7

[45] Mafongoya P, Gubba A, Moodley V, Chapoto D, Kisten L, Phophi M. Climate change and rapidly evolving pests and diseases in Southern Africa. In: Ayuk ET, Unuigbe NF, editors. New Frontiers in Natural Resources Management in Africa. Cham, Switzerland: Springer International Publishing: 2019. pp. 41-57

[46] Ngowi AVF, Mbise TJ, Ijani ASM, London L, Ajayi OC. Pesticides use by smallholder farmers in vegetable production in northern Tanzania. Crop Protection (Guildford, Surrey). 2007;**26**(11):1617. DOI: 10.1016%2Fj.cropro.2007.01.008

[47] Agboyi LK, Djade KM, Ahadji-Dabla KM, Ketoh GK, Nuto Y, Glitho IA. Vegetable production in Togo and potential impact of pesticide use practices on the environment. International Journal of Biological and Chemical Sciences. 2015;**9**(2):723-736. DOI: 10.4314/ijbcs.v9i2.13

[48] Horna D, Falk-Zepeda J, Timpo SE. Insecticide Use on Vegetables in Ghana. Would GM Seeds Benefit Farmers; International Food Policy Research Institute (IFPRI). Accra, Ghana: Environment and Production Technology Division; 2008

[49] Ntow WJ, Gijzen HJ, Kelderman P, Drechsel P. Farmer perceptions and pesticide use practices in vegetable production in Ghana. Pest Management Science: Formerly Pesticide Science. 2006;**62**(4):356-365. DOI: 10.1002/ps.1178

[50] Amoabeng BW, Gurr GM, Gitau CW, Stevenson PC. Cost: Benefit analysis of botanical insecticide use in cabbage: Implications for smallholder farmers in developing countries. Crop Protection. 2014;**57**:71-76. DOI: 10.1016/j.cropro.2013.11.019

[51] Thomson LJ, Hoffmann AA. Ecologically sustainable chemical recommendations for agricultural pest control? Journal of Economic Entomology. 2007;**100**(6):1741-1750. DOI: 10.1093/jee/100.6.1741

[52] Hill MP, Macfadyen S, Nash MA. Broad spectrum pesticide application alters natural enemy communities and may facilitate secondary pest outbreaks. PeerJ. 2017;**5**:e4179. DOI: 10.7717/peerj.4179

[53] Abdel-Raheem M. Effect of insecticides on natural-enemies. In: Insecticides-Impact and Benefits of Its Use for Humanity. London, UK: IntechOpen; 2021. DOI: 10.5772/intechopen.100616

[54] Stapel JO, Cortesero AM, Lewis WJ. Disruptive sublethal effects of insecticides on biological control: Altered foraging ability and life span of a parasitoid after feeding on extrafloral nectar of cotton treated with systemic insecticides. Biological Control. 2000;**17**(3):243-249. DOI: 10.1006/bcon.1999.0795

[55] Bommarco R, Miranda F, Bylund H, Björkman C. Insecticides suppress natural enemies and increase pest damage in cabbage. Journal of Economic Entomology. 2011;**104**(3): 782-791. DOI: 10.1603/ec10444

[56] Gross K, Rosenheim JA. Quantifying secondary pest outbreaks in cotton and their monetary cost with causal-inference statistics. Ecological Applications. 2011;**21**(7):2770-2780. DOI: 10.1890/11-0118.1

[57] Hardin MR, Benrey B, Coll M, Lamp WO, Roderick GK, Barbosa P. Arthropod pest resurgence: An overview of potential mechanisms. Crop Protection. 1995;**14**(1):3-18. DOI: 10.1016/0261-2194(95)91106-P

[58] Venkatesh J, Priya S, Balasubramaniam M, Aarthy C, Thenmozhi S, Balasubramanie P. Continuing issues in the use of pesticides for procuring life in developing countries. Life Science Journal-Acta Zhengzhou University Overseas Edition. 2012;**9**(4):304-308

[59] Sánchez-Bayo F. Indirect effect of pesticides on insects and other arthropods. Toxics. 2021;**9**(8):177. DOI: 10.3390/toxics9080177

[60] Downing AL, DeVanna KM, Rubeck-Schurtz CN, Tuhela L, Grunkemeyer H. Community and ecosystem responses to a pulsed pesticide disturbance in freshwater ecosystems. Ecotoxicology. 2008;**17**:539-548. DOI: 10.1007/s10646-008-0211-3

[61] Schmolke A, Thorbek P, Chapman P, Grimm V. Ecological models and pesticide risk assessment: Current modeling practice. Environmental Toxicology and Chemistry: An International Journal. 2010;**29**(4):1006-1012. DOI: 10.1002/etc.120

[62] Bernardes MFF, Pazin M, Pereira LC, Dorta DJ. Impact of pesticides on environmental and human health. In: Toxicology Studies-Cells, Drugs and Environment. London, UK: IntechOpen; 2015. pp. 195-233

[63] Furlan L, Kreutzweiser D. Alternatives to neonicotinoid insecticides for pest control: Case studies in agriculture and forestry. Environmental Science and Pollution Research. 2015;**22**(1):135-147. DOI: 10.1007/s11356-014-3628-7

[64] Neumann M, Schulz R, Schäfer K, Müller W, Mannheller W, Liess M. The significance of entry routes as point and non-point sources of pesticides in small streams. Water Research. 2002;**36**(4):835-842. DOI: 10.1016/S0043-1354(01)00310-4

[65] Banaee M. Adverse effect of insecticides on various aspects of fish's biology and physiology. Insecticides—Basic and Other Applications. 2012;**6**:101-126

[66] Anderson SE, Meade BJ. Potential health effects associated with dermal exposure to occupational chemicals. Environmental Health Insights. 2014;**8**:EHI-S15258. DOI: 10.4137/EHI.S15258

[67] Rani L, Thapa K, Kanojia N, Sharma N, Singh S, Grewal AS, et al. An extensive review on the consequences of chemical pesticides on human health and environment. Journal of Cleaner Production. 2021;**283**:124657. DOI: 10.1016/j.jclepro.2020.124657

[68] El-Nahhal Y, Radwan A, Radwan AM. Human health risks: Impact of pesticide application. Journal of Environment and Earth Science. 2013;**3**(7):199-209

[69] Hoppin JA, LePrevost CE. Pesticides and human health. In: Environmental Pest Management: Challenges for Agronomists, Ecologists, Economists and Policymakers. 2017. pp. 249-273. DOI: 10.1002/9781119255574.ch11

[70] Kalyabina VP, Esimbekova EN, Kopylova KV, Kratasyuk VA. Pesticides:

Formulants, distribution pathways and effects on human health–a review. Toxicology Reports. 2021;**8**:1179-1192. DOI: 10.1016/j.toxrep.2021.06.004

[71] Oerke EC. Crop losses to pests. The Journal of Agricultural Science. 2006;**144**(1):31-43. DOI: 10.1017/S0021859605005708

[72] Manosathiyadevan M, Bhuvaneshwari V, Latha R. Impact of insects and pests in loss of crop production: A review. In: Dhanarajan A, editor. Sustainable Agriculture towards Food Security. Singapore: Springer; 2017. DOI: 10.1007/978-981-10-6647-4_4

[73] Razaq M, Mensah R, Athar H-ur-R. Insect pest management in cotton. In: Jabran Kh, Chauhan BS, editors. Cotton Production. USA: John Wiley & Sons Ltd; 2020. pp. 85-107. DOI: 10.1002/9781119385523.ch5

[74] Binns MR, Nyrop JP. Sampling insect populations for the purpose of IPM decision making. Annual Review of Entomology. 1992;**37**(1):427-453

Chapter 17

Insecticides and Natural Enemies: Applications in Integrated Pest Management Programs – Challenges, Criteria, and Evaluation for Recommendations

Fatemeh Yarahmadi and Ali Rajabpour

Abstract

Integrated Pest Management (IPM) has emerged as a crucial approach in sustainable agriculture, aiming to minimize the impact of pests while reducing reliance on synthetic pesticides. This paper delves into the evolution of IPM over the past six decades, highlighting its goals of sustainable agricultural practices and decreased pesticide application. It also addresses the challenges posed by socioeconomic, environmental, and health factors associated with pesticide application. The discussion explores the weaknesses in current IPM practices, such as multiple definitions leading to confusion and inconsistencies between theory and implementation. Moreover, inadequate farmer engagement and ecological awareness are brought to light. Although chemical control remains central in plant health programs, its limitations and repercussions are considered within the realm of IPM. An alternative concept, Agroecological Crop Protection, is proposed as a more sustainable approach integrating agroecology into crop protection strategies. This interdisciplinary concept operates at various levels, spanning from field to landscape, emphasizing the importance of ecological balance. Quantitative evaluation methods for assessing IPM effectiveness are examined, encompassing metrics, data collection techniques, and statistical analyses, incorporating economic and ecological indicators. Furthermore, qualitative evaluation methods, such as surveys and stakeholder interviews, emphasize the significance of social and cultural factors influencing IPM practices.

Keywords: Integrated Pest Management (IPM), sustainable agriculture, synthetic pesticides, agroecology, chemical control, socioeconomic factors, ecological balance, quantitative and qualitative evaluation

1. Introduction

1.1 Overview of Integrated Pest Management (IPM)

Integrated Pest Management (IPM) represents a holistic approach to pest regulation aimed at diminishing the dependency on chemical pesticides. Widely recognized as the most efficient method for managing pest populations within agricultural systems, IPM integrates a variety of control strategies to achieve sustainable pest management. These strategies encompass biological control, habitat manipulation, cultural practices, resistant varieties, and minimized pesticide application [1–3]. The key principles of IPM (**Figure 1**) highlighted by Torres and Bueno [3] include regular monitoring and identification of pest populations to assess the need for intervention; implementation of cultural practices like crop rotation, sanitation, and habitat management to reduce pest populations and minimize pesticide application; employing biological control methods such as predators, parasitoids, and pathogens to regulate pest populations and maintain ecological balance; using mechanical and physical control techniques like barriers, traps, and mechanical means to manage pests without relying solely on chemical insecticides; and, ultimately, resorting to chemical control judiciously and selectively as a last resort when other control methods are insufficient in managing pest populations.

In the context of this chapter, IPM serves as a strategic framework for harmonizing the application of insecticides and natural enemies in pest management initiatives. By underscoring the need to strike a balance between pest regulation and preservation of beneficial organisms, IPM advocates for sustainable pest management practices that curtail environmental impacts and encourage enduring pest suppression.

Throughout the chapter, discussions on challenges, standards, and assessment criteria for offering recommendations concerning insecticides and natural enemies will be contextualized within the fundamental principles of IPM. The chapter will underscore the pivotal role of IPM in informing decision-making processes pertinent to selecting insecticides, evaluating their effects on natural enemies, and devising efficient pest management strategies that prioritize ecological sustainability and pest control efficacy.

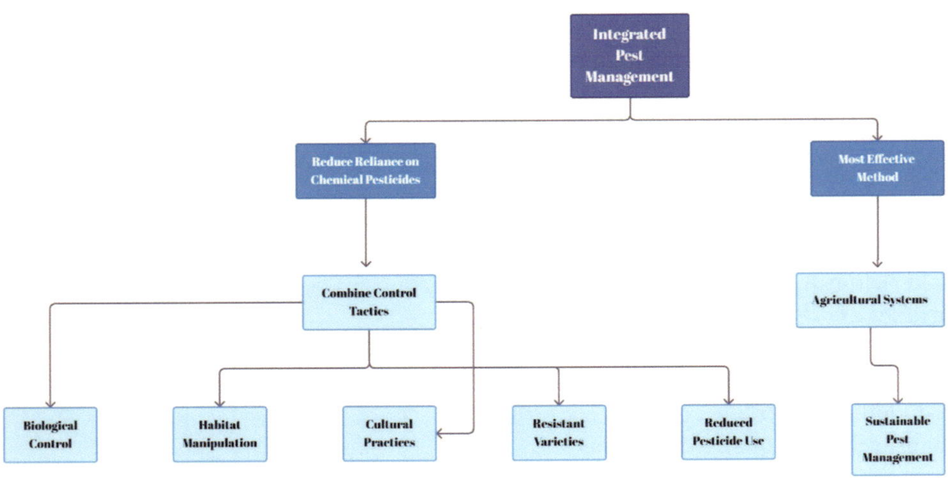

Figure 1.
Strategies for Integrated Pest Management in agriculture.

1.2 Importance of balancing insecticide application and conservation of natural enemies

Implementing practices that balance disturbance and conservation in agroecosystems can create a more resilient and ecologically balanced system, supporting natural enemies of pests for improved biological control and reduced reliance on chemical insecticides. By integrating insecticide application with the conservation of natural enemies in IPM, pest populations can be effectively controlled while minimizing harm to beneficial organisms. This approach not only reduces reliance on chemical pesticides but also promotes a sustainable and environmentally friendly pest management strategy in agricultural systems [2, 4–7].

1.3 Objectives and scope of the chapter

- Exploring the challenges associated with integrating insecticide application and the conservation of natural enemies in IPM programs.
- Discussing the criteria and guidelines for evaluating the compatibility of insecticides with natural enemies in agricultural systems.
- Providing recommendations for effectively incorporating insecticides into IPM programs while minimizing negative impacts on beneficial organisms.
- Evaluating the effectiveness of different insecticides in managing pest populations while maintaining a balance with natural enemy populations.
- Addressing the need for sustainable pest management strategies that prioritize the conservation of natural enemies and reduce reliance on chemical insecticides.
- Offering insights into the evaluation methods and criteria used to recommend insecticide applications in IPM programs, considering both pest control efficacy and environmental sustainability.

Overall, the chapter likely aims to provide a comprehensive overview of the challenges, criteria, and evaluation processes involved in utilizing insecticides within the framework of IPM, with a focus on promoting sustainable pest management practices that support natural enemy populations in agroecosystems.

2. Applications of insecticides and natural enemies in Integrated Pest Management programs

2.1 Evolution of IPM

Over the past 60 years, IPM has undergone significant evolution, transitioning from a predominantly chemical-focused methodology to a more comprehensive and eco-friendly pest control strategy. Originating in the 1950s as a response to the adverse effects of widespread pesticide application, IPM was designed to amalgamate diverse pest management techniques, reduce dependence on synthetic pesticides, and advocate for sustainable agricultural practices. The objectives of IPM have broadened to not only include efficient pest regulation but also encompass environmental preservation, biodiversity conservation, and safeguarding human health and safety [5, 8].

A key aim of IPM is to curtail the utilization of synthetic pesticides by integrating alternative pest control methods such as biological control, cultural practices, and the cultivation of pest-resistant crop varieties. By diminishing reliance on pesticides, IPM seeks to address the adverse socioeconomic, environmental, and health repercussions linked to pesticide application. These challenges comprise the development of pesticide-resistant pest populations, environmental contamination, harm to nontarget organisms, and health risks from pesticide exposure [9–11].

The evolution of IPM has underscored the significance of adopting sustainable pest management strategies that prioritize long-term ecological equilibrium and minimize adverse impacts on the environment and human health. By integrating a spectrum of pest control tactics and emphasizing the preservation of natural enemies, IPM has emerged as a cornerstone of contemporary agricultural sustainability endeavors [4, 6, 8, 12]. Continued advancement and adherence to IPM principles are pivotal in addressing the intricate challenges associated with pesticide application and advancing a more sustainable and eco-conscious approach to pest management in agriculture [13].

2.2 Weaknesses in current IPM practices

Weaknesses within the current framework of IPM stem from various factors that pose challenges to its successful implementation and efficacy. A critical issue revolves around the plethora of divergent IPM definitions in existence, causing confusion among practitioners, policymakers, and stakeholders. The absence of a standardized definition can lead to varying interpretations and strategies related to IPM, impeding consistent application across diverse agricultural systems and regions [8, 14].

Inconsistencies in the concepts, implementation, and policies of IPM exacerbate the challenges faced in current IPM practices. While the core principles of IPM emphasize the integration of diverse pest control methods and minimizing pesticide application, discrepancies in their actual application and depiction in policy frameworks can lead to confusion, impeding the successful implementation of IPM programs. Misalignments between theoretical principles and practical implementation can result in inadequate pest management outcomes and limit the overall effectiveness of IPM strategies.

Another weakness lies in the limited involvement of farmers and inadequate ecological comprehension within current IPM practices. The successful implementation of IPM relies on active farmer participation and the adoption of pest management techniques on their farms. Nonetheless, inadequate farmer engagement, restricted access to information, and resource constraints can impede the proper adoption of IPM. Without a robust ecological knowledge base and active farmer involvement, the full potential of IPM in promoting sustainable pest control and reducing environmental impact may not be fully realized [13, 15, 16].

Mitigating these weaknesses in existing IPM practices necessitates collective efforts to define terms clearly, bridge the gap between theory and practice, enhance farmer participation and education, and cultivate a comprehensive understanding of ecological principles in pest management. By tackling these challenges, stakeholders can collaborate toward fortifying IPM practices and unlocking its complete potential as a sustainable and efficient approach to pest management in agriculture [8, 17, 18].

2.3 Criteria for pest management recommendations and challenges in implementing IPM recommendations

Effective recommendations for pest management are vital for the success of IPM programs. When evaluating pest control methods, several key criteria must be taken into account, including efficacy, safety, environmental impact, and cost-effectiveness. Efficacy pertains to the ability of a pest control method to effectively manage target pests and achieve desired outcomes. Safety considerations involve assessing the potential risks of the method on human health, nontarget organisms, and the environment. The environmental impact assesses the overall effects of the pest control method on ecosystems, biodiversity, and natural resources. Cost-effectiveness evaluates the economic feasibility and practicality of implementing the pest management recommendation (**Table 1**) [13, 19, 20].

These criteria play a crucial role in the decision-making process when selecting and deploying pest management strategies. Stakeholders need to balance the advantages and disadvantages of various control methods based on these criteria to determine the most suitable approach for their specific pest management requirements. By considering efficacy, safety, environmental impact, and cost-effectiveness, decision-makers can make informed choices that prioritize sustainable pest management practices and minimize adverse effects on the environment and human health [12, 13, 21].

Despite the advantages of IPM, there are practical challenges that farmers and practitioners encounter when implementing pest management recommendations. Challenges related to the adoption, education, and policy aspects may impede the effective incorporation of IPM practices into agricultural systems. Farmers may resist change due to concerns about the perceived effectiveness or complexity of IPM methods, leading to hesitancy in embracing new pest management strategies. Limited access to education and training on IPM practices can also hinder implementation efforts, as farmers may lack the necessary knowledge and skills to effectively apply IPM principles on their farms [13, 17, 18].

Policy impediments, such as a lack of supportive regulations or incentives for IPM adoption, can further impede the widespread implementation of sustainable

Criteria for pest management recommendations	Challenges in implementing IPM recommendations
Efficacy: Ability to effectively manage target pests and achieve desired outcomes.	*Adoption resistance:* Concerns about effectiveness or complexity of IPM methods may hinder acceptance.
Safety: Assessing risks on human health, nontarget organisms, and the environment.	*Education barriers:* Limited access to training and knowledge may impede effective application of IPM practices.
Environmental impact: Effects on ecosystems, biodiversity, and natural resources.	*Policy impediments:* Lack of regulations or incentives for IPM adoption can obstruct implementation efforts.
Cost-effectiveness: Evaluating economic feasibility and practicality of pest management recommendation.	*Interdisciplinary collaboration:* Engagement among researchers, extension agents, policymakers, and farmers is necessary to develop tailored solutions for challenges in IPM implementation.

Table 1.
Criteria and challenges in implementing Integrated Pest Management (IPM) recommendations.

pest management practices. Without clear policy frameworks that promote and stimulate IPM strategies, farmers may find it challenging to prioritize and invest in sustainable pest management approaches. Overcoming these challenges necessitates interdisciplinary collaboration among researchers, extension agents, policymakers, and farmers to develop tailored solutions that address the specific needs and constraints of agricultural systems. By fostering partnerships and sharing knowledge across disciplines, stakeholders can collaborate to overcome practical challenges and promote the successful implementation of IPM recommendations in agriculture [15, 19].

2.4 Agroecological Crop Protection

Agroecological Crop Protection is an innovative approach that integrates principles of agroecology with crop protection strategies to promote sustainable and environmentally friendly agricultural practices. This concept proposes a holistic and interdisciplinary approach to managing pests and diseases in crops while enhancing ecosystem health and resilience [1, 4, 22–26].

Here is how Agroecological Crop Protection optimally integrates agroecology for crop protection and emphasizes interdisciplinary strategies at different levels (**Figure 2**) [9, 25, 27]:

Field level: At the field level, Agroecological Crop Protection focuses on enhancing biodiversity and ecosystem services to naturally regulate pest populations. This includes practices such as intercropping, crop rotation, and cover cropping to create a diverse and resilient agroecosystem that supports beneficial organisms and suppresses pests.

Utilizing agroecological principles, farmers can implement techniques like habitat manipulation, conservation biological control, and agroforestry to enhance natural pest control mechanisms within the field, reducing the reliance on synthetic pesticides.

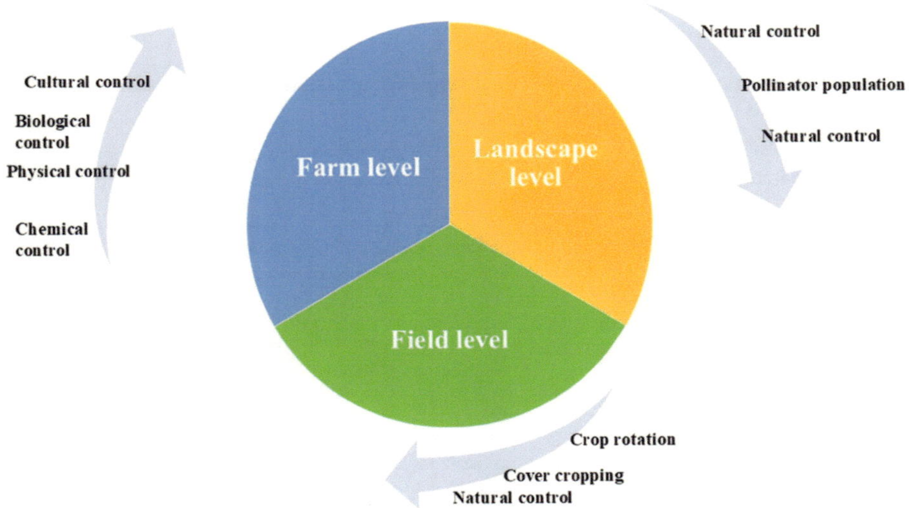

Figure 2.
Elements of Agroecological Crop Protection.

Farm level: Agroecological Crop Protection promotes the implementation of IPM strategies that combine biological, cultural, physical, and chemical control methods in a synergistic manner. This approach emphasizes the use of nonchemical options as the first line of defense against pests, with minimal reliance on synthetic inputs.

Farmers are encouraged to adopt agroecological practices that enhance soil health, promote beneficial insect populations, and improve crop resilience to pests and diseases. By fostering a healthy and balanced agroecosystem, farmers can reduce the need for external inputs and enhance the sustainability of crop protection measures.

Landscape level: Agroecological Crop Protection recognizes the importance of landscape-level management in promoting biodiversity and ecosystem services that benefit crop protection. This involves creating functional agroecological landscapes that support natural enemies of pests, enhance pollinator populations, and improve overall ecosystem health.

Interdisciplinary collaboration at the landscape level involves engaging with stakeholders such as farmers, researchers, policymakers, and conservationists to design and implement landscape-scale strategies for sustainable crop protection. This may include initiatives like buffer strips, hedgerows, and agroforestry systems that enhance biodiversity and connectivity across agricultural landscapes [4, 5].

By integrating agroecology for crop protection at the field, farm, and landscape levels, Agroecological Crop Protection offers a comprehensive and sustainable approach to managing pests and diseases while promoting ecological resilience, biodiversity conservation, and long-term agricultural sustainability [4, 5, 27].

2.5 Evaluation methods in Integrated Pest Management (IPM): quantitative and qualitative perspectives

The evaluation of Integrated Pest Management (IPM) programs relies heavily on quantitative methodologies to gauge their effectiveness. Metrics such as pest population levels, crop yield, pesticide application, and economic costs are commonly utilized to assess the impact of IPM strategies. Data collection techniques, including field observations, sampling, and monitoring methods, are employed to gather pertinent information on pest dynamics and management outcomes. Subsequent statistical analyses are then conducted to interpret the data and determine the significance of the findings [20, 28, 29].

Economic indicators, such as input costs, labor expenses, and market prices, serve as crucial benchmarks for quantifying the financial advantages of implementing IPM practices. By comparing the costs and benefits of IPM against traditional pest control methods, stakeholders can make well-informed decisions regarding the economic viability of embracing sustainable pest management strategies. Ecological indicators, such as biodiversity, soil health, and water quality, also play a vital role in assessing the environmental ramifications of IPM programs and evaluating their contributions to ecosystem sustainability [13, 15, 23, 30].

Quantitative evaluation methodologies offer valuable insights into the efficacy of IPM strategies, aiding stakeholders in comprehending the tangible outcomes of their pest management endeavors. Through the use of quantitative metrics and data analysis techniques, researchers and practitioners can evaluate the performance of IPM programs, pinpoint areas for enhancement, and elevate pest control results while fostering sustainable agricultural practices [31].

Apart from quantitative analyses, qualitative methodologies are indispensable for evaluating IPM programs and capturing stakeholder viewpoints on pest management practices. Qualitative methods like surveys, interviews, focus groups, and participatory workshops offer valuable insights into the social, cultural, and behavioral aspects of IPM adoption [13].

Stakeholder feedback and perceptions are pivotal in comprehending the obstacles, possibilities, and motivations that influence the implementation of IPM strategies. By engaging with farmers, extension agents, policymakers, and other stakeholders, researchers can gather valuable insights into the hurdles to IPM adoption, the perceived benefits of sustainable pest management practices, and the social dynamics that shape decision-making processes in agriculture [13, 32].

Social and cultural factors wield considerable influence in shaping attitudes toward pest management practices and impacting the adoption of IPM strategies. Integrating qualitative methods into evaluation procedures enables researchers to grasp the human dimensions of pest control and customize IPM recommendations to address the distinct requirements and preferences of various stakeholder groups [24]. Appraising the effectiveness of IPM programs and understanding stakeholder perspectives on sustainable pest management practices have crucial role in sustainable agriculture. By amalgamating both approaches, researchers can generate comprehensive insights into the impacts of IPM strategies, pinpoint avenues for improvement, and advocate for the adoption of environmentally friendly and economically viable pest management solutions [13, 30].

2.6 Balancing act: insecticides and natural enemies in IPM

IPM embodies a holistic approach to pest control, underscoring the significance of striking a harmonious balance between the application of insecticides and the preservation of beneficial natural enemies within agricultural systems. This equilibrium is pivotal for fostering sustainable and eco-friendly pest management practices. While insecticides play a crucial role in managing pest populations and mitigating crop damage, their excessive application can pose risks to valuable natural enemies like predators, parasitoids, and pollinators, which are instrumental in regulating pest numbers and upholding ecological equilibrium [12, 23, 33].

To optimize pest control outcomes while safeguarding natural enemies, it is imperative to prioritize the adoption of selective and environmentally benign insecticides within IPM frameworks. By opting for insecticides that target specific pests while sparing beneficial insects, farmers can minimize adverse impacts on natural enemies while efficiently controlling pests [12]. Moreover, cultural practices such as crop rotation, habitat diversification, and the preservation of natural enemy populations can bolster the efficacy of natural pest control methods [1, 3–5].

By maintaining a delicate equilibrium between insecticides and natural enemies in IPM, farmers can institute sustainable pest management approaches that safeguard crop vitality, diminish environmental footprints, and foster biodiversity in agricultural settings. This integrated methodology not only ensures effective pest control but also develops the long-term resilience and sustainability of agroecosystems [7, 18, 25, 33].

Key components of successful IPM strategies encompass monitoring pest populations to determine action thresholds, employing selective insecticides that spare beneficial organisms, timing insecticide applications to reduce harm to natural enemies, enriching habitats for beneficial insects, and incorporating cultural practices that boost biodiversity in agricultural ecosystems. These strategies play a pivotal role in establishing a symbiotic relationship between pest management measures and natural enemy conservation, ultimately yielding more sustainable and efficient pest control outcomes [1, 4, 20].

2.7 Synergy and trade-offs: navigating insecticide-natural enemy interactions

The intricate interactions between insecticides and natural enemies in agro-ecosystems involve both synergy and trade-offs. Synergy refers to the cooperative effects of insecticides and natural enemies in enhancing pest control, while trade-offs occur when insecticides harm beneficial organisms. Synergy and trade-offs play a critical role in navigating the intricate interactions between insecticides and natural enemies within agroecosystems. While insecticides are effective at controlling pest populations, their impact on beneficial natural enemies must be carefully considered to maintain ecological balance and sustainable pest management practices [25, 34, 35].

When used strategically, insecticides can complement the work of natural enemies by targeting specific pest species that may be resistant to natural control methods. By leveraging the synergistic effects of insecticides and natural enemies, farmers can achieve more robust and effective pest management outcomes [7, 10].

However, there are also trade-offs to be mindful of when using insecticides in the presence of natural enemies. Some insecticides may inadvertently harm beneficial organisms, disrupting the delicate balance of predator-prey interactions and leading to unintended consequences in the ecosystem. For example, broad-spectrum insecticides can indiscriminately affect both pests and natural enemies, potentially reducing the overall efficacy of natural pest control [33, 34].

To navigate these interactions effectively, it is essential to consider the potential synergies and trade-offs between insecticides and natural enemies in pest management decisions. Adopting IPM strategies that prioritize the use of selective insecticides, minimize nontarget effects, and enhance habitat and biodiversity conservation can help mitigate the negative impacts of insecticides on natural enemies while maximizing their complementary effects on pest control [3, 21, 33].

By considering the balance between insecticides and natural enemies, farmers can optimize pest management practices, protect beneficial organisms, and promote sustainable agriculture. Further research, education, and policy efforts are essential to advance our understanding and implementation of these complex interactions in agroecosystems [13, 17, 18, 25].

2.8 Evaluating insecticides: criteria for effective pest management

When it comes to selecting insecticides for pest management within the framework of IPM, careful evaluation is essential to ensure effectiveness while minimizing negative impacts on the environment and nontarget organisms. The process of evaluating insecticides involves considering several key criteria to align with IPM goals and promote sustainable pest management practices [34].

- *Efficacy:* The primary criterion for selecting an insecticide is its efficacy in controlling the target pest species. It is crucial to select insecticides that are proven to be effective against the specific pest problem at hand while considering factors like application method and timing for optimal results [36].

- *Selectivity:* Selectivity refers to the insecticide's ability to target pests while minimizing harm to beneficial organisms, such as natural enemies, pollinators, and other nontarget species. Selecting insecticides with lower nontarget toxicity helps preserve ecological balance and biodiversity within agroecosystems [23].

- *Residual activity:* The duration of insecticidal activity on treated surfaces or crops, known as residual activity, plays a crucial role in sustained pest control. Balancing the need for long-lasting efficacy with the potential risks of environmental persistence is important in selecting insecticides that maintain pest suppression without undue harm [3].

- *Mode of action:* Understanding the mode of action of insecticides is essential for managing resistance development in pest populations. Rotating or alternating insecticides with different modes of action can help prevent or delay the emergence of resistant pests and prolong the effectiveness of pest control strategies [37–40].

- *Environmental impact:* Evaluating the environmental impact of insecticides involves considering factors such as toxicity to nontarget organisms, potential for environmental persistence, and effects on water quality and beneficial insects. Choosing insecticides with lower environmental impact reduces risks to ecosystems and human health [23, 39].

- *Application safety:* Ensuring the safe and effective application of insecticides is crucial for protecting applicators, bystanders, and consumers from exposure risks. Adhering to proper application practices, including equipment calibration, personal protective equipment use, and adherence to label instructions, promotes safe handling and application of insecticides [23].

By evaluating insecticides based on these criteria for effective pest management, farmers and pest control professionals can make informed decisions that align with the principles of IPM. Selecting insecticides that balance efficacy with selectivity, residual activity, mode of action diversity, environmental impact, and application safety contributes to sustainable pest management practices that minimize the negative impacts of chemical control methods on agroecosystems and promote long-term pest control solutions [25].

2.9 Challenges ahead: integrating insecticides and natural enemies

As agriculture continues to evolve toward more sustainable and environmentally-friendly practices, the integration of insecticides and natural enemies presents a promising approach for pest management. However, this harmonization of chemical and biological control methods comes with its own set of challenges that must be addressed to maximize efficacy while minimizing negative effects on beneficial organisms and ecological balance (**Figure 3**) [13, 34, 36].

Compatibility of methods: One of the primary challenges in integrating insecticides and natural enemies is ensuring the compatibility of these two approaches. Some insecticides may harm or disrupt the activity of natural enemies, leading to unintended consequences such as resurgence of pest populations or disruption of ecological balance. Finding insecticides that are selective toward pests while sparing beneficial insects is a key consideration in this integration process [3, 33, 35, 36].

Timing and application: Coordinating the timing and application of insecticides and natural enemies can be complex, as both methods may have specific requirements for optimal effectiveness. Factors such as the life cycle of the target pest, the activity period of natural enemies, and environmental conditions must be

DOI: http://dx.doi.org/10.5772/intechopen.1005830

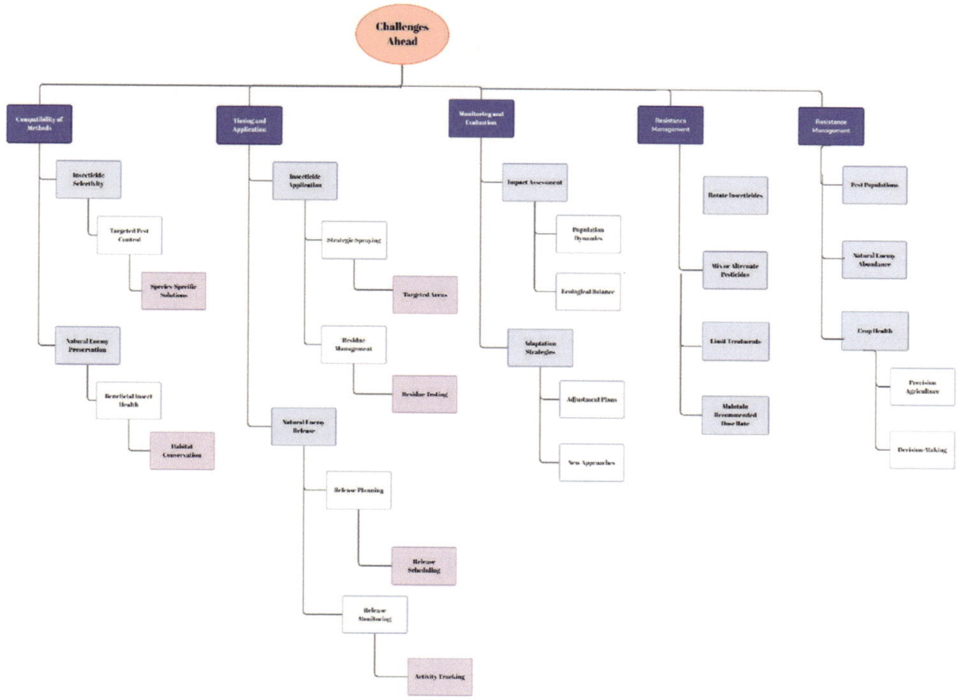

Figure 3.
Challenges ahead in integrating insecticides and natural enemies.

carefully considered to ensure that both control measures work synergistically rather than antagonistically [10, 36].

Resistance management: The risk of pest resistance to insecticides is a significant concern in Integrated Pest Management strategies. Overreliance on chemical control can accelerate the development of resistance in pest populations, reducing the long-term efficacy of both insecticides and natural enemies. Implementing resistance management practices, such as rotating insecticides with different modes of action and using biological control agents strategically, is essential to address this challenge [10, 37–41].

Monitoring and decision-making: Effective integration of insecticides and natural enemies requires regular monitoring of pest populations, natural enemy abundance, and crop health. Making informed decisions about when and how to implement control measures based on this monitoring data is crucial for maintaining the balance between pest control and conservation of beneficial organisms. Implementing precision agriculture techniques and advanced monitoring tools can help improve decision-making in Integrated Pest Management [20, 21].

Education and awareness: Building awareness and providing education to farmers, growers, and pest control professionals about the benefits and challenges of integrating insecticides and natural enemies is essential for successful implementation. Training programs, workshops, and extension services can help foster a better understanding of the complexities involved in this approach and promote adoption of sustainable pest management practices [13, 17, 18, 32].

Addressing these challenges in integrating insecticides and natural enemies requires a holistic and adaptive approach that considers the complexities of agroecosystems and the interactions between pests, natural enemies, and control methods. By overcoming these hurdles through innovative research, practical solutions, and stakeholder collaboration, agriculture can move toward more sustainable pest management practices that harness the synergies between chemical and biological control methods while safeguarding ecosystem health and biodiversity [8, 16].

2.10 From lab to field: implementing IPM strategies

The journey from laboratory research to practical implementation of IPM strategies, which encompasses a collaborative, multidisciplinary approach bridging the gap between scientific discoveries and effective pest control solutions, is eloquently detailed by Rajabpour and Yarahmadi [42]. This comprehensive process encompasses the translation of research findings into field trials, validation of efficacy in real-world settings, and the adoption of sustainable practices that harmonize insecticides and natural enemies for effective pest management [8, 13, 32].

Laboratory research: The foundation of implementing IPM strategies lies in rigorous laboratory research that explores the biological, chemical, and ecological aspects of pest control. Scientists study the efficacy of insecticides, the behavior of pests and natural enemies, and the interactions between different control methods to develop innovative solutions for sustainable pest management. Laboratory studies provide valuable insights into the mechanisms of action of insecticides, the selectivity toward target pests, and the impact on beneficial organisms [2, 43, 44].

Field trials: Translating promising research findings from the laboratory to field trials is a critical step in validating the efficacy of IPM strategies under real-world conditions. Field trials allow researchers to assess the performance of insecticides and natural enemies in diverse agroecosystems, considering factors such as crop type, pest pressure, climate variability, and farmer practices. By conducting field trials, scientists can evaluate the practicality, scalability, and economic viability of IPM approaches and fine-tune control strategies for optimal pest management outcomes [19, 27, 30].

Monitoring and evaluation: Continuous monitoring and evaluation of pest populations, natural enemy abundance, crop health, and control measures are essential for the successful implementation of IPM strategies. Researchers use advanced monitoring tools, such as pheromone traps, remote sensing technologies, biochemical and molecular tools, field surveys and sampling protocols and predictive modeling, to track pest dynamics and make informed decisions about the timing and intensity of control interventions. This data-driven approach enables proactive pest management and adaptive responses to changing environmental conditions [11, 14, 20].

Technology and innovation: Advancements in technology and innovation play a key role in enhancing the implementation of IPM strategies in agriculture. IPM platforms, precision agriculture tools, smart farming technologies, and digital solutions provide farmers and pest control professionals with real-time data, decision support systems, and customized recommendations for sustainable pest management practices. By leveraging technology, stakeholders can optimize resource use, minimize environmental impact, and maximize the efficacy of insecticides and natural enemies in pest control [14, 26, 28, 41].

Capacity building and extension: Building capacity and promoting extension services are essential components of successful IPM implementation. Training programs, workshops, policy supports, and knowledge-sharing initiatives help farmers, growers, and industry stakeholders adopt IPM practices, understand the benefits of integrating insecticides and natural enemies, and implement sustainable pest management solutions. By fostering partnerships between researchers, extension agents, policymakers, and practitioners, the adoption of IPM strategies can be accelerated and mainstreamed across agricultural landscapes [13].

By bridging the gap between laboratory research and on-farm application, the implementation of IPM strategies offers a holistic and science-based approach to pest management that balances efficacy, sustainability, and environmental stewardship. Through collaboration, innovation, and a shared commitment to sustainable agriculture, stakeholders can work together to address pest challenges, support biodiversity conservation, and promote resilient agroecosystems that benefit both farmers and the environment [16, 25, 42].

2.11 Recommendations and future directions in Integrated Pest Management (IPM)

Based on the findings and challenges identified, several recommendations can be proposed to enhance the effectiveness and sustainability of IPM programs [13, 17, 18, 20, 42, 43]:

- *Education and training:* Providing farmers, extension agents, and other stakeholders with training on IPM principles, techniques, and best practices can improve knowledge and capacity for sustainable pest management.

- *Stakeholder engagement:* Engaging with farmers, researchers, policymakers, and industry stakeholders in collaborative decision-making processes can promote the adoption of IPM strategies and address the diverse needs and perspectives of different stakeholders.

- *Integrated approach:* Integrating diverse pest management methods, including cultural, biological, and chemical controls, can enhance the resilience of IPM programs and reduce reliance on pesticides.

- *Monitoring and evaluation:* Implementing robust monitoring and evaluation systems to assess the impact of IPM strategies, track pest populations, and measure economic and ecological outcomes is essential for optimizing pest control efforts.

- *Research and innovation:* Investing in research on emerging technologies, pest biology, and ecological interactions can drive advancements in IPM practices and address evolving challenges in pest management.

Emerging technologies and research areas hold great potential for advancing IPM practices and addressing current and future pest management challenges. Some key areas of focus include [10, 17, 21, 25, 26]:

- *Precision agriculture:* Utilizing technologies such as drones, sensors, and data analytics for precision pest monitoring, targeted pesticide applications, and decision support systems can optimize pest control efforts and reduce environmental impacts.

- *Biological control:* Research on new biological control agents, such as parasitoids, predators, and pathogens, can expand the toolbox of IPM strategies and enhance natural pest suppression in agroecosystems.
- *Genetic resistance:* Developing crop varieties with genetic resistance to pests and diseases through breeding and biotechnology can reduce the need for chemical pesticides and enhance crop resilience to pest pressures.
- *Climate change adaptation:* Studying the impacts of climate change on pest populations, crop phenology, and ecosystem dynamics can inform adaptive IPM strategies that mitigate climate-related risks and promote agricultural sustainability.

In conclusion, by implementing the recommended strategies and embracing emerging technologies and research areas, the future of IPM holds great promise for advancing sustainable pest management practices, enhancing agricultural productivity, and promoting environmental stewardship in agriculture. Continuous collaboration, innovation, and knowledge exchange will be essential for driving the evolution and adoption of effective IPM programs in the years to come.

3. Conclusions

The evolution of Integrated Pest Management (IPM) has demonstrated a significant shift toward a more holistic and sustainable approach to pest management over the past six decades. Initially developed in response to the negative impacts of indiscriminate pesticide use, IPM has expanded its goals to encompass environmental protection, biodiversity conservation, and human health considerations. By integrating various pest control methods and reducing reliance on synthetic pesticides, IPM aims to address challenges such as pesticide resistance, environmental pollution, and risks to human health.

Despite the progress made in IPM practices, weaknesses persist in the current implementation of IPM programs. Challenges include the lack of standardized definitions, inconsistencies between theory and practice, and insufficient farmer engagement and ecological understanding. Addressing these weaknesses requires efforts to clarify definitions, bridge gaps between theory and practice, enhance farmer engagement, and promote a holistic understanding of ecological principles in pest management.

Effective pest management recommendations are essential for successful IPM programs, with criteria such as efficacy, safety, environmental impact, and cost-effectiveness playing a crucial role in decision-making processes. While quantitative evaluation methods provide valuable insights into the effectiveness of IPM strategies, qualitative approaches are essential for capturing stakeholder perspectives and understanding the social and cultural factors influencing IPM adoption.

Looking toward the future, recommendations for improving IPM programs include education and training, stakeholder engagement, an integrated approach to pest management, robust monitoring and evaluation systems, and investment in research and innovation. Emerging technologies such as precision agriculture, biological control, genetic resistance, and climate change adaptation hold promise for advancing sustainable pest management practices and addressing evolving challenges in pest control.

Eventually, by implementing these recommendations and embracing emerging technologies and research areas, the future of IPM looks promising for promoting sustainable pest management practices, enhancing agricultural productivity, and fostering environmental stewardship in agriculture. Collaboration, innovation, and knowledge exchange will be crucial for driving the continued evolution and adoption of effective IPM programs in the years to come.

Acknowledgements

The authors express their gratitude for the support of the Agricultural Sciences and Natural Resources University of Khuzestan in writing this chapter of the book. The authors acknowledge the use of ChatGPT-3.5 for language polishing of the manuscript.

Conflict of interest

The authors declare that there is no conflict of interest.

Acronyms and abbreviations

IPM	Integrated Pest Management
BC	biological control
NE	natural enemies
SA	sustainable agriculture

Author details

Fatemeh Yarahmadi and Ali Rajabpour*
Agricultural Sciences and Natural Resources University of Khuzestan,
Bavi, Khuzestan, Iran

*Address all correspondence to: a_rajabpour2000@yahoo.com

References

[1] Landis DA, Wratten SD, Gurr GM. Habitat management to conserve natural enemies of arthropod pests in agriculture. Annual Review of Entomology. 2000;**45**:175-201. DOI: 10.1146/annurev.ento.45.1.175

[2] Safari N, Rajabpour A, Seraj AA. Evaluation of various diets and oviposition substrates for rearing *Orius albidipennis* Reuter. Journal of Entomological Society of Iran. 2016;**35**:29-37

[3] Torres JB, Bueno ADF. Conservation biological control using selective insecticides—A valuable tool for IPM. Biological Control. 2018;**126**:53-64. DOI: 10.1016/j.biocontrol.2018.07.012

[4] Letourneau DK. Conservation biology. In: Conservation Biological Control. Netherland: Elsevier; 1998. pp. 9-38. DOI: 10.1016/B978-012078147-8/50048-7

[5] Bianchi FJJA, Booij CJH, Tscharntke T. Sustainable pest regulation in agricultural landscapes: A review on landscape composition, biodiversity and natural pest control. Proceedings of the Royal Society. B, Biological Sciences. 2006;**273**:1715-1727. DOI: 10.1098/rspb.2006.3530

[6] Grijalva I, Skidmore AR, Milne MA, Olaya-Arenas P, Kaplan I, Foster RE, et al. Integrated pest management enhances biological control in a US midwestern agroecosystem by conserving predators and non-pest prey. Agriculture, Ecosystems & Environment. 2024;**368**:109009. DOI: 10.1016/j.agee.2024.109009

[7] Tooker JF, O'Neal ME, Rodriguez-Saona C. Balancing disturbance and conservation in agroecosystems to improve biological control. Annual Review of Entomology. 2020;**65**:81-100

[8] Deguine J-P, Aubertot J-N, Flor RJ, Lescourret F, Wyckhuys KAG, Ratnadass A. Integrated pest management: Good intentions, hard realities. A review. Agronomy for Sustainable Development. 2021;**41**:38. DOI: 10.1007/s13593-021-00689-w

[9] Kogan M. Integrated Pest Management: Historical perspectives and contemporary developments. Annual Review of Entomology. 1998;**43**:243-270. DOI: 10.1146/annurev.ento.43.1.243

[10] Bielza P. Insecticide resistance in natural enemies. In: Horowitz AR, Ishaaya I, editors. Advances in Insect Control and Resistance Management. Cham: Springer International Publishing; 2016. pp. 313-329. DOI: 10.1007/978-3-319-31800-4_16

[11] Gul H, Gadratagi BG, Güncan A, Tyagi S, Ullah F, Desneux N, et al. Fitness costs of resistance to insecticides in insects. Frontiers in Physiology. 2023;**14**:1238111. DOI: 10.3389/fphys.2023.1238111

[12] Dean AN, Niemi JB, Tyndall JC, Hodgson EW, O'Neal ME. Developing a decision-making framework for insect pest management: A case study using Aphis glycines (Hemiptera: Aphididae). Pest Management Science. 2021;**77**:886-894. DOI: 10.1002/ps.6093

[13] Bueno AF, Pannizi AR, Hunt TE, Dourado PM, Pitta RM, Gonçalves J. Challenges for adoption of Integrated Pest Management (IPM): The soybean example. Neotropical Entomology. 2021;**50**:5-20. DOI: 10.1007/s13744-020-00792-9

[14] Janssen SJC, Porter CH, Moore AD, Athanasiadis IN, Foster I, Jones JW, et al. Towards a new generation of agricultural system data, models and knowledge products: Information and communication technology. Agricultural Systems. 2017;**155**:200-212. DOI: 10.1016/j.agsy.2016.09.017

[15] Grasswitz TR. Integrated Pest Management (IPM) for small-scale farms in developed economies: Challenges and opportunities. Insects. 2019;**10**:179. DOI: 10.3390/insects10060179

[16] Kaur T, Kaur M. Integrated Pest Management: A paradigm for modern age. In: Pests, Weeds and Diseases in Agricultural Crop and Animal Husbandry Production. London, UK: IntechOpen; 2020. DOI: 10.5772/intechopen.92283

[17] Dara SK. The new integrated Pest management paradigm for the modern age. Journal of Integrated Pest Management. 2019;**10**(1):1-9. DOI: 10.1093/jipm/pmz010

[18] Baker BP, Green TA, Loker AJ. Biological control and integrated pest management in organic and conventional systems. Biological Control. 2020;**140**:104095. DOI: 10.1016/j.biocontrol.2019.104095

[19] Khan MA, Khan Z, Ahmad W, Paul B, Paul S, Aggarwal C, et al. Insect Pest resistance: An alternative approach for crop protection. In: Hakeem KR, editor. Crop Production and Global Environmental Issues. Cham: Springer International Publishing; 2015. pp. 257-282. DOI: 10.1007/978-3-319-23162-4_11

[20] Dent D, Binks RH. Insect Pest Management. 3rd ed. UK: CABI; 2020. DOI: 10.1079/9781789241051.0000

[21] Tiwari AK. Insect pests in agriculture identifying and overcoming challenges through IPM. Archives of Current Research International. 2024;**24**:124-130. DOI: 10.9734/acri/2024/v24i3651

[22] Letourneau DK, Jedlicka JA, Bothwell SG, Moreno CR. Effects of natural enemy biodiversity on the suppression of arthropod herbivores in terrestrial ecosystems. Annual Review of Ecology, Evolution, and Systematics. 2009;**40**:573-592. DOI: 10.1146/annurev.ecolsys.110308.120320

[23] Kremen C, Miles A. Ecosystem services in biologically diversified versus conventional farming systems: Benefits, externalities, and trade-offs. Ecology and Society. 2012;**17**(4):40. DOI: 10.5751/ES-05035-170440

[24] Wezel A, Casagrande M, Celette F, Vian J-F, Ferrer A, Peigné J. Agroecological practices for sustainable agriculture. A review. Agronomy for Sustainable Development. 2014;**34**:1-20. DOI: 10.1007/s13593-013-0180-7

[25] Akanmu AO, Akol AM, Ndolo DO, Kutu FR, Babalola OO. Agroecological techniques: Adoption of safe and sustainable agricultural practices among the smallholder farmers in Africa. Frontiers in Sustainable Food Systems. 2023;**7**:1143061. DOI: 10.3389/fsufs.2023.1143061

[26] Gurr GM, Liu J, Pogrebna G. Harnessing artificial intelligence for analysing the impacts of nectar and pollen feeding in conservation biological control. Current Opinion in Insect Science. 2024;**62**:101176. DOI: 10.1016/j.cois.2024.101176

[27] Heibatian A, Yarahmadi F, Lotfi J-AA. Field efficacy of biorational insecticides, azadirachtin and Bt, on *Agrotis segetum* (Lepidoptera: Noctuidae) and its carabid predators in the sugar beet fields. Journal of Crop Protection. 2018;**7**:365-373

[28] Ya-Ali P, Yarahmadi F, Mehrnia MA. Efficacies of two nano-formulations of Tasmanian blue gum essential oil to control *Callosobruchus maculatus*. Journal of Economic Entomology. 2020;**113**:1555-1562. DOI: 10.1093/jee/toaa069

[29] Zhou X, Yang C, Yesmin S, Islam MA, Sarkar A. Bibliometric analysis of integrated pest management practices. Horticulturae. 2023;**9**:852. DOI: 10.3390/horticulturae9080852

[30] Rejesus RM, Jones MS. Perspective: Enhancing economic evaluations and impacts of integrated pest management farmer field schools (IPM-FFS) in low-income countries. Pest Management Science. 2020;**76**:3527-3536. DOI: 10.1002/ps.5912

[31] Tonle FBN, Niassy S, Ndadji MMZ, Tchendji MT, Nzeukou A, Mudereri BT, et al. A road map for developing novel decision support system (DSS) for disseminating integrated pest management (IPM) technologies. Computers and Electronics in Agriculture. 2024;**217**:108526. DOI: 10.1016/j.compag.2023.108526

[32] Lane DE, Walker TJ, Grantham DG. IPM adoption and impacts in the United States. Journal of Integrated Pest Management. 2023;**14**:1. DOI: 10.1093/jipm/pmac028

[33] Yarahmadi F, Salehi Z, Lotfalizadeh H. Compatibility of insecticides and *Elachertus inunctus* Nees (Hymenoptera: Eulophidae) for controlling *Tuta absoluta* Meyrick (Lepidoptera: Gelechiidae) in greenhouse condition. Journal of Plant Diseases and Protection. 2018;**125**:357-364. DOI: 10.1007/s41348-018-0164-6

[34] Abdel-Raheem M. Effect of Insecticides on Natural-Enemies. In: Eduardo Rebolledo Ranz R, editor. Insecticides—Impact and Benefits of Its Use for Humanity. IntechOpen; 2022. DOI: 10.5772/intechopen.100616

[35] Zandi-Sohani N, Rajabpour A, Yarahmadi F, Ramezani L. Sensitivity of *Bemisia tabaci* (Hemiptera: Aleyrodidae) and the generalist predator *Orius albidipennis* (Hemiptera: Anthocoridae) to vapors of essential oils. Journal of Entomological Science. 2018;**53**:493-502. DOI: 10.18474/JES17-113.1

[36] Cloyd RA. Compatibility of insecticides with natural enemies to control pests of greenhouses and conservatories. Journal of Entomological Science. 2006;**41**:189-197. DOI: 10.18474/0749-8004-41.3.189

[37] Casida JE, Durkin KA. Neuroactive insecticides: Targets, selectivity, resistance, and secondary effects. Annual Review of Entomology. 2013;**58**:99-117. DOI: 10.1146/annurev-ento-120811-153645

[38] Sparks TC, Nauen R. IRAC: Mode of action classification and insecticide resistance management. Pesticide Biochemistry and Physiology. 2015;**121**:122-128. DOI: 10.1016/j.pestbp.2014.11.014

[39] Khan Mirza F, Yarahmadi F, Lotfi Jalal-Abadi A, Meraaten AA. Enzymes mediating resistance to chlorpyriphos in *Aphis fabae* (Homoptera: Aphididae). Ecotoxicology and Environmental Safety. 2020;**206**:111335. DOI: 10.1016/j.ecoenv.2020.111335

[40] Madesh K, Komala G, Chandralekha R, Tripathi P. Mode of action of novel insecticides. In: Advanced Trends in Plant Protection. India: Shri Vaishnav Institute of Agriculture; 2024. pp. 255-291

[41] Siddiqui JA, Fan R, Naz H, Bamisile BS, Hafeez M, Ghani MI, et al.

Insights into insecticide-resistance mechanisms in invasive species: Challenges and control strategies. Frontiers in Physiology. 2023;**13**:1112278. DOI: 10.3389/fphys.2022.1112278

[42] Rajabpour A, Yarahmadi F. Decision System in Agricultural Pest Management. Singapore: Springer Nature; 2024

[43] Shahpouri A, Yarahmadi F, Zandi SN. Functional response of the predatory species *Orius albidipennis* Reuter (Hemiptera: Anthocoridae) to two life stages of *Bemisia tabaci* (Genn.) (Hemiptera: Aleyrodidae). Egyptian Journal of Biological Pest Control. 2019;**29**:14. DOI: 10.1186/s41938-019-0119-7

[44] Mousavi K, Rajabpour A, Parizipour MHG, Yarahmadi F. Biological and molecular characterization of *Cladosporium* sp. and *Acremonium zeylanicum* as biocontrol agents of *Aphis fabae* in a tri-trophic system. Entomologia Experimentalis *et* Applicata. 2022;**170**:877-886. DOI: 10.1111/eea.13217